LEARNING RESOURCES CTR/NEW ENGLAND TECH.
GEN   TK7895.A8 D38 1994
Davis, Brend   The economics of automatic

3 0147 0001 9829 4

TK 789

Davis, Brendan.

The economics of automatic
testing

NEW ENGLAND INSTITUTE
OF TECHNOLOGY
LEARNING RESOURCES CENTER

The Economics
of Automatic Testing

Second Edition

The Economics
of Automatic Testing

Second Edition

Brendan Davis

McGRAW-HILL BOOK COMPANY

London · New York · St Louis · San Francisco · Auckland
Bogotá · Caracas · Lisbon · Madrid · Mexico · Milan
Montreal · New Delhi · Panama · Paris · San Juan
São Paulo · Singapore · Sydney · Tokyo · Toronto

NEW ENGLAND INSTITUTE
OF TECHNOLOGY
LEARNING RESOURCES CENTER

11/94

28633990

Published by
McGRAW-HILL Book Company Europe
Shoppenhangers Road, Maidenhead, Berkshire, SL6 2QL, England
Telephone 0628 23432
Fax 0628 770224

British Library Cataloguing in Publication Data
Davis, Brendan
  Economics of Automatic Testing. –2Rev.ed
  I. Title
  620.0044
  ISBN 0-07-707792-X

Library of Congress Cataloging-in-Publication Data
Davis, Brendan
  The economics of automatic testing/Brendan Davis. — 2nd ed.
    p. cm.
  Includes bibliographical references and index.
  ISBN 0-07-707792-X: £45.00
  1. Automatic checkout equipment–Cost effectiveness.
  2. Electronic industries–Quality control I. Title.
  TK895.A8D38   1994   93–2633
  621.3815′068′5–dc20   CIP

Copyright © 1994 McGraw-Hill International (UK) Limited.
All rights reserved. No part of this publication may be reproduced, stored in a retrieval system, or transmitted, in any form or by any means, electronic, mechanical, photocopying, recording, or otherwise, without the prior permission of McGraw-Hill International (UK) Limited.

12345   HW   97654

Typeset by P&R Typesetters Ltd, Salisbury, UK
and printed and bound in Great Britain by
BPCC Hazell Books Ltd
Member of BPCC Ltd

To Maggie, Emma, James, Jennifer and Tessa

Contents

Preface

The first edition of this book was prepared in 1980 and 1981 and published in 1982. Since then there have been many changes in the electronics industry and in its testing arena. Some of these changes have a significant impact on the way that we should view the economics of test. Because of this, and because of the encouragement of some of my colleagues, I finally decided to put pen to paper, or fingers to keyboard, and prepare this second edition. However, not everything has changed. Many of the key issues and concepts of the early eighties are the same today. The following paragraph is reproduced verbatim from the preface to the first edition.

Electronics technology is growing rapidly in terms of its use and its complexity. It is no longer easy to define the 'electronics industry' since the technology is penetrating into just about every industry you can think of. The applications for 'semiconductor-based' products continue to multiply—think of your car, your television set, your watch, and your children's toys. But while semiconductor technology has substantially lowered the cost of microprocessors, memories, and supporting chips in the last decade, the complexity of electronic equipment has increased substantially. In 1970, a typical complex semiconductor chip contained the equivalent of 1000 transistors; by 1980 chips containing the equivalent circuitry of 100 000 transistors were fairly common. At the same time, this accelerating technology has reduced product lifetimes drastically, with three years being fairly typical. Customers buying these high-technology products are also demanding greater reliability. Electronic equipment manufacturers are therefore caught in a squeeze—business success depends on getting more complex, but more reliable, products to market faster than ever before. To make matters even worse there is a shortage of skilled and experienced electronics engineers.

The situation and the trends outlined here are still valid. They are simply bigger and more important than they were twelve years ago. Component and board complexity continues to rise, customers continue to demand higher quality and the competition forces the need to bring new products to market faster than ever. Sadly the only comment in the paragraph reproduced above that is not so valid today is the last comment about the shortage of skilled and experienced engineers. The downturn in the electronics industry that took place in 1985 resulted in cutbacks and streamlining of operations. Recovery was slow and in many cases did not return to the levels of early 1985 before the next recession hit. The second half of 1989 saw the start of the 'outbreak of world peace'. This was a major benefit to mankind but the reduced military spending that followed added to the problems of the electronics industry. The commercial benefits, however, may come from the opening up of markets in Eastern Europe and what was formerly the Soviet Union as they slowly rebuild their economies. In the longer term, these countries will simply add to the numbers of competitors fighting for a share of the world electronics market, and their lower labour costs will give them an advantage for a while.

The recession in the electronics industry led to a deeper recession in the automatic test equipment (ATE) industry. Not only was there little need to increase testing capacity but a lot of surplus testers also became available on the second user market. Under such conditions there are only two reasons to buy new test systems. The first of these is when the new equipment can result in cost savings and the second is when the existing equipment cannot cope with the test requirements of new technologies. The effect of operating in times of recession leads quite naturally to an increase in the importance of good economic analysis. Indeed, in the twenty-two years that I have been involved with ATE the interest in test economics has varied in a direct relationship with economies. When the economy of the country is buoyant nobody is interested, but when money starts to tighten up it becomes most important. Naturally I believe that it is important at all times. There is no need to waste good money even if the economy is good.

Apart from the recessions there has been another important reason for the relatively lower sales of ATE. The eighties saw a change in the awareness and the understanding of the importance of quality that has been more revolutionary than evolutionary. The degree of improvement in component defect rates and the yield of defect-free boards has been quite dramatic, especially when you factor in the increase in complexity that has also taken place. The higher component quality has led to less use of incoming inspection and the higher board yields has led to the need for fewer board testers. Fewer failing boards means fewer re-tests after repair and therefore increased tester capacity. This quality revolution has arguably been the most important change since the first edition of this book was published. As a result I have included a completely new chapter to cover the impact this has on how we should think about test economics.

Other major changes that have taken place include the increased degree of integration between the design and test function, the increased use of custom devices, the changes in manufacturing to the use of surface mount technology and JIT (just in time) processes, the introduction of new types of testers, the adoption of testability disciplines such as SCAN and Boundary Scan and the adoption of new test strategies. The impact of these and other trends are covered in this new edition. The net effect of all of the changes that have occurred since the first edition was published is that the cost of test, as a percentage of the total product cost, has risen dramatically even though some of the changes are aimed at lowering test costs. This means that even in the absence of any recessionary problems, the economics of test are more important today than they have ever been. In the past, test economics has been used mainly to compare alternative test strategies, alternative testers or to justify the purchase of equipment. Today the more enlightened companies are using test economics in the planning stages of a new product to determine the life cycle cost of the product. Various alternative design approaches and test strategies can be compared to find the optimum approach to meeting the main objectives. These objectives will have been driven by the four main market forces of cost, quality, time and technology. Survival in a competitive market is heavily dependent on introducing new products that cost less, that have better quality, that are brought to market on time and that incorporate the latest component and manufacturing technologies.

My main personal hopes for this book are that it should broaden the use of economic analysis techniques as applied to testing decisions and that it should help to reduce the amount of time needed to perform the analysis. To this end the book has been structured to be both tutorial and a reference work. Most of the chapters contain numerous examples as well as reference tables and charts. The more useful reference material is also contained in appendices for more convenient access.

Many people have aided and abetted in the preparation of this second edition. Some have provided ideas and information, some have reviewed and critiqued the manuscript and some have helped in other ways. They are too numerous to mention individually but they know who they are and they know that I am grateful to them all.

Brendan Davis

1. Introduction and overview

It is widely recognized that automatic testing of electronic components and sub-assemblies has a number of distinct advantages over more manual test methods. Probably the biggest advantage is in the area of economics, since although automatic test equipment (ATE) has a relatively high capital cost its overall operating cost is usually significantly lower than the alternatives of manual testing or no testing.

This alternative of 'no testing' is what most people would like to achieve. If we could guarantee that the design was error-free, that every component purchased would meet all of its specifications and be assembled and soldered correctly and that boards in a system were interconnected properly, we could eliminate testing. Assuming that the design required no adjustments or calibration, the product should be shipped to a customer directly off the production line or from the system integration area. In the real world, if we did this, our field-service costs would be astronomical—they are probably high enough even with all the testing we do anyway—and we would rapidly go out of business because of a terrible reputation for poor quality and reliability. Testing, then, can be thought of as a *cost-avoidance* activity.

Within the production cycle of an electronic product there are a number of stages at which testing or inspection can be performed. It is generally accepted that the earlier in the process a fault is found, the less it costs.

It should be clear, however, that if all the testing that could possibly be done is in fact performed, then probably too much would be done, resulting in a less than optimum operation. Economically speaking, more testing than is necessary is just as costly as not doing enough. Quite naturally there will be differences between available test equipment: not only the differences between the products of the various vendors but also differences within the product range of an individual vendor. Deciding which product is best suited to your requirements can be a very difficult task, especially when some of the more important differences—such as software support—and the credibility of the vendor are difficult to measure.

The major purpose of this book is to:

1. Outline the advantages of automatic testing relative to the alternatives of manual testing or not testing.
2. Highlight the major cost and savings areas, showing how to calculate potential savings.
3. Highlight the important areas for consideration when evaluating automatic test equipment.

4. Discuss the trade-offs to be considered when trying to optimize the testing strategy for a product.
5. Show how to perform a financial analysis of an intended investment in ATE and present the results in a way that financial and upper management can understand.

Throughout the book the emphasis will be on economics rather than on the technical aspects of ATE. Inevitably, however, some reference needs to be made to technical matters since they affect the economic performance of a given type of system. In particular, the emphasis will be on minimizing costs and maximizing productivity rather than simply lowering costs and increasing productivity. Any ATE system, no matter how poor and no matter how poorly used, should perform more economically than the manual testing alternative, but is not necessarily the optimum solution. To optimize the testing operation, four major objectives need to be accomplished:

1. The development of testable designs.
2. Selection of the optimum testing strategy.
3. Selection of the most cost-effective equipment.
4. The effective management and control of the testing department.

The first three objectives are addressed by this book. The fourth objective is outside the scope of this book and probably deserves a book all to itself to do justice to the subject.

Another objective to be kept in mind when considering the others is the maintaining, or better still the improvement, of quality and reliability of the product. The effects of the rapid growth in the use and complexity of semiconductor devices referred to in the Preface is having another effect. In some areas, products are becoming more and more alike. The chips are available to everyone *and* it is easy to copy an innovative design. The result is that companies have to search for other ways to differentiate their products from those of their competitors. An increasing form of differentiation is that of quality and of after-sales support such as field service. More and more, it will be the company with the better reputation for quality and service that will gain market share over its rivals. The testing strategies selected must therefore make every attempt to improve quality and reliability to make the field-service job easier to perform in an efficient manner.

Another issue that has gained substantially in its importance over the past ten or twelve years is that of 'time-to-market'. Getting new products to market on time, and hopefully ahead of the competition, has become increasingly important. This has been driven by the intensification of world-wide competition that has tended to shorten product lifetimes as companies try to leap-frog their competitors with better products at lower prices. Possibly one of the best examples lies in the personal computer industry. Technological changes have been rapid and price reductions have been dramatic. The 'laptop' computers that enjoyed premium prices just two years ago (1990) are now virtually obsolete as the smaller and lighter 'notebook' PCs have caught up and surpassed them in terms of performance. My own laptop, just two years old, is currently available for one third of its original (discounted) price. A colour notebook launched in March 1992 at a list price of £5500 is already available for less than £3000 and it is only July 1992 as I write this.

I recently read an article about Michael Dell, founder of the Dell Computer company. In the article it mentioned that their new product design cycle is already down to eight months and the oldest product in their catalogue is just fourteen months old. A number of direct selling PC manufacturers have recently discontinued all models using the 386 series of processors. They only make 486 based units with a starting price below £1000 and the 586 chip is expected soon.

By the time this book is available this will all be stale news. The prices I have quoted will seem very high and the 486 will possibly be on the way out. This pattern, to one degree or another, is common throughout the electronics industry—shorter product life cycles, increased use of technology, lower prices and better quality. This is all good news for the end user, but the cause of many headaches for the engineering, manufacturing and test engineering people.

1.1 What cost quality?

More often than not quality, or more particularly high quality, conjures up thoughts of a Rolls-Royce, a Mercedes or a Cadillac. As a result, many people equate high quality with high cost. However, if you think of quality as 'fitness for use' or as 'conformance to specifications', it becomes clear that it is possible to have a high-quality model of a much lower priced vehicle, or indeed a poor-quality example of one of the more exotic machines mentioned above. This concept of 'quality', epitomized by promotional slogans such as 'this is the Rolls-Royce[1] among dishwashers', is really only part of the overall meaning of quality. This use of the word is really referring to the 'grade', 'class' or 'degree of excellence' of a product or service. More generically, this can be referred to as the 'quality of design' since the grade selected will be a conscious decision of the design of the product. Looked at from the point of view of 'fitness for use', the quality of design will be a function of how well the marketing people have interpreted the real needs of the market-place, the concept of their product offering and the quality of the specifications laid down for the product.

Once the product is being manufactured, the important issue will be the 'quality of conformance' to the specifications. It is with the quality of conformance that the test department and quality assurance departments will be most concerned.

Looked at in this way it is clear that, most of the time, higher 'quality of design' can only be achieved at higher cost. What is perhaps less clear is that higher 'quality

---

[1] The Rolls-Royce Car Company object, quite rightly, to the use of their name to promote other products. I object to it also but for other reasons. The 'Rolls-Royce' definition of quality is totally incorrect. There is always confusion between high-grade or luxury products and the correct 'conformance to requirements' definition of quality. Chapter 2 explains this more completely.

of conformance' can often be achieved with a *reduction* in costs. This can be achieved because it is usually less costly to prevent a defect at an early stage than to locate it and rectify it at a later stage when value has been added and it is more difficult to identify the cause of the problem.

This concept is extremely important. It is the key to the whole process of determining the optimum testing strategy for a particular product. Earlier and better testing will lead the way to increased productivity, quality and lower costs.

There are other elements to the overall concept of quality. One of these is sometimes referred to as 'the abilities'. Availability of a product to the user is obviously an important parameter. This, in turn, is determined by the 'reliability' and the 'maintainability'. The other element is that of field service since inevitably some of these complex products will go wrong at times. When they do, the efficiency of the field-service operation, in terms of its promptness, competence and integrity, will have a major influence on the customer's perception of the quality of the product.

1.2 Why economics of automatic testing?

I believe that a careful economic analysis is the only effective way to compare alternative testing strategies. Comparing such alternatives is a complex problem involving many factors, and as such it is necessary to reduce these factors to some common denominator. The only common denominator available is cost and, as stated earlier, testing is a cost avoidance operation. Automatic test equipment (ATE) is therefore the solution to what is primarily an economic problem. There are, however, more fundamental reasons to perform an economic analysis for engineering decision making. Most university courses on all kinds of engineering subjects offer an 'engineering economics' or 'engineering economy' course module. Unfortunately these modules are not always mandatory and can often be left out of the total credits needed for a degree. There is some evidence that this is changing and that more universities are making the economics module mandatory. Equally, however, there is evidence that some economics modules cover the subject in a far greater degree of detail than is normally used in industry. The danger here is that the complexity of the course material may well put some students off the subject. This is a pity because the subject is really an essential part of the education of a good engineer. The reason is simple. Almost all engineering decisions are economic decisions. That may sound like a sweeping statement, but consider the following examples.

A new design project has gone through the 'top level' and the 'partitioning' stages. It is now necessary to decide if a part of the design should be implemented with an application specific integrated circuit (ASIC). It is agreed that this should be done so now it is necessary to decide on which ASIC technology should be used. Next, a decision has to be made about including some 'design for test' (DFT) circuitry in the ASIC. If it is agreed that this is worth while the next decision will be about the form that this should take, and so on, and so on. Decisions, decisions—all of them need to be based on economics because most of these decisions will have a significant impact on the life cycle cost of the project.

Similar decisions have to be made by the test engineering department when a new design is started. If it is a major new product it will be necessary to decide on the test strategy to adopt. A smaller project will probably have to fit in with the current test strategy: in-circuit, functional, combinational, manufacturing defects analyser (MDA), hot mock-up, final test, fault coverage targets, rework routing, fixture preparation. These and many other issues have to be considered, and cost is the common denominator. That is how it is with all engineering projects.

An excellent definition of engineering was presented by Arthur Wellington in a book published in 1887. He defined engineering as:

> The art of doing something well with one dollar that any bungler can do with two, after a fashion.

The book was *The Economic Theory of the Location of Railways*, in which he described how he used economic analysis techniques to determine the optimum layout for the railway (railroad) system in the United States. His definition neatly sums up the difference between good and bad engineering.

1.3   The comparison of alternatives

If you are not convinced that performing an economic analysis is the best way to compare engineering alternatives, there is another reason to do it that is hard to refute. None of us are in business to make products or services. We are in business to make money for the shareholders and the employees. Successful businesses make more money than unsuccessful businesses. The products and services are a means to an end. We make money by doing what we do best. That is the essence of economics in its broader sense. We do what we do best, make some money and then buy what we need from someone who has specialized in that. For example, farmers sell their produce and buy PCs from electronics companies to enable them to run their businesses more efficiently. For these reasons it is logical that most important decisions should be based on an economic or financial analysis, but there is a more fundamental reason.

Most of our engineering decisions involve comparing relatively similar alternatives such as one ASIC technology versus another or one in-circuit board tester versus another. Making the best decision is still a difficult task, but consider the following situation. The senior management who have to approve the spending of the money are usually comparing very different alternatives. They may have to decide between funding a new test system, a new CAD (computer aided design) system, a numerically controlled machine or some new vehicles. With such diverse alternatives to choose from, assuming they are not able to fund all of them, their relative financial performances will be the only sensible way to make a choice. This becomes obvious when you consider the main objectives of senior management. Their job is to maximize the return on the shareholders' investment in the company.

Most companies are owned by one or more stockholders or shareholders. The majority of these stockholders do not work for the company. They invest their money in companies in the hope of a greater return than they can get by investing their money in a bank. A board of directors is appointed by the shareholders to look after their interests. In turn the directors employ a general manager, who also usually sits on the board, to run the company in such a way that the shareholders get a good return on their investments.

Many of these shareholders have no loyalty to the company at all. To them it is just an investment and if the company does not perform well they will sell their stock and put their money somewhere else. If a lot of shareholders try to sell their stock, then the price goes down and so does the value of the company. In order to keep a good market for its stock, the company therefore needs to perform well and 'performance' here means *profitability* and *growth*.

The company also needs to obtain money from other sources—usually in the forms of loans. In order to obtain such financing at a good rate of interest, the company must show a good 'track record' and show that its future prospects are good. To a large degree this is dependent on its management being seen as doing a good job and capable of continuing to do so by making correct decisions.

Any business enterprise exists to fulfil a specific mission and a social function. The mission is primarily one of economic performance and the social function is to contribute to the economy of the country and provide employment and a good standard of living for a number of people. If a business does not perform economically it will be unable to achieve this social function fully. The inevitable result will be redundancies, high unemployment and, possibly, the closure of factories. For this reason, business management must always put economic performance first in any decision it takes. Overall their decision making falls into three areas:

1. What business to be in.
2. Marketing strategy—products, promotion, pricing, distribution, etc.
3. The use of capital.

The primary aim is long-term profitability and growth so that the company is attractive to shareholders and is able to obtain other sources of funds such as loans. If this is achieved, then other parts of its mission such as fulfilling social needs will fall naturally into place.

So far in this section there have been several mentions of profit and profitability. To many people 'profit' is a dirty word. It conjures up images of the fat mill owner making vast sums of money at the expense of his poorly paid workers. Perhaps if everyone were to switch to the American terminology of *earnings*, there would be less stigma attached. For some reason, 'earnings' seems much more acceptable than 'profit'. Whatever we call it, however, does not change the fact that it is necessary for the continued well-being of a company and its employees.

The investment in ATE falls into the 'use of capital' area of management decision making. Figure 1.1 shows the flow of money within a company. The 'capital employed' is made up of 'fixed assets' such as buildings and equipment (including ATE) and

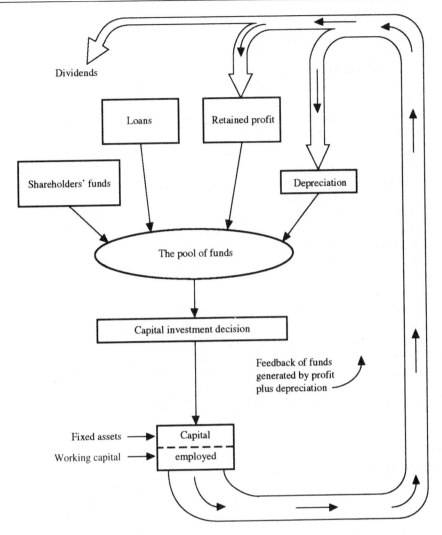

Figure 1.1 The flow of funds

'working capital'. The decision to add to fixed assets needs a lot of thought, especially since management and directors are judged on the basis of return on capital employed (ROCE), so that for them the lower the assets are the better.

Making correct decisions on major capital investment is vital to maintaining profitability. Therefore, someone in the company should always be looking at major purchases from a purely financial point of view. If two totally different investment proposals are competing for some of the funds, a well-managed company will pick the one that gives the highest return on investment (ROI). The answer then to the question 'Why is a financial analysis necessary?' is that it is good (necessary) management practice to do so; the only way top management can decide between dissimilar investments is on the basis of financial performance.

Unfortunately, the economic performance of ATE is not always given the importance it deserves when equipment is evaluated. Research has shown that in many cases it is treated as an afterthought—a necessary evil to be performed because 'upper management likes that kind of thing'. The steps leading to a purchase often look like those in Fig. 1.2. Based upon technical performance, the support capabilities of the suppliers and the overall credibility of the suppliers, a decision is taken in favour of one piece of equipment. Having made this decision the evaluation team will then perform some economic analysis to see if this system will 'pay its way'. Quite often at this stage it is compared with the alternative of 'not using this ATE' and so just about any system will show a good economic performance. Since the whole point of purchasing the system is to achieve a better economic performance it would be more appropriate to include economics as one of the primary parameters at the stage when the short list of possibilities is being evaluated, as in Fig. 1.3.

Other problems that research has shown to exist are that the economic analysis and financial appraisal are sometimes performed by a financial person who has little knowledge of the applications; that the most common method used is simple payback analysis, which is very poor for comparing alternatives; and that the person responsible for the running of the equipment was often unaware of the criteria used for measuring the financial performance of the equipment.

The usual result of poor or inadequate economic and financial appraisal is that management will reject the proposal with a request for more information. This results in a delay in getting the purchase approved, which could have a major impact on a new product introduction or the efficiency of present operations.

1.4 Other factors

The way in which an ATE system will perform economically is obviously a function of how it is designed. The philosophy of the design, the importance given by the manufacturer to different aspects of what it should be and the way the various capabilities are handled from a hardware and software point of view will all have a bearing on the efficiency and, therefore, economic performance of the equipment. Naturally, the equipment must meet the minimum technical requirements of the job it is to perform in order for it to have any form of economic performance at all. Equally, if the equipment is capable of doing things we do not need to do—and these features cost money—then a portion of our cash will have been invested in a useless asset. A common mistake made when purchasing ATE is to buy more technical capability than is really required. The two main reasons for this are:

1. As an 'insurance' for the future. 'We may need that feature at some time'.
2. The 'since we are asking management for a pile of money anyway, we may as well go the whole hog' syndrome.

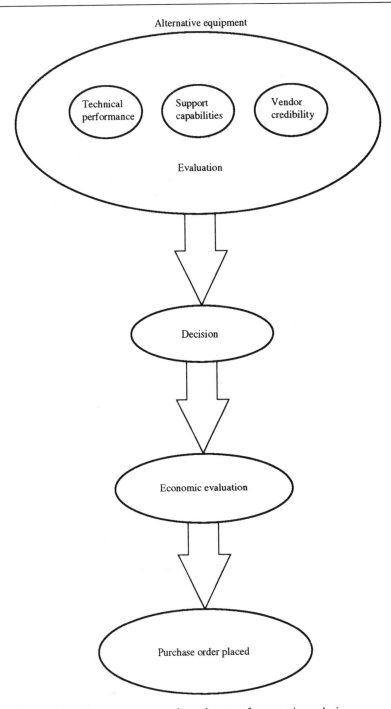

Figure 1.2 The wrong approach to the use of economic analysis

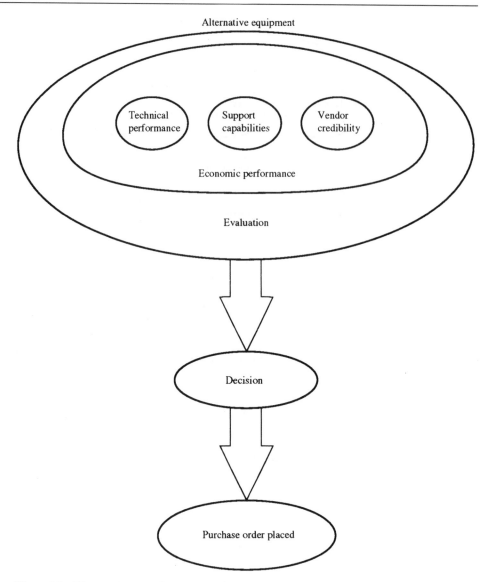

Figure 1.3 The correct use of economic analysis

In any event it will be necessary to evaluate the technical capabilities of available equipment. However, this should be done with economics in mind so that we only buy what we need now and for the foreseeable future, and that specified accuracies, etc., are not out of court relative to the real requirement.

Another important consideration will be the evaluation of the vendor (supplier). This again should be done with economics in mind. It is, however, rather difficult to assign a value to having a system out of operation, or not getting some other form of support when it is needed. It is a fact of life that very few, if any, companies purchasing a large software-based tester can be totally independent of the vendor. Such systems need regular updating to keep track with advances in technology, and the user will be dependent on the vendor for that support. Buying a large ATE system can be analogous to getting married. You develop a partnership and if you do not get the support and cooperation you need you end up living in a state of simply tolerating the partner or getting involved in an expensive divorce. The evaluation of the vendor in terms of credibility—'Will the vendor be around tomorrow?'—and the support that can be provided are, therefore, as important as the other selection parameters. The system with the best technical performance and the best economic performance will be of little use if the vendor goes out of business or is unable to support the products, since without this support the expected economic performance will not be achieved.

Throughout the book, the more detailed analysis of cost savings and economics generated by ATE apply to the testing of electronic components and assembled printed circuit boards (PCBs). It is hoped that the principles discussed will also be applicable to other forms of automatic testing. The techniques covered in the financial analysis section will, of course, apply to any capital investment decision.

1.5 Test strategy defined

It is unfortunate but the term 'test strategy' can mean different things to different people. There are to my knowledge three major uses of the term. The first of these relates to the nature and the order of the tests performed at a specific test stage. For example, the test program generated for an in-circuit test system (usually automatically) typically has the following sequence:

1. Short test.
2. Passive component tests.
3. Apply power.
4. Test linear devices.
5. Test non-bussed digital devices.
6. Test the busses.
7. Test the bussed digital devices.
8. Special tests.

This is a fairly logical sequence and many programmers will refer to it as the *test strategy*, but I prefer to call this the *test plan* or simply the *test program*. The second use of the term relates simply to the testing technology that is used. You can therefore have an in-circuit, a functional or a boundary scan test strategy. This use of the term is characterized by the name 'strategy independent tester' or 'multi-strategy tester' applied to some of the more sophisticated board testers that can perform a very wide range of tests. These are really different testing techniques, methods or tactics rather than strategies.

The third use of the term, and the one used throughout this book, relates to the overall philosophy of test and the routing of the units under test. This routing is important because you have two very different strategies based upon the same set of test stages, as is indicated in Fig. 1.4. The only difference here is the route that the boards take. However, these diagrams do not show the complete test strategy. They simply show what happens after the boards are assembled and soldered. There will be other elements of the test strategy prior to this stage. The amount and the nature of any design for test (DFT) or built-in self-test (BIST) in the product design is a part of the test strategy, as is the decision to perform or not perform any incoming inspection on selected devices. This will be the definition used throughout the book. Of the three common uses of the term 'test strategy' this is the only one that can

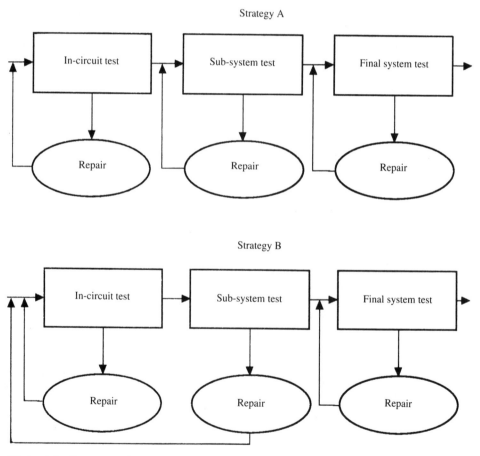

Figure 1.4   Two strategies with the same set of test stages

really be called a 'strategy' since a strategy is the overall plan that is put in place to reach a specific objective. Everything else is the tactics, plans or methods that are employed to implement the strategy.

1.6 Optimization versus sub-optimization

An emphasis throughout this book will be the need to determine optimum solutions to engineering problems. Whether these are design related, manufacturing related, test related, quality related or whatever, the analysis should consider all the issues. Too many of the problems in the industry are caused by sub-optimal decisions. In most cases such decisions are taken because of a narrow viewpoint in terms of the scope or the time horizon. Put more simply, sub-optimal decisions are usually the result of the organizational structure or of taking a short-term view. The departmental structure of most companies, with their attendant departmental budgets and objectives, forces many decisions to be made that have an adverse effect on other departments and therefore on the company as a whole. In the past, and still today in many companies, the engineering department could make decisions about a new design that could cause major problems for the test department and for manufacturing,

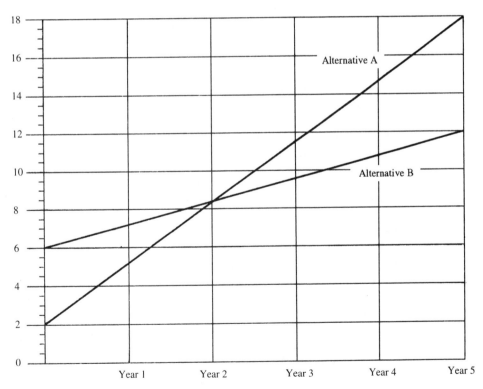

Figure 1.5   The optimal choice depends on your time horizon

test engineering could make decisions about the test strategy that could cause major headaches for the field service department and so on.

The growth in the importance of quality that took place in the eighties led to a large amount of education on the subject. This has resulted in many individuals taking a much broader view of things with a resultant reduction in the amount of sub-optimal decision making. However, it is not perfect yet. Many companies still only pay lip-service to the quality ethic and many sub-optimal decisions are still made.

Taking a short-term view of things is another reason for poor decisions being made. The graph in Fig. 1.5 illustrates this clearly. The two lines represent the cost of two alternative courses of action. A time horizon of less than two years would result in the alternative with the highest long-term costs being selected. This problem is also related to the use and the misuse of financial *return on investment* (ROI) calculations. All too often methods are used to compare the financial attractiveness of alternatives that are completely inappropriate or just plain wrong. Many major decisions are based upon a simple payback analysis which is not a measure of ROI at all. To make matters worse the payback time calculation is performed incorrectly in about eight cases out of ten. These problems are addressed in Chapter 11 on financial appraisal.

Yet another reason for poor decision making is the simple fact that, in some cases, not all of the relevant costs are included. This leads us into the next subject.

1.7 Life cycle costing

The obvious way to make optimum decisions that result in the long-term well-being of a company is to consider as much as possible the life cycle cost implications. The military and other government agencies have done this for many years and the concept is becoming more widely used in industry.

When the ATE industry was becoming established many potential users of the equipment were put off by the high purchase prices. The ATE vendors had to convince people that the overall cost of ATE was less than the cost of manual testing and that the cost of ownership of their particular product was lower than that of the competition. The cost elements that were usually considered in such an exercise were:

1. The purchase price.
2. The maintenance costs.
3. The cost of programs and fixtures.
4. The cost of testing and the diagnosis of defects.

Occasionally the cost of repairs or re-work of the units under test would be included, as it should be, with the test and diagnosis costs. Typically these costs would be calculated for a five year period and compared to the cost of manual testing or the costs for an alternative ATE system. Figure 1.6 shows the results of such an exercise, with the maintenance cost being around 50 per cent of the cost of the tester and the other cost areas being between 1.5 and 4.5 times the cost of the tester. Thus the 'cost

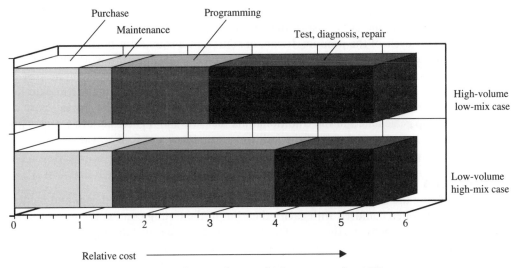

Figure 1.6   Traditional 'cost of ownership' cost areas for ATE

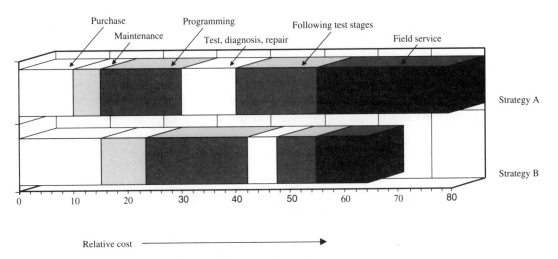

Figure 1.7   Basic 'life cycle' cost areas for ATE

of ownership' over the five year period could be between 3 and 6 times the purchase price.

This approach was very logical and quite correct but it did not really go far enough. The testing process involves more than just the ATE system. There are usually one or two stages of testing following the board tester, and then there is the field-service operation. The costs at each of these following test stages will be very dependent on the performance of the ATE. Therefore, for a true measure of the cost of different test strategies or different testers these other costs should be included in the analysis. This then is the life cycle cost of the test strategy, which in turn forms a part of the overall life cycle of the product. It can, of course, be a significant portion of the total life cycle cost. To be complete, any additional design engineering work that relates directly to the proposed test strategy should also be included in its life cycle cost.

Figure 1.7 shows how the life cycle testing costs for a product might compare for two different test strategies. In this example the strategy with the lowest 'cost of ownership', calculated in the traditional manner, has the highest life cycle cost. This is due to the increased cost of the following test stages and field service.

1.8 Strategies and tactics

In defining the term 'test strategy' I drew a distinction between strategies and tactics. These are often confused with each other, but in the testing context there is a simple differentiation between the two. The test strategy is the overall approach taken to reach the objectives and the test tactics are the specific testers, test stages or test technologies chosen to implement the strategy. A fact that is often overlooked is that a good strategy implemented with poor tactics can fail to meet its objectives. Sometimes a poor strategy implemented with good tactics may have a better chance of success. Ideally, if we want to optimize the results we need to select the best strategy and the best tactics, as is illustrated in Fig. 1.8. This implies that any economic model

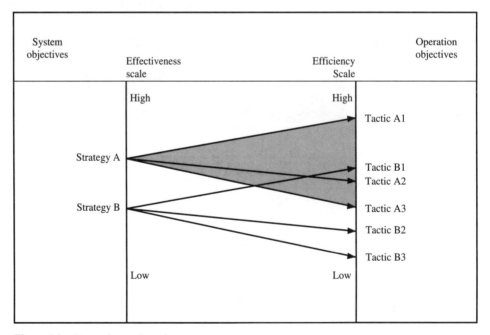

Figure 1.8 Strategies and tactics

that is developed, or obtained commercially, to analyse different test strategies should be capable of modelling the subtle differences between different testers so that the tactics can also be analysed. The modelling approach outlined in Chapter 8 shows how this can be done.

1.9 Does testing add value?

It has become quite fashionable in recent years to profess that testing adds no value to the process or to the product. This concept has almost certainly emanated from the quality revolution of the eighties. During that time, and still today, many thousands of people received large amounts of training on quality. Much of this included the attendance of seminars held by the leading quality experts and the reading of their books or the viewing of their videos. Most of these experts talk about the need to eliminate test and inspection because these activities add no value. Do everything right the first time, every time, and there will be no need for test. They are quite right, of course. It is the correct concept and what we should all be striving for. However, we should not forget that these quality experts are talking about quality in general for all kinds of manufacturing and service industries. The electronics industry is a little different to the average of everything else. That may sound like a bit of an excuse but the electronics industry really is different. We have to live with a situation where complexity is constantly increasing at the same time that physical dimensions are constantly decreasing. This presents problems for design, manufacturing, test and quality, factors that are unmatched by any other industry. The complexity problems are enormous to a degree that people outside the industry find hard to comprehend. I occasionally lecture on quality issues to audiences made up mainly from mechanical and automotive companies. To give them a feeling for the complexity of electronics testing I use the following example.
'You have a VLSI tester that can perform one hundred million tests per second.' This simple fact, though not exactly state of the art, usually causes mutterings of disbelief to begin with. 'Now you need to test a digital device that has 64 input pins by applying every possible combination of "ones" and "zeros" and monitoring the results of the tests on the output pins. At the one hundred million tests per second rate this will take 5850 years!' It takes a bit of believing but it is true. Two raised to the power of sixty-four is about eighteen million million million. Of course we would not attempt to test it in this way, but it does illustrate the problem.

In the mid-eighties I attended one of the quality seminars referred to earlier. This one was presented by Dr Deming, generally considered to be the guru of all quality gurus. He had been talking about eliminating test and inspection so I asked him if he felt that this would be possible in electronics. He said that he thought that electronics was possibly the only exception to the rule because of the constant increase in complexity and the frequent technology changes. He countered with a story of a visit he made to a shoe factory where he talked to one of the senior inspectors. This man was testing the strength of the leather by hanging up the complete hide, punching a hole near the bottom edge and then hanging a weight on a hook through the hole.

'How long have you been doing this?' asked Deming. 'Twenty-eight years', came the reply. 'How often do the hides fail the test?' asked the quality man. 'Never had one go yet', was the reply. This rather silly sounding example was apparently quite true. It does illustrate how easy it can be to retain testing procedures that do absolutely nothing.

There have been dramatic improvements in defect rates in electronic devices and in the manufacturing process. However, the closer you get to the ideal of zero defects the harder it becomes to improve further or even to maintain the gains you have already made. With complexity increases adding to the number of opportunities for obtaining a defect it becomes harder still to improve yields. A 100 parts per million (ppm) defect rate (0.01 per cent) for solder shorts may sound quite good but a fairly average board may have 2000 solder joints on it. Statistically this would result in over 18 per cent of the boards having a solder short, and this is only one possible source of defects.

We are still a long way from the point where testing could be eliminated. The graph in Fig. 1.9 illustrates the problem. This shows the overall fault coverage of the test strategy that is required in order to ship boards with a given defect rate, for different levels of yield out of manufacturing. As an example, if you want to ship boards with fewer than 100 ppm defects when the yield out of manufacturing is 70 per cent then the fault coverage of the test strategy will have to be 99.97 per cent or better. For a yield of 90 per cent out of manufacturing the same output quality could be achieved with a fault coverage of 99.91 per cent. Even if you could achieve a yield out of manufacturing of 99.9 per cent, you will still require a fault coverage of 90 per cent in order to meet the 100 ppm target. These figures are for a single board. For multi-board products you may well need to do much better than this. The obvious conclusion here is that we need *both* high yields from manufacturing and high levels of fault coverage in order to meet the quality goals that are necessary in order to remain competitive.

The answer to the question 'Does testing add value?' I believe is therefore yes. Testing has always been a cost-avoidance activity. We test because it is cheaper to test than to not test. Only when we reach a point where the manufactured quality is so high that the cost of testing is higher than the cost of not testing will testing become a non-value-adding process. We may well have reached that point when we consider the value of performing an incoming test on certain classes of components, but there are still other situations where an incoming test can be justified. We need to remember Dr Deming's story about the testing of the shoe leather and recognize when a test or an inspection stage can be realistically eliminated. Regular analysis of the present and alternative test strategies using detailed modelling is probably the best way to do this.

I usually argue that testing adds value because increased costs are equivalent to negative value. Therefore avoiding increased costs is the same as adding value, and that is precisely what testing does. In a keynote address at the First International Workshop on 'The Economics of Design and Test' (Ambler *et al.*, 1992), R. L. Campbell of AT&T Bell Labs put forward a similar argument. His keynote was entitled 'Creating

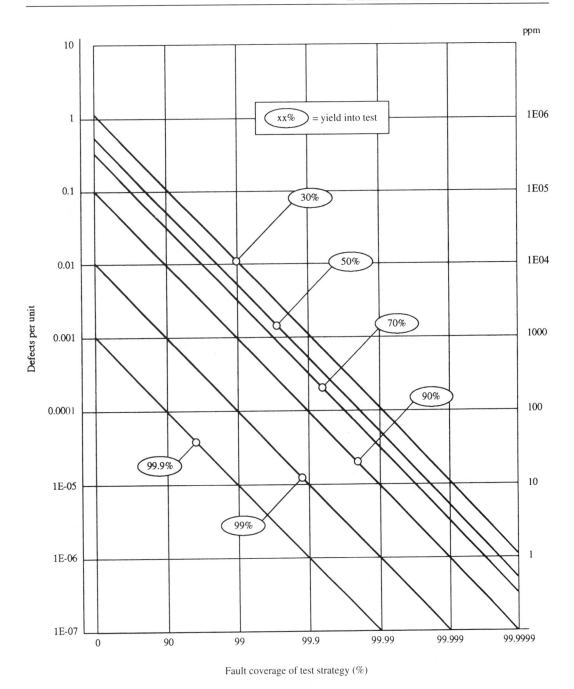

Figure 1.9 Fault coverage requirements versus output quality (DPU) for various levels of process yield

wealth—through testing' and in this he stated that: 'We create wealth when we incur a cost (C) to create a value (V), if $C < V$. The amount of wealth we create is $V - C$, and the relevant measure of our wealth generating engine is V/C—the amount of value created per unit cost incurred.' He went on to argue that value is determined by the customer in terms of what they will pay for the product and their perception of its value. A major part of this perception is the quality and reliability elements of the product. Testing adds value to the *manufacturing process* if it reduces the overall cost of the operation, and testing adds value to the *product* if the customer is happy with the increased quality and reliability that the testing helped to achieve.

1.10   Six-sigma quality

Another way to look at this problem of complexity counteracting the low defect rates that are being achieved is to calculate the board failure rate, assuming that all processes are at the six-sigma quality level. Six-sigma quality is a target that many companies have set for themselves, so what exactly does it mean? Sigma is the Greek letter used in probability and statistics to denote a *standard deviation*, which is generally used as a measure of dispersion about the mean value of a set of measurements. For example, you set up a business to produce 1000 ohm resistors and you take a sample of two from one of the production machines and measure their values to be 900 ohms and 1100 ohms. Your conclusion might be that the mean value is 1000 ohms so you are on the right track. If you now take two from another resistor-making machine and these measure 990 ohms and 1010 ohms, the mean value would also be 1000 ohms, but the sample has much less 'dispersion about the mean'. Unless this is an unrepresentative sample then clearly this second machine will produce fewer defects. There is less variation in its output.

The calculation of the standard deviation is fairly straightforward:

1. Find the arithmetic mean of the values.
2. Find the deviation (difference) of each value from the arithmetic mean.
3. Square each deviation.
4. Add the squared deviations.
5. Divide the sum of the deviations by the number of measurements taken (this result is called the *variance*).
6. Find the square root of the variance and this is the *standard deviation*.

Plotting a reasonably large number of measured values of the results of a process in the form of a histogram will usually result in the familiar *normal distribution* or *Gaussian* curve. This bell-shaped curve is common for a wide range of natural phenomena (such as the height of a person) and also for many manufacturing processes. A few processes can be biased to the high side or the low side of the mean value and these are referred to as *skewed*. A more detailed discussion is outside the scope of this book but you can find this topic in any good book on quality control or statistics. For our purposes the important issue is that a narrow distribution of values will result in a low value for the standard deviation and as the process tends towards making every unit the same then the standard deviation tends towards zero.

Most technical things have an engineering specification and it is usually this that determines what is a good unit and what is a defective unit. Figure 1.10 shows the distribution curves for two similar processes, one of which is under tighter control than the other. The engineering specification is the same in both cases and it is clear that the process with the least variation will produce fewer defects. It should also be clear that for the better process there will be more standard deviations sitting between the engineering specifications. Mathematically, regardless of the narrowness of the curve, a range of ± 1 standard deviation will include 68.26 per cent of the measured values. A range of ± 2 standard deviations will include 95.46 per cent of the measured values. These are the examples shown in Fig. 1.10. In the left-hand curve the 1 sigma points are at the engineering specification points and so this process will produce a yield of 68.26 per cent or 31.74 per cent defects. In the process on the right the 2 sigma points line up with the engineering specification and so this process will have a yield of 95.46 per cent or produce 4.54 per cent defects. For processes with these levels of variation this is the best you can ever expect. Most of the time it will be much worse than this. The above example assumes that the mean value for the process lines up with the centre value of our engineering specification and is stable. In practice the mean value of most processes will shift or drift away from this nominal value because of variations in materials, equipment, temperature, humidity, the phase of the moon or any number of other factors.

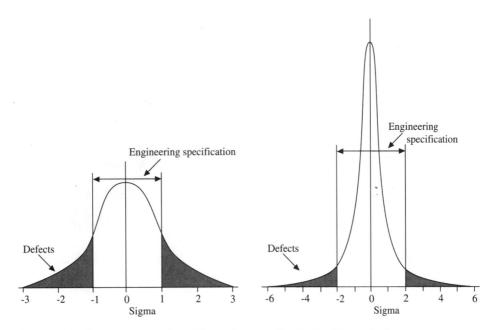

Figure 1.10 Two processes with different degrees of variation (dispersion)

If a process is said to be at six-sigma then the engineering specification is effectively at the ± 6 sigma points and there will be very few defects produced. The six-sigma concept is in fact a lot more than this simple mathematical explanation. Many companies are using the six-sigma concept as a philosophy for their quality improvement process. It is an extremely effective system to teach everyone about the common enemy of variation and an easily understood measure of progress towards superior quality performance and all that it can result in. This aspect of six-sigma is outside the scope of this book. What we need to understand from a test strategy planning point of view is what it means in practical testing terms.

First of all, what does six-sigma mean in terms of the number of defects that we are likely to see if we reach this goal? The example shown in Fig. 1.10 indicated that a process at ± 2 sigma would have a yield of 95.46 per cent. The figure for ± 6 sigma is 99.999 999 8 per cent, which is very good. That is equivalent to 0.002 defects in a million (0.002 ppm). However, this again assumes that the mean value for the process matches the nominal value of the requirements. Because of the inherent shifts and drifts in most processes the six-sigma standard is modified to allow for shifts and drifts of ± 1.5 sigma. This may sound quite a lot, but bear in mind the fact that at this level sigma is a very small number. The result of this recognition that shifts and drifts will occur results in the 6 sigma points being equivalent to a yield of 99.999 66 per cent, which is in turn equivalent to 3.4 ppm. This is illustrated in Fig. 1.11.

Is six-sigma good enough? If we reach this point can we scrap all of our test systems? Can we forget about developing optimum test strategies? Can we tell all of the field-service engineers to go home? The answer is no. Six-sigma effectively refers to the defect rate for one part or one process step. In a complex electronic product there are many opportunities for defects or nonconformities. 'One process step' means

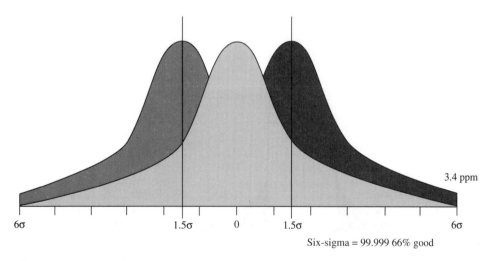

Six-sigma = 99.999 66% good

3.4 ppm

6σ   1.5σ   0   1.5σ   6σ

Figure 1.11   The six-sigma quality standard assumes that shifts or drifts of ± 1.5 sigma can occur in the mean value

just that. It should not be confused with one process stage. If your board contains 2000 soldered joints, that counts as 2000 process steps. There are 2000 opportunities for a defective solder joint and this will result in a yield of 99.32 per cent not 99.999 66 per cent. This calculation is made by raising the probability that one item will be defect-free, to the power of the number of opportunities for nonconformance. In this example this is

$$0.999\,996\,6^{2000} = 0.9932$$

If these boards are assembled into a system containing, say, 20 boards there would be 40 000 solder joints and the yield with the process at six-sigma would be 87.28 per cent. This is obviously a fairly complex product but bear in mind that we are only addressing one source of defects. The point I am making is that, wonderful though it may be, six-sigma will not eliminate the need for testing. Imagine building the space shuttle with all parts and process steps at six-sigma and not performing any testing. There is no way it would ever get off the ground.

Defect rates are now commonly expressed in parts per million (ppm) rather than in per cent (parts per hundred) but we need to be clear about what this means. *Parts per million* is an abbreviation of 'nonconformities per million opportunities for nonconformance' (npmo). Every component and every individual process step is an opportunity for a nonconformance, or a defect, to occur. That includes every step in the design process as well as in the manufacturing process and your supplier's processes.

I hope I am not giving the impression that six-sigma is not a good thing. On the contrary, I believe that we should do everything possible to improve quality to the

Table 1.1 A complex PCB with all elements of the fault spectrum at six-sigma quality (99.999 66 per cent good)

Defect	Power	P (good)	DPU (defects per unit)
Design errors	350	0.998 810 7	0.001 189
Board defects (joints)	2500	0.991 536 0	0.008 464
Board defects (tracks)	500	0.998 301	0.001 699
Wrong component	350	0.998 810 7	0.001 189
Wrong value	200	0.999 320	0.000 680
Misplaced	350	0.998 810 7	0.001 189
Component defects	350	0.998 810 7	0.001 189
Solder shorts	2500	0.991 536 0	0.008 464
Open circuits	2500	0.991 536 0	0.008 464
Thermal shock	350	0.998 810 7	0.001 189
			0.033 716

0.0337 FPB (faults per board), 96.68 per cent yield, 3.32 per cent of boards defective, 1.0169 FPFB (faults per faulty board). For a ten-board product the yield would be 71.38 per cent (50 boards, 18.53 per cent).

six-sigma point and beyond. However, we do have to be realistic about the continuing need to test high-complexity products. The example in Table 1.1 shows how the yield would turn out for a board with all parts and process steps at six-sigma.

1.11   A simple defect occurrence model

One of the most important pieces of information needed in test strategy analysis is the fault spectrum. This is a list of defect types, and the quantity of each, that you typically have or expect to have for a given unit under test (UUT). This knowledge, or at least a good estimate, is needed in order to determine the right kind of test stages and the likely escapes of defects from stage to stage. The more accurate you want your answers to be the more accurate your knowledge of the fault spectrum will need to be. Each expected defect type will need to be separately listed in the fault spectrum because different tester types will have varying degrees of success at detecting each different type of defect. This level of detail is required for the detailed analysis of test strategies and test tactics that will lead to major decisions such as changing your strategy or purchasing a new test system. However, for much of the general discussion of strategy decisions this level of detail is not required so I will

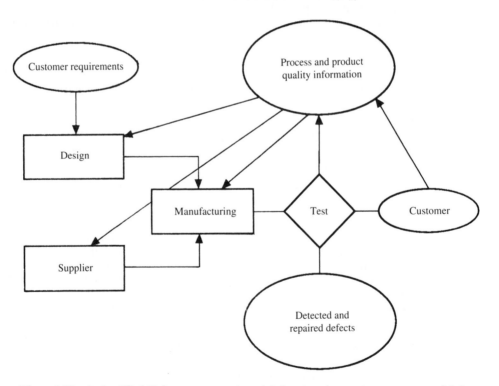

Figure 1.12   A simplified 'defect occurrence' model showing three primary sources of defects

refer from time to time to a simpler fault spectrum that is produced from a simplified defect occurrence model. This model is depicted in Fig. 1.12 and essentially says that there are three basic sources of defects. Defects can be generated by your design process, by your supplier's processes and by your manufacturing process. In other words, your test process (test strategy) will need to deal with design induced defects, supplier induced defects and manufacturing induced defects. The more detailed fault spectrum needed for analysis work is effectively an expansion of this simple three-part fault spectrum.

The overall yield coming out of the production process will be the result of the sum of the defects in these three categories. The job of the production test strategy will be to prevent as many of these defects as possible from escaping to the field by detecting the defects and providing the diagnostic information to enable an effective repair. At the same time the various test stages should collect defect data and repair data for your quality improvement process.

1.12   Forecasting, estimating and guesswork

The economic analysis of test strategies will typically fall into two areas:

1. The analysis of present methods to see if any changes would be worth while.
2. The analysis required to determine the best alternative for some future period or some future product.

The first of these situations is usually relatively straightforward because there will be a lot of real valid data available. This is not the case for the second situation, which is also likely to be the more common of the two. Most of the time this type of analysis will require a forecast of the likely situation in the future, possibly up to five years ahead. This will be particularly true when calculating a *return on investment* for the purchase of some new equipment, since this is most commonly done over a three to five year period. Virtually all analyses involving design or design to test issues will be for some future time and so require a lot of prediction. Even if the project is similar to something that has already been done the inevitable change in component, manufacturing or test technology that will be implemented in the new design will mean that some of the key parameters will change.

As a minimum it will be necessary to predict the number of new designs, their complexity, the production volumes, the process yield and the fault spectrum—quite a tall order. However, it has to be done if you are to have any chance of making the right decisions. If you think about it, almost all business decisions involve a lot of forecasting, estimating, guesstimating or just plain guesswork. The successful managers and the successful companies are the ones that guess correctly more often than they guess wrongly. Forecasting is usually more scientific than pure guesswork and we have to use whatever techniques we can to get close to the right numbers. It is also necessary to look at a range of outcomes and to determine which of the variables have the biggest impact on the results. In this way we can devote more time to determining what these high sensitivity variables might be. It can be tempting to leave something out of the equation because we are not sure of its value, but there are two simple rules that need to be uppermost in the mind of any analyst:

1. If you wait to get all of the data for an accurate result, it is probably too late.
2. If you leave a parameter out of the equation you are effectively making an assumption about its value and its importance.

The first of these rules is particularly valid in the electronics industry where the only thing that is constant is change.

1.13   Business perspectives

Selecting the optimum test strategy by performing a thorough economic analysis is only a part of a much bigger scheme of things. Decisions about the inclusion of 'design for test' features in a new design, the production test strategy, the test tactics that will be employed, the field-service strategy and so on cannot be taken in isolation. These decisions have to be made with due consideration and understanding of the wider business issues. What is driving the market for our products? What are the customers' expectations? How long will the product last? These and other issues have to be understood by everyone in the organization who can make or influence a decision that will affect the performance of the product in the market-place. Understanding the primary market forces that drive the electronics industry is fairly easy because they have not really changed for some years. They have not changed but they have intensified and they continue to intensify. It is very simple (see Fig. 1.13):

> All you have to do, product after product, is to lower the *costs*, increase the *quality*, include the latest *technology* and get it to market on *time*.

If these are the driving forces then they are also the goals that you have to meet, and this is where the problems start. It is generally felt that these goals are in conflict with each other:

1. Lowering costs may cause a lowering of quality.
2. Reducing time to market may lower quality.
3. Introducing a new technology may lengthen the time to market and also increase costs.
4. Pressure to improve quality may slow down time to market and increase production costs.

This can be thought of as the conventional wisdom that is based upon years and years of experience. However, conventional wisdom has been challenged many times in recent years and proved to be wrong. New ways of looking at problems and new ways of solving them have emerged in many areas. The quality revolution discussed

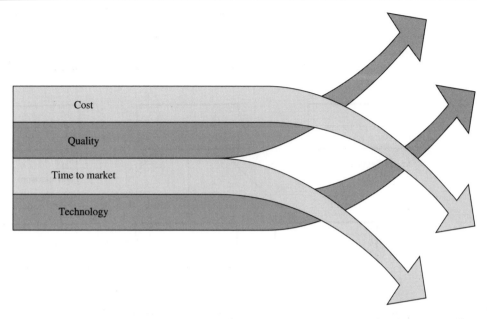

Figure 1.13 The challenge of the four primary market forces—reduce costs, increase quality, reduce time to market and implement new technologies

in the next chapter is perhaps one of the better examples. The conventional wisdom used to be that productivity was all important, but that had a negative effect on quality. The new wisdom says that if you improve quality in the right manner then productivity will increase automatically. The conventional wisdom used to say that you should have many suppliers for a given part so that you do not run out of stock and you can play one off against the other to get the best prices. The new wisdom says that you should have only one supplier of a given part and that you should develop a closer relationship with them to get the best quality. The list can go on and on.

I believe we have a similar situation when it comes to meeting the four market driven goals simultaneously:

1. Cost and quality are not in conflict. The lowest overall life cycle cost for a product will generally be achieved when the quality is at or near the maximum. If we could achieve zero defects by doing everything right the first time, every time, then we would have minimum costs.
2. Reducing time to market, done the correct way, should have a positive impact on quality. Increasing the degree of integration between design and test while taking advantage of the widely available design for test techniques will result in shorter design cycles, more rapid test program generation and higher fault coverage test programs. Concurrent engineering saves time and improves quality.

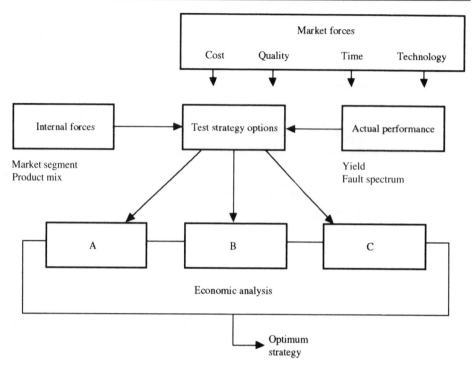

Figure 1.14 The four primary market forces influence the choice of test strategy because they are the objectives that we must reach in order to survive

3. Incorporating a new technology into a product should improve the performance and lower the cost. It need not increase the time to market if the design and test integration is done well with the right tools in place.
4. Your quality improvement drives should result in lowered costs and improved time to market. Doing things right first time costs less and takes less time. Since the constantly increasing complexity of electronic products means that we cannot yet eliminate testing then we also have to do the *right testing right*, first time and every time.

There is therefore no need for there to be any conflict between the four major market forces (see Fig. 1.14), but to reach this situation will require careful planning and implementation of all the pieces that are needed to operate an electronics operation in today's highly competitive market.

A change of emphasis

The four forces of cost, quality, time and technology have been the main driving forces for the past twenty to thirty years, but not always with the same degree of importance. In the late sixties and throughout the seventies the main driving force for many companies was cost reduction. Thanks to the teachings of the quality experts we now know that this drive to reduce costs and increase productivity was a major contributor to the low levels of quality that were common in those times. Component defect rates of between 1 and 5 per cent were commonplace, and the process yield of boards out of manufacturing varied between 50 per cent and zero—and these boards were incredibly simple by today's standards. The eighties were without doubt the decade of quality awareness. The quality driving force became number one in many organizations, and rightly so. The improvements made by the really committed companies were astounding, and they all proved that the good Dr Deming and his peers were right. Increase quality and productivity does go up, and so does profitability and market share. Unfortunately a lot of the progress that was being made came grinding to a halt in 1985 when a major recession hit the electronics industry with a speed and ferocity that had never been experienced before. The ATE industry took a double hit. Not only did they suffer from the effects of the recession but they also began to suffer the effects of the industry's quality improvement. When production yields increase the world needs fewer testers. When you reduce the amount of re-testing after repair, the capacity of the testers increases considerably.

We now have a mixed situation with regard to the quality revolution. The companies who implemented the new ideas early on have tended to stick to the program because they have benefited from it and continue to benefit. Unfortunately there are also a large number of companies who only pay lip-service to the new quality ethic. They talk about it a lot because they know that their customers will disappear if they do not, but internally the efforts are often only superficial. The recessions we have had in the past seven years are partly to blame for this. There is so little money to spare that some companies cannot spend any of it on major quality improvement processes even if they fully believe in the longer term benefit. There is never a good time to implement a good idea.

However, there is still a feeling in the industry that if you do not already have your quality act together then it is probably too late. The big issue for the nineties is *time to market*. This has now risen to the top of the pile of forces. Time to market, or for some companies *time to volume*, is now receiving much attention. The competitiveness of the industry and the rapid changes in technology are the forces behind the force. Every company is trying to leap-frog its competitors by being first to market with new features brought about by new technologies at lower prices. What about *technology*? Where does that sit in the hierarchy of the four forces? Some would argue that it is really on top. Others would say that it sits in the background. It depends on your point of view, but there is little doubt that this is the force that drives the other forces. The quality improvements we have seen have come partly from the education we have all received either directly or indirectly from the quality gurus, but many of the improvements have only been made possible by technological advances. Advances in EDA/CAE (electronic design automation/computer aided engineering) have made it possible to design bigger devices that improve overall quality by replacing many smaller devices. Advances in manufacturing automation and robotics have led to fewer assembly mistakes. Surface mount technology has resulted in a more automated, more consistent process. Advances in soldering technology for both the 'through hole' and surface mount boards has led to far fewer solder defects and testing technology has advanced to cope with the changes and increases in complexity. Technology has driven down the costs alongside new manufacturing approaches such as JIT (just in time), and it is technology that is making it possible to reduce design cycle times and test program generation times to shorten the time to market.

The strategy hierarchy

These are the forces that should shape the thinking behind every major strategic decision. They should remain embedded in our brains in such a way that each time we are faced with a decision to make we always question its impact on the four forces. This is just as valid if we are developing a new test strategy or the corporate strategy for the company. There is a hierarchy of strategies within any organization that should fit together in a cohesive manner such that the corporate strategy is supported properly by all of the strategies beneath it. The development of the corporate strategy should be the starting point. I mentioned earlier in this chapter that the job of the senior management is to ensure that the owners of the company, the shareholders, get a good return on their investment. They do this by making the right decisions about:

1. What business to be in.
2. Marketing strategy—products, promotion, pricing, distribution, etc.
3. The use of capital.

The corporate strategy is essentially about what business to be in and the markets to pursue. Typically the corporate group will define the business they are in and then analyse such things as the market opportunities, the overall capabilities of the company, their competitive position and the strengths and weaknesses of the company. They will need to determine if they want to be a leader or a follower in their chosen markets and so on. None of this analysis will be worth anything unless it is done with the four forces in mind. For example, if a market opportunity is identified it will be necessary to determine the customer's expectations in that market and then determine if the company's capabilities and strengths match those expectations. 'Can we really meet the cost, quality and time to market requirements of this market?' 'Do we have the capability to implement the technologies needed in the time available?' 'Can we support the customers properly in the geographical areas we have targeted?' These and other searching questions all relate to the four forces.

Once the corporate strategy is finalized it will be communicated to the various business units within the organization. They should then go away and develop their own individual strategic plans, which should support, or add to, the corporate plan. These business unit plans will include the specific marketing plans defining the products that need to be developed, the price they should sell for, when they should

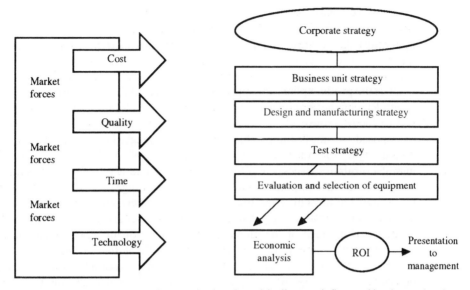

Figure 1.15 The hierarchy of strategic planning with all stages influenced by the market forces

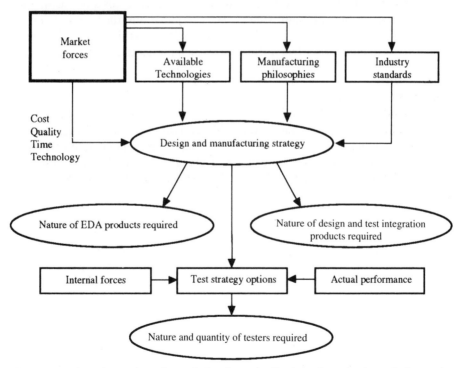

Figure 1.16 An alternative view of the factors affecting the selection of the optimum test strategy

be available, what their profit contribution should be, what the support plans are and so on. Again this kind of specific planning needs to be done with the four forces uppermost in the minds of the business managers of the business units. The same questions have to be asked for each part of the plan. 'Can we do it at the right price, with the right quality, at the right time'. 'Can we handle the technologies involved?'

Once the product plans are approved the design and manufacturing strategy needed to implement the plan can be established (see Fig. 1.15). An important part of this will be the test strategy. Here I am using the term 'test strategy' in its broader sense as opposed to simply the production testing strategy, including decisions about the design style, the inclusion of DFT, the use of ASICs, other testability and manufacturability issues, etc. Out of this will come a set of requirements for CAE (computer aided equipment) and design to test integration tools. At this point the future and the present come together. All of the new product plans and the design and manufacturing strategy now have to be brought together with the reality of the current performance of the manufacturing operation in terms of the yield and the fault spectrum. These and the internal forces of the *nature of the markets* served and the *mix of products* will all have to be taken into account before deciding on the optimum test strategy. Once again all of this decision making has to be constantly tested with the four forces. We have also reached the point where economic analysis really comes into its own. By analysing the life cycle costs of the alternatives we can zero in on the optimum strategy. We can make sure that we do the right testing the first time and every time. This is essentially what this book is all about. Hopefully the techniques described will also be useful for other engineering economics problems.

Reference

Ambler, A. P., M. Abadir and S. Sastry (eds) (1992) *Economics of Design and Test—For Electronic Circuits and Systems*, Ellis Horwood Limited Chichester. This publication contains extended versions of the papers presented at the First International Workshop on 'The Economics of Design and Test' which took place in Austin, Texas, in September 1991. The Workshop was sponsored by ACM/SIGDA.

2. The quality revolution

2.1 Background

Since the first edition of this book was published in 1982, changes have taken place in the understanding and the importance of quality that have been nothing short of revolutionary. There was nothing new about the concepts. Quality experts had been preaching about the right way to view and to achieve quality for years, but relatively few companies were willing to listen. Probably the main reason for this was the poor understanding that senior managers had about what quality really was. The comments in Sec. 1.1 headed 'What cost quality?' in Chapter 1 remain completely unchanged from the first edition. The problem was that most managers believed that better quality could only be achieved at higher cost. The revolution began when senior management became enlightened about quality and what it could do for them. The changes that took place were revolutionary rather than evolutionary because of the sheer speed with which they occurred.

In the electronics industry the revolution was a two-stage process in that the main driving force behind the movement changed. It began as a direct response to the competitive threat from Japan. In the late seventies it had generally been assumed that the Japanese copies of semiconductor devices whose designs had originated in the United States were rather second rate. The shock that started the quality revolution in the electronics industry came when several major manufacturers published the results of some internal test programs. The initial reports stated that the Japanese copies were between 5 and 50 times better than their American counterparts in terms of both the initial defect levels and the longer term reliability as shown by accelerated burn-in procedures. Following these revelations other manufacturers conducted their own tests and came up with similar results. Confidence in the suppliers was naturally shaken. To make matters worse there were several highly publicized cases where leading component suppliers had made 'mistakes' and shipped virtually untested devices as being high reliability military grade components. Some of these 'mistakes' had been occurring for several years so many of the suspect devices were already in active service. Independent test houses also fell into disrepute for having inadequate controls to prevent the shipping of untested components. In one case a greedy test house was found to be shipping far more devices than they had the capacity to handle. Components were simply marked to indicate that they had been tested and then shipped a week or two later to the customer who had paid heavily for the service. This particular case resulted in prison sentences for several managers at the test house concerned.

The American semiconductor industry responded with amazing speed and effort. They quickly recognized the seriousness of the challenge they were facing but they were still too late to prevent the Japanese companies from taking the lion's share of some parts of the market. The response of the US semiconductor industry was both practical and verbal. They started to improve their processes but they also started talking about quality. Major advertising and promotional campaigns were run. You could not open an electronics magazine without finding articles on quality and promotional material that explained how the challenge from Japan was being met.

This then was the first stage of the quality revolution: a response to a major competitive threat. It led to a demand for education about quality at all levels and a desire to find the secret behind the Japanese ability to produce high-quality products at a lower price.

Shortly before the revelations about the quality of Japanese semiconductors were made public, a book was published that was to have a major impact on stage two of the revolution. This was *Quality Is Free* by Philip Crosby (1979). Crosby had been director of quality for ITT for fourteen years and had held various other quality positions before that. He had developed a very common-sense view about the importance of quality to the performance of a company, and how you should go about achieving better quality. His book was a best seller. The main contribution that it made, in my opinion, was to show senior level management that quality was indeed free and that it is cheaper to produce high-quality products than to produce poor-quality products provided that you use the correct definition of quality and that you use the right approach to achieving it. That approach is 'defect prevention' as opposed to the 'detect and fix' approach that was widely used. In this respect Crosby and all of the other quality gurus are in full agreement. The concept applies to all products and services. At the time ITT was a major conglomerate and Crosby was responsible for quality in areas as diverse as electronics, cosmetics and hotels.

Quality Is Free and supportive articles in business magazines showed senior management the direct positive relationship that exists between quality, productivity and profitability. This was in complete contrast to many earlier beliefs that high quality costs more and takes longer, so having a negative impact on profitability and productivity. The revelation that the cost of the poor quality being achieved could be as high as 20 to 25 per cent of annual revenues led to the obvious conclusion that quality improvement could increase profits.

This then started stage two of the quality revolution. What had begun as a direct response to a competitive threat quickly changed to a profit-driven activity based on good business sense. At the same time, however, management did not lose sight of the fact that quality improvement was very necessary for survival. Quality quickly became a major basis for competition and regularly scored higher than such issues as price or performance in surveys about purchasing decisions.

2.2 The Japanese secret

Early on in the quality revolution many people in the United States tried to discover the secret behind Japan's quality success. After all it had only been some twenty years earlier that the expression 'that's a Japanese copy' had been synonymous with 'that's cheap rubbish'. What they discovered was that the improvement was due to a large extent to the teachings of a small number of US based quality experts—people like W. Edwards Deming, Joseph Juran and Armand Feigenbaum. There were others also but these are the three that are most often cited for their early work in Japan. Of the three Dr Deming is usually given the most credit and Japanese industry gave his name to their most prestigious quality award.

All of these experts had been active in the United States but very few listened to their ideas. Productivity was the name of the game, not quality, and if you drive productivity up quality will inevitably fall. The same was true in Europe. Almost all of the industrial disputes in the previous thirty years or so had some sort of productivity agreement included in the settlement terms. Dr Deming and the others were trying to show that if you try to drive productivity up, quality will be driven down. In contrast, if you drive quality up then productivity will rise automatically. This becomes very obvious if you look at the two production lines illustrated in Fig. 2.1. The lower line has the potential to create defects at each stage of the process. To correct for this we have test or inspection stages and repair or re-work stages. A repaired product

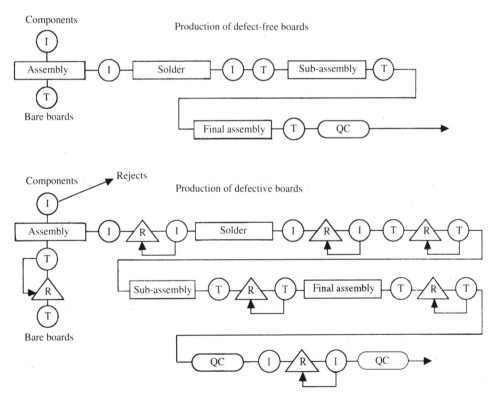

Figure 2.1 The two production lines in most factories

is re-tested before passing on to the next stage. If the re-test fails the product goes back to the repair station. The top line makes no mistakes. No defects are added and many of the inspection and test stages could be eliminated. It should be fairly obvious which of these lines will be the most productive. This may appear to be an extreme example today but this is how it was in the seventies and the early eighties. The more controlled process of today would tend to have fewer inspection and test stages but if the manufacturing process can still generate defects this simply means that some will be detected later rather than immediately. It is therefore clear that we all effectively have both of these lines in our factories. If you have a yield of 80 per cent good boards then 80 per cent of your production is manufactured on the upper line and 20 per cent of your production is manufactured on the lower line. If you could move all of your production to the upper line, through quality improvement, then your productivity would increase automatically.

Dr Deming was invited to Japan in the early fifties to lecture on his ideas and concepts. Joseph Juran, possibly best known as the editor of McGraw-Hill's *Juran's Quality Control Handbook* (Juran and Gryna, 1988), also worked in Japan around this time. Armand Feigenbaum is perhaps less well known than Deming and Juran but he has been particularly influential in one very important concept. When most other quality experts were concentrating their efforts on the quality of the generic product or service being offered, Feigenbaum was taking a much broader view. His book *Total Quality Control* was first published in 1951 by McGraw-Hill and was later translated into Japanese. It is now generally accepted that TQC, sometimes also referred to as 'total quality commitment', is the right philosophy on which to base your quality process. Feigenbaum describes the need for TQC in the following way: 'The goal of competitive industry is to provide a product and service into which quality is designed, built, marketed, and maintained, at the most economical costs which allow for full customer satisfaction.' He goes on to define TQC as an 'effective system for integrating the quality development, quality maintenance, and quality improvement efforts of the various groups in an organisation so as to enable marketing, engineering, production, and service at the most economical levels which allow for full customer satisfaction'.

The inclusion of marketing is one area where Feigenbaum's concepts differed from many others. 'What has marketing or sales go to do with quality' was a typical response. Well, that all depends on how you define quality.

2.3   What is quality?

Looking up the word 'quality' in a dictionary is of no help. Most dictionaries offer similar nebulous definitions that relate mostly to what is generally accepted by quality management professionals to be the wrong meaning of the word—expressions such as 'a distinguishing characteristic or attribute', 'the basic character or nature of something or someone', 'the degree or standard of excellence, especially a high standard', 'musical tone colour or timbre', etc. Examples of common usage include 'she has many excellent qualities' which probably should be written as 'she has many excellent attributes' and 'this is a high-quality product' which usually means 'this is a high-grade or luxury product'. You cannot really blame the compilers of dictionaries for this. They are simply reflecting the common use of the word in the English language.

It is the last definition quoted above that leads to the most common of the misunderstandings about quality. 'Degree of excellence or grade level' implies that high quality is usually synonymous with high price. This leads to the 'Rolls-Royce' image of quality referred to in Chapter 1. Other misunderstandings include the equating of an attribute of the product to its quality even when the attribute has no bearing on its functionality. A common example of this is the weight of an object. Heavy products are frequently perceived to have better quality than lighter ones, even when weight is not a necessary attribute to the performance of the product. In fact, excess weight may even be detrimental to the product's performance or the result of poor design.

The most commonly accepted correct definitions of quality, for our purposes in electronics manufacturing, are all based upon the concept of 'conformance to requirements'. Earlier definitions used the term 'fitness for use' but essentially they mean the same thing. If a product meets all of the requirements of the user it is fit for the use it was intended for and is therefore a quality product. If a Rolls-Royce motor car conforms to the requirements of its owner, then it is a *high-quality* product as well as being a *high-grade* product. Similarly, if a relatively inexpensive family saloon car conforms to the requirements of its owner then it, too, is a *high-quality* product even though it is lower in *grade* or *luxury level* compared to the Rolls-Royce.

Although 'conformance to requirements' is the generally accepted correct definition, there is another way of saying this that I personally feel suggests a more complete definition. This is simply that if a product or a service 'meets the customers expectations' then it is a quality product. It has to conform to requirements in order to meet expectations, but this definition broadens the view we should have about quality and clarifies why Feigenbaum is right about including marketing and service is his definition of TQC. Feigenbaum defines product and service quality as 'The total composite product and service characteristics of marketing, engineering, manufacture, and maintenance, through which the product and service in use will meet the expectations of the customer'.

The conformance to requirements definition includes marketing to some degree because the product should be specified by marketing to engineering based upon their market research. If they fail to include a feature in their specification that the customer needs, then the product cannot conform fully to the customer's requirements. The product will therefore be of poor quality no matter how well it is designed and manufactured. Maintenance is not considered to influence product quality by some definitions. This probably stems from an assumption that the product quality has been established by the time you ship it. However, the customer does not see it that way. The customer's expectations are not limited to the generic product or service. He or she will also expect to get complete and effective documentation with the product. If training is required an adequate amount at a standard necessary to operate the product efficiently will be expected, as will other forms of customer support as and when needed. In total the customer's view of your quality goes well beyond that of the generic product itself. Indeed, it will even extend to such seemingly mundane issues as how your telephonist answers phone calls, how well typed and laid out your letters are, the style of your advertising, etc.

This then is the concept of TQC. It is indeed *total*. It includes everything you do and how you do it. If we return for a moment to the subject of marketing and how it can effect quality, I think I can show why I feel that the *meeting customer's expectations* definition encompasses more than the *conformance to requirements* version. If you want to meet your customer's expectations then it is vital that you set those expectations correctly in the first place. We generally set out customer's expectations with the promotion and the selling of our products. Advertising, brochures, direct mail, exhibitions, seminars and many other methods are used to promote our products and therefore set the expectations of the customer. If, as a result of this promotion or a conversation with one of your sales people, the customer's expectations are set higher than they should be then the customer will be disappointed when the product is received. The perception will be that the product is of poor quality because the customer did not get what was expected. If this happens there are usually only two courses of action open to you. Either you take back the offending product, even though it may conform fully to its engineering specification and have been manufactured to the highest possible standards, or you modify the product to give the customer what was expected. Both of these courses of action are expensive, but the alternative would be to lose the customer altogether and that could cost even more.

This example should clarify how sales and marketing can influence the customer's perception of quality, even when the product has been specified correctly by marketing, designed correctly by engineering and built correctly by manufacturing—zero defects but poor quality. Do not forget that when it comes to *perception of quality* the customer has all of the votes.

Similar examples can easily be developed to show how the various elements of customer support, not simply the maintenance function, can affect the customer's perception of quality and your ability to retain the loyalty of your customers. We should never forget that the customer usually has a choice. The customer can almost always go somewhere else for products or services like yours. An example from my own personal experience involved a major world-wide chargecard company. I had used this particular card for about sixteen years for all of my business travel expenses and had always paid my bills reasonably promptly. One day I received a rather curtly worded letter from them chasing a payment. I checked my records and found that I had paid the bill two weeks earlier so I telephoned them. I was put through to the 'customer services' department, which I thought was a rather curious name for a department whose sole function was to chase overdue accounts. However, I explained the situation, the computer was consulted, and I was told that they had not received the payment. I was also told *why* they had not received the payment. Apparently there was a localized postal strike which was preventing some mail from reaching them but which did not seem to be affecting their mail getting out. I was told, 'Don't worry, we understand the problem, just ignore the letter.' So I asked why, if they were aware of the problem, they had sent out the letter in the first place. 'Oh, the computer sends out the letters automatically', came the cheery response. I said, 'That's OK, just make sure it doesn't send any more of its nastygrams to me.' 'Yes, we can put a stop to it', came the reply. Over the course of the next week I received two more letters and a telegram. As instructed I ignored the letters but I could not ignore the Western Union telegram. I telephoned again. A different but equally cheery voice said, 'Yes, the postal strike is still causing a problem, just ignore the letters.' 'What about the telegram.' 'Ignore it, the computer sends those out as well.' 'But I was told that the nastygrams would be stopped.' 'Yes, we can do that.' 'I'm glad to hear that because I don't like receiving threatening letters. If I get another one you can have your card back.' 'Don't worry, I'll see to it right away.' Five days later I received another letter, this time threatening legal action. I have never used their card since. Despite letters telling me how difficult my travelling would be without their card I have never had a problem using the card I got to replace it. I will not name the offending card company but I did prove one thing. You can leave home without it!

This is a good example of setting the customer's expectations incorrectly. If they had told me that the computer was programmed to send out the letters and there was nothing they could do about it, I would probably have accepted the situation and still be using their card today. However, I was told on two occasions that they could stop the letters and they failed to do so. I got annoyed and cancelled the card because there were several alternatives for me to choose from.

2.4 The lessons learnt

Philip Crosby defines five stages of managerial maturity with regard to quality. Stage 1 is a state of 'uncertainty'. At this stage quality problems are severe but nobody knows why. The cost of quality, or rather the cost of the lack of quality, is probably around 20 per cent of sales, but no one knows this.

Stage 2 is 'awakening'. People question the need to suffer the quality problems but management are unwilling to spend any money to get their quality act together. The main emphasis is on appraisal and getting the product shipped. The reported cost of quality is between 3 and 5 per cent, but is actually between 16 and 20 per cent.

Stage 3 is 'enlightenment'. Someone has worked it out and has convinced management about the need for quality improvement. Problems are being identified and fixed at source. Cost of quality is reported to be around 8 to 10 per cent, but is actually about 12 per cent.

Stage 4 is 'wisdom'. Management now understand all of the quality issues. They actively participate in the program and recognize their personal role in establishing a culture of continuous improvement. Defect prevention has become a routine part of all processes. The cost of quality is reported to be around 6 per cent, but is actually 1 or 2 per cent higher than this.

Stage 5 is 'certainty'. Everyone in the organization knows why they have little or no real quality problems. Cost of quality is reported to be between 2 and 3 per cent and that is what it actually is.

By the time an organization reaches the 'wisdom' stage there are four lessons that will have been learnt. Crosby refers to these as the four absolutes of quality:

1. The definition. The definition of quality is *conformance to requirements*, or meeting your customer's expectations.
2. The system. The system established to achieve quality should be based upon the *prevention of defects*—do it right the first time (DIRFT).
3. The standard. The standard to aim for is *zero defects*. Continuous improvement is essential.
4. The measurement. The measurement of progress in your continuous improvement (CI) efforts is the *cost of non-conformance*—the cost of not meeting your customer's expectations.

The prevention of defects requires constant monitoring of the various processes involved in defining, designing, manufacturing and supporting your products. The processes have to be analysed and understood in great detail so that you can effectively determine the root cause of defects. Statistical process control methods are widely used to monitor a process and make judgements about how well under control it is, but this does not usually indicate the cause of any process variations. A set of new tools emerged from Japan in the early eighties, which are usually referred to as the 'seven new tools for synthesis'. These are now quite widely used in the United States to determine the cause of quality problems. A description of these tools is beyond the scope of this book but what we do need to understand is what they are trying to achieve. These, and the more conventional tools such as Pareto diagrams, control charts, scatter diagrams, etc., are all used to determine the amount and the cause of variation in a process. Variation is the enemy. If we can eliminate variation then we are in control of our quality.

According to Crosby's 'absolutes of quality', the standard to aim for is zero defects. However, Dr Deming says that 'zero defects is not good enough—we have to do better than that'. When I attended one of his four-day seminars in the mid eighties the great man kept repeating this statement. Eventually after two days someone in the audience plucked up the courage to ask him how we could achieve better than zero defects. His response was quick and to the point. 'What is a defect?' he growled. He went on to make the point that we usually define defects ourselves. We define the engineering specifications. Defects occur when the capability of the process is not up to meeting the engineering specification consistently. Many defects are failures by degree. A resistor is outside its 5 per cent tolerance, the leakage current of an op-amp is out of spec, the propagation delay of a digital device is too long, etc. Defects of this type usually follow a normal or Gaussian distribution pattern, as shown in Fig. 2.2. If the measured values of a parameter are plotted as a histogram it will usually follow this bell-shaped outline. Those devices that fall outside of the engineering specification are deemed to be defective and are discarded.

As indicated above, we have a situation where the process capability is not good

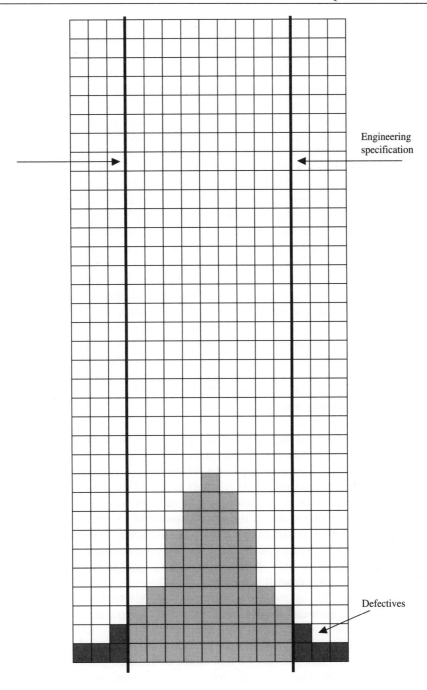

Figure 2.2 A process that cannot match the engineering specification—too much variation

enough to make all of the devices within the engineering specification. The variation in values results in defects. If we could improve this process to the point where the extremities of the variation in values just fits within the engineering specification points we will have reached the zero defects point. This situation is shown in Fig. 2.3. Here we see the results of the measurement of a population of 64 components plotted as a histogram. There are no rejects so we have reached our goal. But have we? Unfortunately any process that exhibits variation will almost certainly also exhibit shifts and drifts in the central or mean value of the measured parameter. This fact of life is accepted in the definition of six-sigma quality, referred to in Chapter 1.

The result of such a shift or drift is shown in Fig. 2.4. Here we see 64 components with the same degree of variation in values as in Fig. 2.3 but with the mean value changed. Now the process has produced three defective devices even though the degree of variation is the same. We can see what Dr Deming meant when he said that we have to achieve better than zero defects. We have to go beyond the zero defect point to such a degree that any normal shifts and drifts in the mean value will still not result in a defective device being produced. This situation is shown in Fig. 2.5 and requires much less variation in the process. The objective is to keep the mean value midway between the upper and lower specification limits and to monitor this closely. If the mean value shifts or drifts this change will be spotted quickly and corrective action can be taken before any defectives are produced. In this example the mean value can shift or drift by plus or minus two cells of the x axis and there will still be no defects. Notice in this example how the 64 items form a taller but narrower curve as the variation reduces, thus implying a low value for the standard deviation. This in turn will mean that there will be more 'sigmas' between the engineering specification limits.

This need for constant monitoring implies a changing role for the testing process. For defect prevention and constant monitoring of the manufacturing process, test systems need to provide data to the quality management system. The automatic test systems are in the best position to do this because they know the quality status of the items being manufactured. Is it good? Is it bad? If it is bad, why is it bad? The three basic issues for manufacturing management are *cost*, *quality* and *schedule*. However, the quality status of the work-in-progress will have a major impact on its cost and its schedule possibilities. Only if you have accurate information about quality can you hope to have accurate information on cost and schedule. The quality status of a printed circuit board will determine where it goes next, how long it will remain in-process and what costs it will incur.

Commercial test systems are beginning to reflect these new requirements by providing the essential quality data. However, generating masses of data is not necessarily going to help. You need the right kind of data and you need it in the right form. If the test system simply provides data about defects it is not going far enough for two main reasons. First of all the defect data generated by the test system may be incorrect. The diagnostic accuracy of test systems is not perfect by a long way. Frequently the repair action that needs to be taken to fix a problem differs from the diagnosis made by the tester. Also, the diagnostic resolution may be poor in some

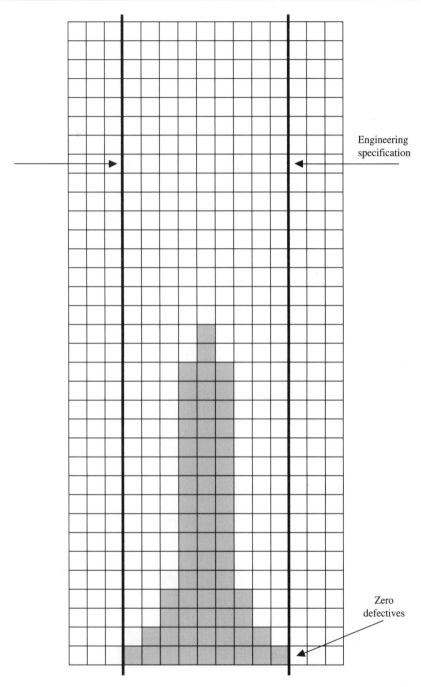

Engineering
specification

Zero
defectives

Figure 2.3 'Zero defects', when all items produced are within the engineering specifications—
less variation

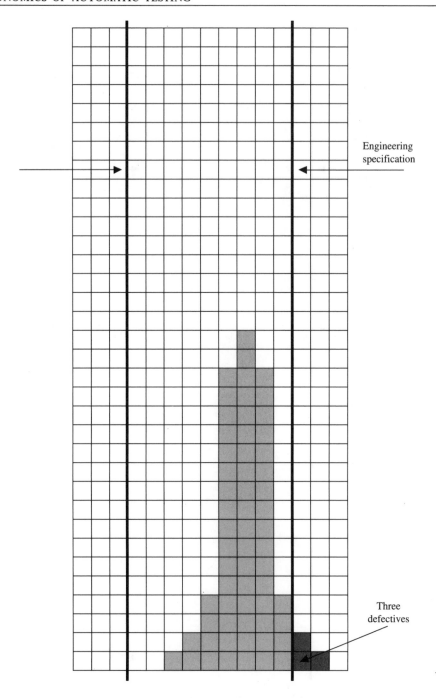

Engineering specification

Three defectives

Figure 2.4 The same degree of variation as in Fig. 2.3 but with a shift or drift in the mean value. Defective items produced

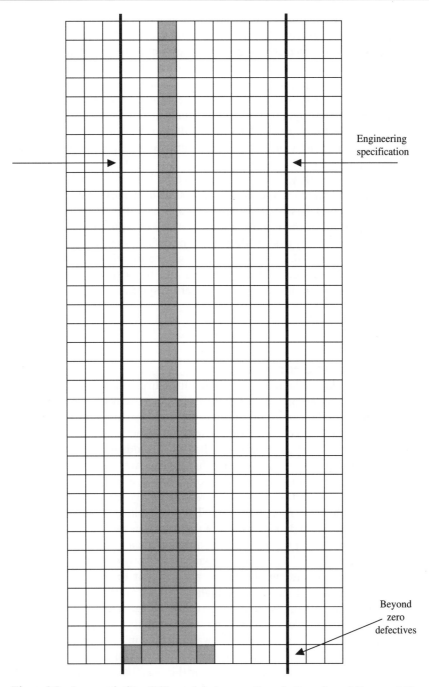

Figure 2.5 Less variation. Still no defects even though there is a shift or a drift away from the mean value. Beyond zero defects

situations. For accurate quality information which can be used to track down the cause of the defects you need to have verified data. You need to know what repair action fixed the problem, so the repair process has to be included in the data collection scheme as well. The second major problem concerns the nature of the data. Simple defect data will help you to determine the cause of defects and, by Pareto analysis, which defects you should track down first. However, as you approach zero defects you get less and less data to analyse. Eventually if you reach the zero defects point there will be no defect data at all. There will still be variation, and shifts and drifts, in the process. At this point we need measured values rather than simple pass/fail or good/bad information.

Some of the quality data will have to come from the process itself because some defects do not manifest themselves as a shift in a measurable parameter once the zero defects point has been reached. A good example is the common solder short. You can plot all of the usual quality charts by counting the number of shorts on a batch, daily or weekly basis to determine the defect rates, the variation and the trends. Once you get to zero defects there is nothing to count any more. There is still variation, shifts and drifts in the soldering process and this can now only be monitored by monitoring the variables within the soldering system.

2.5 Quality and test economics

There is complete agreement among quality gurus on one point in particular. This is the need to base your quality improvements system on the prevention of defects rather than simply adding more quality control, testing and inspection in order to detect more of the defects. This view of test and inspection as theoretically superfluous operations led in part to the concept that testing adds no value to the manufacturing process or the product being made. This concept was discussed in Chapter 1.

The electronics industry had traditionally used a 'detect and fix' approach prior to the quality revolution because it was generally assumed that it would be too expensive to build it right first time. The availability of automatic test equipment that could detect and diagnose most of the defects in seconds provided a cost effective solution to bringing the quality up to par following a relatively sloppy manufacturing process. In a way the ATE helped to make this sloppy manufacturing process a viable operation. Component suppliers were also content to ship relatively simple devices with defect rates of 1 per cent or higher, at least until the Japanese showed what could be done. When the quality revolution began, many of the early 'gains' in component quality were achieved by 'testing quality in' rather than by improving the process. In the early eighties it was not uncommon for users to find that the spread of values followed a 'double-humped curve' rather than the expected 'normal' or 'Gaussian' distribution. The reason for this was that the component suppliers were selecting the better devices to ship to their bigger customers who by this time were already insisting on better quality. The remaining devices, still mostly within specification, were then shipped to the smaller customers. This phenomenon is illustrated in Fig. 2.6.

Unfortunately there are still many companies that have not yet seen the quality light and are still using the defect and fix approach. This worries me greatly because the amount of publicity that quality has had throughout the eighties and the early nineties has been enormous. Indeed, as I pointed out in Chapter 1, for many companies quality is now taking second place to 'time to market' as the most important of the four market forces. This does not reflect a reduction in the importance of quality, rather it reflects a situation where the quality problem is virtually solved so they are now moving on to solve the time to market problems. Among all of the publicity on quality there have been many success stories showing how companies have improved their financial performance as a direct result of their quality process. Therefore, why, with all of this proof around, are there still companies who do not understand or believe in the quality ethic? Some of the answer lies in poor communications or training. When I give training courses or presentations I often ask how many people in the room have heard of Deming, Juran, Feigenbaum or Crosby. All too often the show of hands is pitifully small. It is easy to dismiss this situation by declaring that these are the companies that will not survive, and indeed many of them will not, but

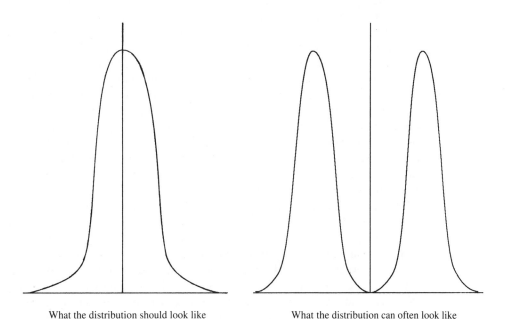

What the distribution should look like   What the distribution can often look like

Figure 2.6 The double humped distribution curve, caused by selecting closer tolerance devices from a large batch in order to offer a choice (e.g. 2, 5 and 10 per cent) with premium prices for the better devices. This can also occur when devices are selected for shipment to favoured customers

perhaps quality has become a victim of its own success. Too much publicity and too much hype can often put some people off. They dismiss it as the latest fad.

I have also had some people tell me that they cannot improve their quality beyond a certain point because of their company's dependence on service revenues for profitability. This kind of comment is symptomatic of a lack of understanding about the degree to which the industry has changed. Quality *is* a vital part of being competitive. Customers *do* demand better quality and lower costs. Poor quality *does* cost more than good quality. These and all the other reasons that are quoted as justification to move to a TQC–CI (total quality commitment–continuous improvement) approach to running a business are all true. The weight of evidence is far too great to be denied.

Another reason why some companies simply pay lip-service to quality, without really doing much about it, stems from the problems of 'short-termism'. The arbitrary way in which a company's financial performance is judged in annual or quarterly periods means that there is never a good time to implement a good idea. Short-term profitability considerations always seem to override the longer term improvements that can be achieved.

Yet another reason for a less than 100 per cent commitment to TQC is that the concept of the business benefits that are available just does not get fully understood. The concept is really quite simple. Dr Deming uses a brilliantly simple analogy to explain it. He says that '...industry has got used to burning the toast and then scraping it'. The brilliance of this simple analogy lies in the fact that it encompasses all the key points about quality, productivity, profitability and market share. The toast is the product and the toaster is the process. The fact that the toast gets burnt implies that the product is spending longer in production than is necessary. This increases costs and reduces the potential capacity of the process. Scraping the toast is a repair or re-work operation. Potentially unnecessary, this increases production costs and capital equipment costs because you need a knife to scrape the toast with. Inventory costs also increase as a result of all the extra work-in-process. Test and inspection costs increase because of all the re-test required for toast that has been re-worked. Scrap costs also increase because there is only a small difference between scraping and scrapping.

Eventually, having scraped all the toast and re-tested it you can give it to the customer. Productivity has clearly suffered and since fierce competition prevents you from passing on the extra costs to your customer, so has profitability. It is worse than that. The toast you deliver is still not as good as toast that was made properly in the first place. It will still retain some of that burnt taste even after the scraping and it will of course have gone cold. The result is that your customers will gradually drift away and buy their toast from someone who makes their toast right the first time. Your market share now begins to decline and the inevitable pressure on profitability will result in the need to lay off some of your staff. You start down a spiral of decline that can only be reversed by fixing your quality. Other approaches will only have temporary effects. You can increase your promotion, cut your prices, offer free entertainment or hire attractive waitresses. However, after a short increase in business the deficiencies of the product will override these incentives and the customers will once again drift away.

The right approach, quite clearly, is to adjust the toaster (the process) to make the toast correctly in the first place. If there is too much variation in the output of the toaster, get it fixed or buy a new one with better process control.

What then is the impact of all this quality emphasis on design and test? What is the impact on test economics? At the design stage there is an obvious need for much better verification of complex designs. Quality has to be designed in and built in. At the custom chip level this requirement has been met for some time by a number of design tools including simulation and synthesis packages. However, the use of simulation techniques for board design has lagged behind for several reasons. The first relates to the availability of simulators powerful enough to cope with complex board designs and the problems of handling a mix of digital and analogue circuitry. Another reason is that many companies are unwilling to take the extra time needed to perform fault simulation on their designs. This is a classical trade-off between cost and quality, on one hand, and time to market, on the other.

Simulation and synthesis technologies are improving all the time, as is the price/performance ratio of the computing power needed to perform the processing. To some degree this will also reduce the time needed to do the job but time to market has become such a big issue that faster processing alone may not be the complete answer to this part of the problem. Some effective method for optimizing the design decisions that involve trade-offs between quality cost and time is essential. The technology issue cannot be left out of this equation either, since that is what drives the verification problems in the first place. Models based upon the time to market model described in Chapter 3 may well form the basis of a design decision model for this kind of analysis.

The most obvious impact that the quality focus should have on test strategy and test tactic decisions is a greater emphasis on fault coverage. Minimize the number of defects by defect prevention and then maximize the detection of the defects you fail to prevent. Strangely, however, the reverse appears to have been the case in some companies. At the time of writing this (mid 1992) there is a strong pressure in the ATE market-place for lower priced testers. This is fine as a concept so long as 'less expensive' does not mean 'less performance'. To some degree this desire for cheaper test systems has been driven by a long and deep recession, but there is also a view that less thorough testing is acceptable now that yields out of manufacturing have improved substantially. 'With fewer faults to find maybe we can get by with less fault coverage'. This kind of thinking can result in throwing the quality baby out with the bathwater. A yield of 88 per cent and a fault coverage of only 85 per cent will result in exactly the same escape rate as a 70 per cent process yield and a 95 per cent fault coverage. The fault coverage has to be maintained or improved in order to benefit from all of the efforts to improve the process yield. If there is an opportunity to improve quality it should be taken unless the cost is prohibitive. Therefore it boils down to a cost analysis to see what makes most sense. In almost all of the analyses that I have been involved in the higher performance tester, even with a higher price

Fault coverage	Escapes	Fault coverage		Escapes
94%	6%	90%		0.6%
97%	3%	90%		0.3%

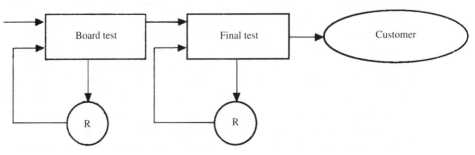

Figure 2.7 If you are serious about quality you should think in terms of 'escape rate' rather than 'fault coverage' (see text)

tag, usually turns out to be the most cost effective solution. The simple example used in Chapter 1 illustrates why this is so. If you compare a test system with a potential fault coverage of 94 per cent with another system that has a potential fault coverage of 97 per cent, then the 97 per cent system will usually have substantially lower life cycle costs (see Fig. 2.7). The amount of this cost difference will depend on volumes, yields, the nature of the product, field-service costs and so on. The reasons for the differences are simple. The 94 per cent system allows 6 per cent of the input faults to escape and the 97 per cent system only allows 3 per cent to escape. Regardless of what testing follows the ATE, this two-to-one difference remains fixed. If the following test stages detect 90 per cent of the remaining defects then the escapes to the field will be 0.6 and 0.3 per cent respectively. If the following test stages detect 99 per cent of the remaining defects then the escapes will be 0.06 and 0.03 per cent respectively. Quite simply this results in a two-to-one difference in the cost of test stages that follow the ATE in the factory and a two-to-one difference in field-service costs. There will also be a two-to-one difference in your quality reputation and customer satisfaction. If you then factor in the cost of losing customers and the increased risk of 'product liability litigation', paying a premium for that extra 3 per cent fault coverage could be the bargain of the year.

2.6 Product liability

In case you are not familiar with 'product liability litigation' this is the process whereby a company can be sued if one of their products causes damage, injury or death. Product liability lawsuits have become very popular in the United States where very large sums of money are awarded if the case is proved. A major turning point in product liability cases occurred some years ago when a very large manufacturer was sued by a lady who had put her dog into a microwave oven to dry it. When the poor animal died she sued, claiming that there were no warnings in the instruction book about the dangers of drying pets in the oven. She was right; there was no mention of it in the manual, so the judge found in her favour. This case caused major problems. If such a ridiculous case could be won then the door was open to all kinds of crazy suits.

2.7 The right approach

The example of the 94 per cent tester and the 97 per cent tester leads to a very important concept. In Chapter 1 I mentioned that the four major market forces of *cost*, *quality*, *time* and *technology* are often thought to be in conflict with each other. At least that is the conventional wisdom. This example supports my assertion that the reverse is true in the case of cost and quality. *The lowest life cycle cost will usually be achieved when the quality is high.* Thus, improving quality primarily by *defect prevention* and then using the best level of *defect detection* to catch the defects that you failed to prevent will result in lower costs. This now brings me back to Dr Deming and the other quality gurus. 'Do it right the first time and test and inspection can be eliminated.' 'Testing adds no value to the product.' 'Better than zero defects is the goal.' 'Reduce variation in the process.' These are all admirable goals and we should all be aiming to reach them, but the reality of the situation in electronics is that the continuous increase in complexity makes it that much harder than it is in most other industries. Remember Dr Deming's response when I asked him if he thought that we could get beyond zero defects in the electronics industry. The example in Chapter 1 showed that six-sigma quality is not good enough to eliminate testing. Maybe seven-sigma will be close (0.019 ppm), but until we can reach these levels the best strategy is to prevent as many defects as possible and then test with as high a fault coverage as possible.

2.8 Quality data collection systems

The quality revolution brought about a change in the role of automatic test systems away from being simply defect detection devices to becoming defect prevention devices. Indeed the ATE industry was somewhat ahead of the game in that the first network systems for linking testers together were actually shipping in 1980, the same year that the agreement was reached to develop Ethernet. The network was a prerequisite for the operation of any automation of the repair loop process and also necessary for any real verification of the defect data needed to improve quality. It is of course possible to collect data directly from a test system without the need for any network.

The data can be transferred via floppy disk to a computer containing the quality management system and the necessary database and report generation software. What gets lost by doing it this way is the verification that the defect reported by the tester really was the cause of the failure. With a fully networked system the repair stations also provide data about the repair action that took place. When the board is re-tested after the repair, the quality system will know if the repair action fixed the problem or not. If it did then the reported defect really was the problem; if not then the repair will have to be performed again. Only when the repair action successfully fixes the fault will the 'defect' be added to the database. This approach prevents the defect database from becoming contaminated with incorrect data, and since this data will be used for defect prevention (quality improvement) actions this can be quite important.

Several ATE vendors offer quality data collection capabilities in some form. If all of the other evaluation criteria are similar then this capability could well be the deciding factor when choosing a tester. In any event, the primary objective of board testers should be viewed as the monitoring of the manufacturing process and a real-time quality monitoring system built into the tester can be invaluable.

2.9 Summary

Quality is one of the four primary market forces that drive the electronics industry today and it has become a major competitive issue. The eighties saw a veritable revolution in quality as US companies began to respond to the threat from the Japanese electronics industry. As more people became better educated about quality, its real meaning and its potential for increased profitability, the revolution gained pace. The early adopters of the new quality ethic soon reaped the benefits, but there are still many companies who for one reason or another just talk about it but do very little real quality improvement.

The increased importance of quality should lead naturally to a greater need for high levels of fault coverage in the test strategy. Unfortunately there seems to be a desire for cheaper testers that may or may not have an adequate coverage. This desire is mainly driven by a need to economize but some people believe that the high yields that are now being achieved means that fault coverage is less important. What is important is that we look at the escape rate from the test strategy to see how this compares with the quality goals, but, in general, the higher the fault coverage, the lower the life cycle costs. The need for constant monitoring of the process presents a new role for the ATE. Board testers are ideally placed and ideally equipped to collect the quality data that is needed to make the best decisions about effective quality improvement.

The ever-increasing complexity of electronic products means that we may never reach the goal of better than zero defects at the high-complexity end of the product spectrum. Quality targets will therefore require a combination of defect prevention, which the testers will help with, as well as the best possible levels of defect detection. This combination will lead to the best quality and the lowest life cycle cost.

References

Crosby, Philip B. (1979) *Quality Is Free*, McGraw-Hill, New York.

Feigenbaum, Armand (1983) *Total Quality Control*, 3rd edn, McGraw-Hill, New York.

Juran, Joseph M. and Frank M. Gryna (1988) *Juran's Quality Control Handbook*, 4th edn, McGraw-Hill, New York.

3. Time to market

The average lifetime of electronic products has decreased considerably over the past ten to twenty years. This has been brought about by increasing global competition and the constant progress in technology. These short lifetimes have made it increasingly important to get new products to market on time. If you are late to market you will have less time to sell your product. As a result you will sell less, you will make less profit and you may also lose market share. Market share is important because research has shown that profitability is directly related to market share in most cases.

A simple analogy may clarify this time to market issue. You can think of the market as being a train carrying money in a series of carriages (Fig. 3.1). The various companies competing for the money on this particular train are the 'passengers'. As soon as the first passenger boards the train it starts to move. The 'product' is a container for the money and a specially designed scoop with which to pick up the money. The first passenger boards the first carriage immediately behind the engine and starts taking the money. Other passengers are able to jump on to the train from a fixed ramp and enter the carriages. Obviously you cannot jump on to a carriage that has already passed by the ramp. Less obviously, you can only pass through the train towards the rear because the doors connecting the carriages will only open in one direction. It would be a waste of time to move towards the engine anyway since the 'passengers' who boarded the train before you will have emptied all of the money from those carriages.

It should be clear that the 'passenger' who boarded the train first has the best chance of collecting the most money. A large company may be able to put more 'passengers' on the train and so collect more money, but they will still collect less than they would have done had they joined the train at the first carriage. The smaller companies who boarded the train first will have taken market share away from the larger company.

You do not get to travel on this train for free. It costs a lot to design the container and the scoop for the money and without these you will not be allowed on board. Two other important things to remember are that the train travels quickly once it starts to move and the train is quite short. If you take too long to develop your scoop you may miss the train completely. You may be better off to design an even better scoop and catch the next train. Assuming that you go on board somewhere near the front of your train and have been collecting money you will notice that once you have moved past the half-way carriage the train begins to slow down and there

The market train

Figure 3.1 Getting to market on time is vital to the profitability of a product. Reproduced by permission of New Vision Technologies

is less money in the carriages. Another train begins to approach, catching you up on a parallel track. You now need to try to jump aboard this train as it moves alongside, but research tells you that the money in this train has been packaged in a different manner. You need a new scoop to collect the money on this train. Your old scoop is worn out and it is also the wrong shape. If your company is on the ball your designers and your manufacturing department will provide you with the right type of scoop for this new train. If they are really on the ball they will get the new scoop to you in time for your heroic leap on to the first carriage of the new train so that you can start collecting as much of the money as you possibly can. And so it goes on. Every time you jump on to a new train there are more 'passengers' all equipped with bigger and better scoops. It gets harder and harder all the time.

3.1 The product life cycle

To understand the sensitivity of the relationship between time to market, revenues and profitability, it is necessary to understand the product life cycle and how it fits into a market. Typically a products life cycle consists of a period of growth, a period of maturity and a period of decline, as depicted in Fig. 3.2. However, the market will continue to exist long after any one product has been discontinued. In general terms a 'market' is a need, or a demand, for a particular type of product. Markets also follow a pattern of growth, maturity and decline, in much the same way as a product but over a longer period of time. Products tend to have shorter lives than the market they serve because design innovation and new technologies result in the development of other products that perform better for a lower price. Over time most markets will be served by products that incorporate different technologies and these technologies

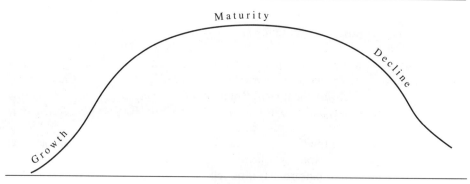

Figure 3.2   The classical product life cycle curve

will also exhibit a life cycle similar in shape to the product life cycle. In essence we have a set of nested life cycles.

3.2   Demand cycles, technology cycles and product cycles

Figure 3.3 shows how marketing managers view things. They usually refer to the overall market cycle as the *demand cycle*. Within the demand cycle there will be several *technology cycles*, and within each technology cycle there will be several *product cycles*. A simple example I use to explain this is that of *hunting weapons*. There has always been a demand for hunting weapons, ever since man appeared on the planet. In those early days the demand was driven by the need for survival. Today this is still the driving force in many undeveloped parts of the world but in the so-called civilized world the demand is for 'sporting' weapons. The demand cycle has been active for thousands of years. In that time the demand has been satisfied by numerous technologies and refinements of these technologies. We moved from stones and clubs to hand-launched pointed projectiles such as the spear. Material technology moved from stone and wood, to bone, and then on to bronze and iron. Eventually

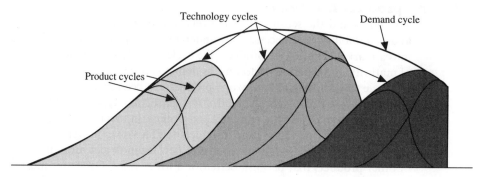

Figure 3.3   Demand cycles, technology cycles and product cycles

a major technological breakthrough saw the development of the bow and arrow. This is turn was eventually superseded by the development of gunpowder and all of the weapons based upon this new technology. The bow still survives to this day, which simply proves that there is often a niche market for older technologies. As an aside it seems hard to believe in this technological age that the old English longbow made famous by the tales of Robin Hood was once regarded as the ultimate deterrent. In the famous battles of Agincourt and Crécy the English were seriously outnumbered, but the superior accuracy and firing rate of the longbow resulted in around 10 000 casualties on the French side and only about 100 casualties on the English side. No wonder the weapon was feared throughout Europe.

The usual pattern of product development in electronics is for continuous improvements in the functionality of the design coupled with minor technological changes. Then a major new technology will be utilized which will tend to make the older technology obsolete very rapidly. There is effectively a period of evolutionary improvements followed by a revolution followed by more evolutionary improvements and so on. Since even the evolutionary changes result in an improved price/ performance ratio it is easy to see why the products tend to have short lifetimes.

3.3 The ideal product life cycle

The product life cycle (PLC) is usually shown as a curve similar to the one in Fig. 3.4. The cycle can usually be subdivided into three to five regions. The simple form would consist of periods of growth, maturity and decline. In the more detailed version the growth period is subdivided into two or three regions. These are usually defined as a period of emergence (sometimes called the embryonic stage), followed by a period of rapid growth, followed by a period of slower growth. Sales and marketing professionals will study the position of a product on its life cycle to make decisions about how to promote and sell the product. Similarly senior management should study the markets they are operating in to determine the corporate strategy of the company. You act very differently in a growth market to the way you would operate in a mature or declining market.

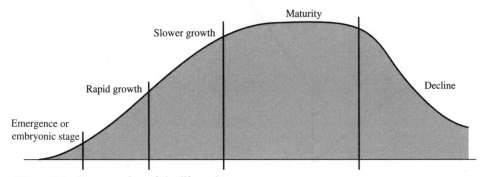

Figure 3.4 Segmentation of the life cycle curve

The life cycle shown in Fig. 3.4 starts at the launch of the product. For a more complete analysis we also need to consider the part of the life cycle prior to the product's introduction to the market, namely the development phase. The ideal product life cycle would look something like the one depicted in Fig. 3.5. The curve has been reduced to straight lines for simplicity, and the basic three stages of growth, maturity and decline are shown. The vertical axis shows development expenses below the horizontal axis and sales revenues above. The horizontal axis indicates time. In this ideal product life cycle the development time is short, and it costs relatively little. After the product introduction the sales grow rapidly, they remain at a high level throughout a long period of maturity and then they decline very slowly. A product life cycle with this kind of an envelope will result in low R&D (research and development) costs, high profits, predictable manufacturing operations and plenty of time to develop the next product. Compare this to the curve in Fig. 3.6 which characterizes the high-technology electronics industry. The product development time is long relative to the overall lifetime. Growth can be rapid or slow depending on

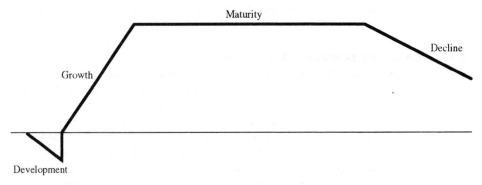

Figure 3.5   The ideal product life cycle

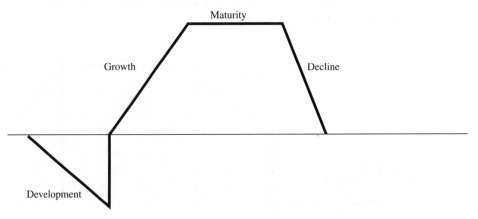

Figure 3.6   The 'high-tech' product life cycle

the nature of the product, its ease of acceptance and the competitive climate. The maturity period is between 'short' and 'non-existent', and the decline is rapid due to the emergence of a new technology. Looking at this curve it is easy to see why the 'high-tech' industry is also a 'high-risk' industry. Obviously the shape of the PLC will vary from project to project and this shape, in the form of a sales forecast, will normally be predicted by the marketing department. Their promotional plans will be defined to try to optimize the shape of the PLC in that they will attempt to gain rapid acceptance (growth) and a long maturity period before competitive products (or their own replacement product) causes the sales rate to decline. If a new product contains new technologies relative to existing market offerings then it may well have a faster growth stage. This is particularly true if the demand curve is itself in the growth stage. However, whether it is working for you or against you at any point in time, it is the high rate of technological change that results in very short windows of opportunity for a specific product incorporating specific technologies. It is this simple fact of life that makes time to market such a critical issue in the electronics industry. Incidentally, the rather nebulous term 'window of opportunity' does have at least one definition. I have seen it defined as being half the expected product lifetime. In other words, you have to introduce your product in the first half of the lifetime for similar competing products using similar technologies in order to have any chance of reasonable sales. If you cannot achieve this then you may as well cancel the development project and start on the next generation product. Bear in mind the fact that the competitors who launched their current generation products six months ago will already be at least six months into the development of their replacement product.

3.4 Time to market and profit

The results of a study performed in the mid eighties by the McKinsey Company raised people's awareness of the relationship between time to market and profits. In their report they gave an example of the impact on profit of several factors based on their findings. This showed that a 50 per cent overrun on development costs would reduce the lifetime profits on a product by only 3.5 per cent. A 9 per cent excess on the manufacturing cost of the product would reduce profits by 22 per cent. However, a six month delay in product availability reduces profits by 33 per cent. This example was for a product with a life of five years, selling into a market with an annual growth rate of 20 per cent and with a 12 per cent per year price erosion. This is hardly typical for the electronics market at present. We would tend to expect a much shorter product life, less market growth and a larger price erosion. It would probably therefore be a lot worse.

Unfortunately, there are very few examples like the McKinsey one available. As a result many people are aware of the importance of time to market but find it difficult to quantify the benefits. In actual fact it is not really that difficult to put some numbers to the problem. It is quite easy to construct a time to market model to determine the impact on revenues and profit using a spreadsheet program.

3.5   Modelling the effects of time to market

A time to market model can be a very useful tool to help make decisions about test strategies and test tactics. Being able to estimate the effects on revenues and profit will enable you to make the best decisions about the alternatives available to you. It will also provide additional justification to go down what may at first sight be a more expensive route. For example, you may determine that each week the product is delayed from reaching the market will result in revenue losses of one hundred thousand dollars and profit losses of between thirty thousand and fifty-five thousand dollars. Armed with this knowledge it will be easier to justify the higher cost of a better tester that can shave a few weeks off the program preparation time. Alternatively, or even additionally, it may make it easier to justify some design for test (DFT) activity that will reduce the test programming effort. How do we determine these effects on revenues and profits?

Figure 3.7 shows how the above example was calculated. I have used a simplified trapezoidal representation of the product life cycle to indicate periods of growth, maturity and decline. *The key assumption in this type of analysis is that the timing of the planned or forecasted decline, and eventual discontinuance, of the product remains*

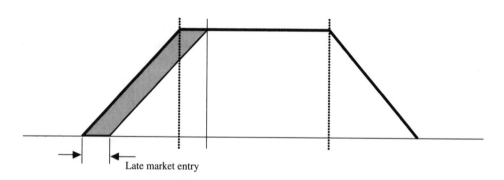

Late market entry

	%	Plan	Actual (S&M fixed)	Actual (S&M variable)
Revenues	100	7200	6800	6800
R&D	10	720	720	720
COGS	45	3240	3060	3060
Sales/marketing (S&M)	25	1800	1800	1700
General/Administrative	10	720	720	720
Profit before tax	10	720	500	600

All amounts in thousands of dollars

Figure 3.7   The reduction of profits as a result of a delayed project

fixed. Since this will usually be the result of competitive action or of the availability of a new technology this is not an unreasonable assumption. For the product under scrutiny it does not matter if it is made obsolete by a competitor or by your own new design.

The product is expected to have a 24 month lifetime, with a growth period of 6 months and a decline period of 6 months. Sales average $200 000 per month during the growth and the decline periods, and $400 000 per month during the period of maturity. From inspection of the life cycle diagram it is clear that the delay in product availability reduces the length of the maturity stage and results in lost revenues at the rate of $400 000 per month. This occurs because it is reasonable to assume that a small delay in product availability will not affect the acceptance of the product and so the length of the growth period will be as originally forecasted. If the delay is substantial, relative to the overall product life, then it would be reasonable to assume that it may take longer to reach the sales rate of the maturity stage because the competition will be well established. Conversely, if you could bring the product to market earlier than planned you may well experience a shorter growth period due to the lack of competition. Because of these possibilities it should be possible to vary the growth period in any model developed to analyse time to market effects.

The profit for the project is calculated using the following assumptions:

1. The R&D costs will be 10 per cent of the expected revenues.
2. The cost of goods sold (COGS) will be 45 per cent of the selling price.
3. Sales and marketing expenses will be 25 per cent of revenues.
4. General and administrative costs will be 10 per cent of revenues.

With these assumptions the forecasted revenues are 7.2 million dollars and the expected profits will be 720 thousand dollars.

The assumptions for calculating the profit when the product availability is late are as follows:

5. Time to market delay is one month so $400 000 are lost from the overall revenues over the life of the product.
6. R&D costs will be the same as the planned amount since this money will already have been spent prior to the product launch. It could be argued that if the product is late due to a development overrun then the R&D costs will be higher than planned. However, the delay may be because the test program was not ready on time. Either way, the added accuracy gained by trying to account for these costs is not usually worth the extra effort.
7. COGS will vary in proportion to the actual revenues (i.e. they will be 45 per cent of the actual revenues). This is effectively saying that this is a JIT (just in time) manufacturing operation. If it is not there may well be inventory surpluses caused by not meeting the shipment plan. However, the added

accuracy of trying to account for this is also not worth the extra effort unless it is likely to be a significant amount of inventory.

8. Sales and marketing costs may remain fixed at the planned rate or they may vary with the actual sales rate. This will depend on your product mix and the organization of the sales and marketing departments. If there is a sales force dedicated to this product then the costs will reflect the plan. If, however, the sales team sell other products as well then they may divert some of their efforts to selling these. This raises another time to market issue. If the product is late and the competitors have become established it will be harder to sell. Human nature being what it is, the sales force may well turn their attention to other easier to sell products so that the lost sales for the new product may be worse than the delay alone would account for. Some of these effects are difficult to predict so a good compromise would be to calculate the revised profits assuming both fixed and varying sales and marketing expenses. For the final results you can either take the average value of the two or, as in my example, quote a range.

9. General and administrative costs will remain at the planned rate.

Note. Large companies with many products will lump many of these costs together. This type of analysis assumes that each product should contribute to the overall performance of the company. Therefore the costs should be the actual costs incurred when these are known and an allocation based upon the revenues contribution when they are not. For example, if the company regularly spends 10 per cent of revenues on R&D then each new product proposal should be expected to produce revenues of ten times the estimated R&D costs. If it seems unlikely that this can be achieved then the proposal should be rejected unless there are good strategic reasons to continue with the development. In practice something more than the tenfold return should be expected to allow for the occasional failure.

The results in the example show profits falling to between 500 thousand dollars and 600 thousand dollars from a plan of 720 thousand dollars. This represents a fall in profits of 17 to 31 per cent for a one month delay in product availability. Why is there such a big impact?

The easiest way to see why the effect is so great is to look at the break-even chart for the project shown in Fig. 3.8. This type of chart shows clearly that profit is the difference between two large numbers, the total revenues and the total costs. If one of those numbers is changed just a little the profit will change a lot. The total revenues line and the total cost line usually cross each other at a fairly acute angle. The profile of the product life cycle for electronics products all too often results in the break-even point being too near the end of life point for comfort.

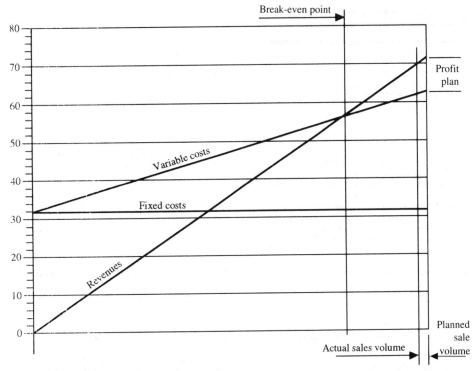

Figure 3.8 The classical break-even chart, showing how profit is the difference between two large numbers

3.6 Other factors to consider

If you decide to develop a spreadsheet model to analyse time to market effects here are a few things you might want to take into account in your model:

1. Make it possible to have an early product introduction as well as on-time and late introductions.
2. Allow for the fact that a late introduction may result in you not reaching the planned sales rate during the maturity period due to competitive pressures. Similarly, an early introduction may result in a higher sales rate during this period as a result of increased market share.
3. Allow for the possibility of price erosion during the latter stages of the product's lifetime. Similarly, allow for the possibility of a price premium if you can bring the product to market ahead of the forecasted time.

Figure 3.9 shows the layout of a time to market model that I developed. For comparison purposes this version also includes a model that represents the product life cycle as a triangle. This model was developed by the ATEQ Corporation of Beaverton, Oregon, and tends to be more applicable to products that have a very short life due to rapid technological changes. The assumption here is that the sales

Time To Market Model

Input parameters for an 'on-time' product availability

Planned Product Lifetime (months)	**A**	24
Peak sales during maturity period (units)	**B**	40
Target sales price (each)	**s**	15
Time to reach peak sales rate	**g**	6
The 'decline' period (months)	**d**	6

Input parameters for a late or an early product availability

		EARLY	EARLY	ON-TIME	LATE	LATE	LATE
Delay time of product availability (months)	**L**	-4	-2	0	2	4	6
[for 'on-time' specify zero (0)]							
[for an early launch specify negative delay time]							
Time to reach peak sales rate	**g**	6	6	6	6	6	6
Price premium for an early availability (%)	**PP**	0	0	0	0	0	0
Price erosion during latter stage of maturity period	**PE1**	0	0	0	0	0	0
Duration of above price erosion period	**T1**	6	6	6	6	6	6
Price erosion during 'decline' period	**PE2**	0	0	0	0	0	0

		EARLY	EARLY	ON-TIME	LATE	LATE	LATE
Revenues for the trapezoidal model		13200	12000	10800	9600	8400	7200
Revenues for the triangular model		16800	13650	10800	8250	6000	4050

Profit and loss calculations (S&M Fixed at planned rate)

			EARLY	EARLY	ON-TIME	LATE	LATE	LATE
Revenues	%	100	13200	12000	10800	9600	8400	7200
Cost of goods sold (COGS)	%	45	5940	5400	4860	4320	3780	3240
Research and Development (R&D)	%	12	1296	1296	1296	1296	1296	1296
Sales and Marketing (S&M)	%	20	2160	2160	2160	2160	2160	2160
General and Administrative (G&A)	%	10	1080	1080	1080	1080	1080	1080
Other expenses	%	2	216	216	216	216	216	216
Profit Before Tax (PBT)			2508	1848	1188	528	-132	-792
Profit Before Tax (PBT)	%	11	19.0	15.4	11.0	5.5	-1.6	-11.0
Change in profit versus plan	%		111.1	55.6	0.0	-55.6	-111.1	-166.7

Profit and loss calculations (S&M varying as planned percentage of actual revenues)

Profit Before Tax (PBT)			2028.0	1608.0	1188.0	768.0	348.0	-72.0
Profit Before Tax (PBT)	%	11	15.4	13.4	11.0	8.0	4.1	-1.0
Change in profit versus plan	%		70.7	35.4	0.0	-35.4	-70.7	-106.1

P & L for the triangular model (S&M fixed relative to plan)

PBT (Triangular/fixed S&M)			4488	2755.5	1188	-214.5	-1452	-2524.5
PBT (Triangular/fixed S&M)	%		26.7	20.2	11.0	-2.6	-24.2	-62.3
Change in profit versus plan	%		277.8	131.9	0.0	-118.1	-222.2	-312.5

P&L for the triangular model (S&M varying as planned percentage of actual revenues)

PBT (Triangular/variable S&M)			3288	2185.5	1188	295.5	-492	-1174.5
PBT (Triangular/variable S&M)	%		19.6	16.0	11.0	3.6	-8.2	-29.0
Change in profit versus plan	%		176.8	84.0	0.0	-75.1	-141.4	-198.9

Figure 3.9 An example of a spreadsheet-based time to market model

grow until a point in time when competition arrives and then they begin to decline. There is no maturity period as such.

Figure 3.10 shows the variable parameters used in this model to determine the revenues. The area under the curve represents the revenues and the loss or gain in revenues is simply the difference in the area of the 'on-time' curve and the 'late' or 'early' curve. The impact on profits is then calculated from a set of typical percentages for the various elements making up the P&L (profit and loss) statement in the same manner as for the example in Fig. 3.7. Figure 3.11 shows a family of profit lines calculated using the model. It is interesting to compare this example with the McKinsey one. Their research showed a reduction of 33 per cent in the profits for a product with a five year lifetime and a six month delay in introduction. The same point in

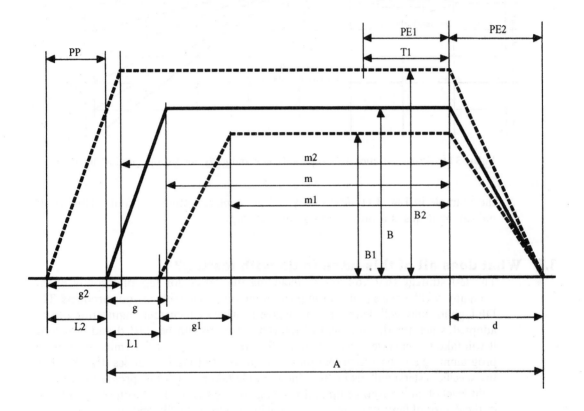

A = expected life of product	L1 = period late, relative to plan
B = sales during the maturity stage	L2 = period early, relative to plan
g = the growth stage	PP = price premium for early period
m = the maturity stage	PE1 = price erosion during latter part of the maturity period
d = the decline stage	PE2 = price erosion during the decline period

Figure 3.10   The variables used in the model shown in Fig. 3.9

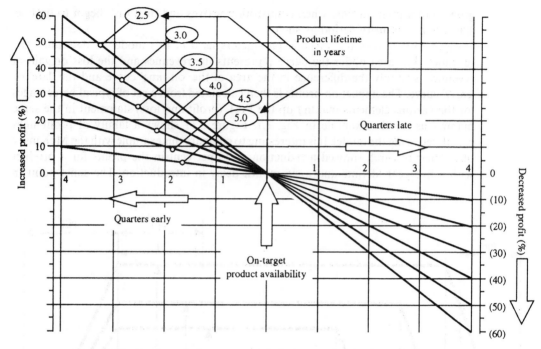

Figure 3.11   Profit variation versus periods, late and early, for various product lifetimes

the example in Fig. 3.11 shows only a 5 per cent reduction in profits. This would indicate that this is a rather conservative example.

3.7   What does all of this have to do with testing?

The test strategy that you adopt, including the design for test philosophy of the company, will have a significant impact on the time to market of your products. The DFT philosophy will determine the degree to which concurrent engineering can be adopted, when the development of the test programs can be started and how long it will take to generate these programs. The test strategy will determine how much programming and fixture preparation is required, and the test tactics (the choice of the specific testers) will also determine the time taken to get test programs with the right level of fault coverage up and running. Test systems vary enormously in terms of the degree of fault coverage they can achieve automatically and in the amount of assistance they give you to improve the fault coverage to acceptable levels. *You cannot trade off fault coverage for time to market. You have to have both in order to be competitive.* The goal relates back to the main driving forces detailed in Chapter 1. To be competitive you have to produce new products that have lower costs, higher quality, the latest technologies and get them to market on time. Two or three out of the four is just not going to be good enough.

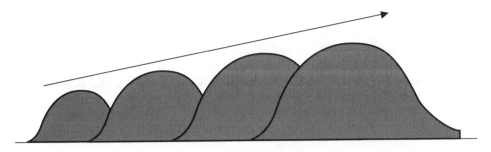

Sustained growth requires on-time product availability

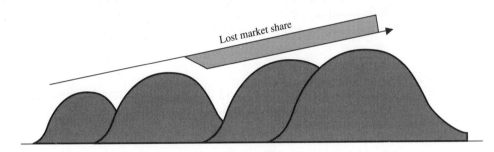

One delayed project can lead to lost market share

Figure 3.12 In a competitive market a single delayed product can result in lost market share that will be difficult to regain

Since the choice of test strategy and the choice of the test tactics will be influential in meeting your time to market goals this has to become a major factor in the analysis to select the optimum strategy and tactics. As a result the financial benefits of meeting your time to market goals can, and should, be included in the cost justification.

I mentioned at the beginning of this chapter that research has shown that there is usually a direct correlation between market share and profitability. The best-known research of this kind was conducted by the Strategic Planning Institute, and is usually referred to as the PIMS study. Figure 3.12 shows in a simplistic manner how a late product availability can cause a loss of market share. Consistent growth requires the introduction of a series of new products that meet the four market force goals. A delayed project can lead to a loss of market share that will be difficult to recover. The same effect will be produced if any of the other market force goals is not met because even if the product is on time lower revenues caused by some other shortfall will result in lost market share.

4. A primer of test economics

This chapter is intended to give a quick overview of some of the basics of test economics and the statistical and mathematical relationships that we need to understand before we can make any reasonably accurate analysis. It is hoped that this will be useful to those that are either new to the world of test or to the problems of economic analysis and the economic comparison of alternatives.

4.1 Jargon

Like most other activities, the world of automatic testing has its own terminology or jargon. Therefore it might be useful to begin with a brief glossary of some of the more commonly used terms.

- *UUT* Unit under test. The component, printed circuit board, backplane, cable, sub-assembly, etc., that is being tested. Sometimes (mainly for components) called the DUT (device under test).
- *Fixture* The adapter that sits between the UUT and the tester. This item interfaces the UUT to the various resources of the test system. Sometimes (mainly for component testers) called a device adapter.
- *Test program* The program, specific to one type of UUT, that gives the test system instructions that define the tests to be performed along with the expected responses of the UUT.
- *Yield* The proportion of defect-free units at some stage in the process. May be expressed as a proportion or as a percentage, e.g. 0.72 or 72 per cent. Most often used to describe the proportion of UUTs that are defect-free when entering the test stage, but a yield can be expressed at any point within a multi-stage process.
- *Fault spectrum* The term used to describe the nature and the distribution of the various different types of defects that can be present on a particular type of UUT.
- *Production mix* Sometimes simply the 'mix'. Describes the nature of the production operation in terms of the number and the variety of UUT types; e.g. a low-mix operation would have relatively few different types of UUT.
- *Diagnosis/repair loop* Refers to the flow of UUTs that fail a test, from the tester (where the diagnosis is made), to the repair or rework area, then back to the tester for re-test after repair.
- *Fault coverage* A figure of merit for a test stage. It is usually stated as a percentage and indicates how many, or what proportion, of the faults that could be present

can be detected by the stage. Fault coverage will be a function of the test method employed (e.g. in-circuit or functional), the capabilities of the tester, the comprehensiveness of the test program and the fault spectrum.

- *Escapes or escape rate*   Effectively the inverse of fault coverage. A measure of the number, or the proportion, of the defects present that escape detection at a test stage. This is actually a better measure of the performance of the test stage than fault coverage because it is a direct measure of the output quality from the stage.

4.2 Economic analysis

All economic analysis involves the *comparison of alternatives*. This is fundamental because the justification of any capital purchase must be based on the potential cost savings that will result from its use. Cost savings can only be estimated when there are at least two possible alternatives because they will be the *difference* between the total costs of these alternatives. You cannot develop an ROI for a test system, or anything else, in isolation. The cost savings that lead to the ROI have to be based upon the cost of some alternative course of action. This alternative course of action would typically be one of the following:

1. Do nothing, i.e. continue to operate in the current way with the current costs.
2. Use some different approach, i.e. solve the problem in another way with the costs of this approach.
3. Use an alternative piece of equipment.

From a testing point of view doing nothing at a particular stage in the process will usually result in higher costs at the following stages. The total (life cycle) costs of the two approaches are compared to determine which is the most cost effective.

The completely different approach could, for example, be having components tested at an outside test lab as opposed to buying a tester and performing the tests in-house. Again the costs of the alternatives are compared to see which is the cheapest, and an ROI can then be calculated.

The third course of action is probably the most common one for ATE because in many cases the complexity of the task makes other alternatives non-viable. However, comparing the economic performance of two relatively similar pieces of equipment is not quite as straightforward as it may seem to be. The correct approach to take for the analysis is really not very obvious. Should the comparison be made directly between the two competing testers or should they each be compared to some baseline alternative of manual testing?

If, as will often be the case, the alternative of manual testing is really not viable then the ROI that would result from comparing this approach with any ATE (no matter how poorly it performs) will be enormous. This has led in the past to many purchasing decisions that have been less than optimum. Probably the best approach to take in a case like this is to assume that one form of ATE or another will be required, and then compare the economic performance of the short-listed alternative

	Alternatives		
	A	B	Difference
Investment	200000	300000	100000
Annual costs	250000	150000	100000

Simple payback for alternative B
3 years?
1 year?

Figure 4.1 Incremental investment. The savings result from the difference in the two investments

testers. Depending on the rules within your company it may be necessary to calculate an ROI for each tester relative to the baseline alternative, and then to compare the testers against each other. In this way you get the total ROI for the investment as well as a clear measure of the economic merits of the viable alternatives.

Even doing it this way, however, you still have to be careful, especially if there is any reasonable difference in the price of the alternative testers. A fairly common situation like this occurs when a so-called 'low-cost' tester is being compared with a higher priced alternative. A simple example will clarify the potential problem.

Let us assume that a $200 000 tester is being compared to a higher performance system costing $300 000, as shown in Fig. 4.1. The annual costs of using the systems are estimated to be $250 000 for the 'cheaper' of the two and $150 000 for the more 'expensive' system. Therefore the higher priced system will save $100 000 per year in operating costs relative to the alternative unit. Taking a simplistic view and calculating a payback time for the $300 000 system, it is quite obviously three years. Or is it? The savings of $100 000 per year will be achieved by investing an extra $100 000 on the tester; therefore the payback will be one year not three. We have to spend a minimum of $200 000 anyway. Therefore the savings result from the difference in the investment and not the absolute amount. As indicated above, the total ROI would have to be determined relative to the alternative of not buying either of the testers.

4.3 Incremental investment

This simple example illustrates the concept of *incremental investment*. The savings are the result of an incremental investment; therefore the ROI should be calculated based on the incremental amount and not the full amount of the investment.

Note. The calculation of a payback time, although very commonly done, is not a measure of ROI and so is not a good method to use to compare alternatives. This is explained in the chapter on financial appraisal. I have used payback in the above example simply because it is an easy way to explain the concept of incremental investment.

4.4   Costs and savings

The economic analysis of alternative courses of action is essentially a three step process.

1. First we determine the costs associated with alternatives.
2. From these costs we determine the savings.
3. From the savings, their timing, the lifetime of the project and the investments required, we determine the financial attractiveness of the alternatives.

The financial attractiveness is usually measured in terms of an ROI which is why the timing of when the savings occur and the expected lifetime of the equipment are important. This is all explained in Chapter 11 on 'financial appraisal'.

Clearly, then, a good understanding of the cost areas is important because this is the starting point for any analysis and the accuracy of the ROI will be heavily dependent on getting the costs right. The first thing we have to do is separate out the one-time costs associated with the purchase of the equipment and the recurring costs associated with using it. The equipment itself, and possibly some other one-time costs, can be *capitalized*. This means that it becomes a fixed asset and can be *depreciated*. The annually recurring running costs would come out of the working capital.

4.5   ATE cost areas

For most automatic test equipment the main costs can be grouped into the following areas:

1. The purchase price (the investment).
2. Any site preparation, training and other initial set-up costs.
3. Test programming and test fixture preparation costs.
4. Testing, diagnosis and repair (TDR) costs.
5. Maintenance costs (hardware and software).
6. The cost of any following test stages.
7. The field-service costs associated with escaping defects.

These will be the main cost areas that will go to make up the total cost of ownership, or more accurately the life cycle cost, of the test strategies and the testers that are being considered.

Let us now take a closer look at these cost areas.

The purchase price and other up-front costs

The purchase price is a fairly obvious one, but in addition to actually buying the tester there can often be other set-up costs involved. These are usually a one-time cost and may include such things as constructing a special area, installing air-conditioning and so on. It might be possible to capitalize some of these up-front costs, in which case they would be added to the purchase price and so form part of the investment. Other costs, such as training for programmers, operators and possibly maintenance engineers, would not be capitalized. These costs would form part of the first year operating costs. Some companies also include the cost of the evaluation process itself in these up-front costs.

Test programming and test fixture preparation costs

These are all of the recurring costs associated with the preparation and the debugging of test programs and the test fixtures needed to interface the unit under test (UUT) to the tester. The costs in this area are usually quite high. How high will obviously depend on the number of test programs that have to be prepared each year. The complexity of the UUT will also influence these costs, as will the degree of integration between design and test. For any company with a large variety of UUTs and many new designs each year, not to mention a large number of ECOs (engineering change orders), this will usually be the biggest cost area. It will easily exceed the cost of the tester in a fairly short time.

Test, diagnosis and repair costs

Test and diagnosis costs will mainly be a function of the test and diagnosis times, the UUT handling times and the labour rate for the operator. They will also be dependent on the efficiency of the repair operation because incorrect repair actions will cause boards to pass unnecessarily around the *diagnosis/repair loop*. This not only adds to the cost but it also decreases the capacity of the tester. For most test stages there can usually be four possible outcomes of a test (see Fig. 4.2):

1. The UUT passes the test because it is defect-free.
2. The UUT passes the test even though it contains a defect.
3. The UUT fails the test because it contains a defect.
4. The UUT fails the test even though it is defect-free.

Obviously only two of these outcomes are valid. The defective UUT that passes the test does so because of fault coverage limitations of the tester or of the program written for it. The hardware of the tester sets the limit on its fault coverage capabilities, the test program generation system (software) sets the limit on what coverage might actually be achievable and the amount of skill and effort applied by the programmer

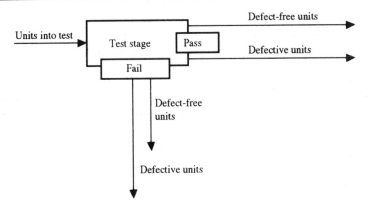

Figure 4.2 The four possible outcomes of a test

determines what is actually achieved. Just how much fault coverage can be achieved in a given time is also heavily dependent on the test systems software and the degree of automation and operator information that it provides. This is a very important measure of the performance of a commercial tester because time is usually one of the major constraints placed on the test program generation process. A tester that enables you to reach a 90 per cent plus coverage in, say, one week is much more powerful than one that can only get you to 85 per cent coverage in the same time.

The defect-free UUTs that fail the test do so because of limitations in the performance of the tester or shortcomings of the test program. A simple example of this would be the in-circuit test of a passive component to some high and low limits. If the actual value is near one of the limits, but inside the specification, and the value being measured is near one end of the tester's measurement capability, it is possible that the tester will measure the component as being outside the limits. Another possible reason for this effect might be that the circuit configuration around the component being measured makes it difficult for the guarding system to work effectively.

TDR costs are also dependent on the *yield* of good units reaching the test stage since high yields result in less diagnosis and fewer repairs. The *fault spectrum* may have a small impact on TDR costs because some fault types may take longer to diagnose than others and some defects are more likely to result in an incorrect repair action.

Maintenance costs

Maintenance costs include the cost of service contracts or on-call service charges and also the cost of keeping the system up to date. Some companies offer higher levels of support for an annual subscription fee. The cost of any additional training, beyond the initial courses that are usually included in the system price, can also be included in this category.

The cost of following test stages

As indicated in Chapter 1, the most accurate way to view test economics is on the basis of life cycle costs as opposed to simply comparing the direct costs of programming and operating the tester. The decision to adopt a particular test strategy essentially determines where in the overall test process each type of defect is most likely to be detected. The decision as to which specific test systems to use to implement the chosen test strategy will determine the effectiveness of the individual test stages. Defects not detected at a given stage will escape to the following stage where, in general, it will cost more to detect them. It may well happen that some of the escaping defects cannot be found at the following stage, so they move on to the next and more costly stage before they can be detected.

The ability to detect defects is a function of the nature of the tester, its theoretical performance, its actual performance and the effectiveness of the test program that is prepared for the specific UUT.

1. *Nature of the tester.* Testers can usually be grouped into certain categories, e.g. in-circuit testers, functional testers, manufacturing defect analysers, combinational testers, etc. For each type of tester there are classes of defects that are difficult or impossible for the tester to detect.
2. *Theoretical performance.* The maximum possible defect detection capability as determined by the nature of the tester.
3. *Actual performance.* The maximum defect detection capabilities of a specific system limited by the actual measurement capabilities, the software performance and trade-offs in the design of the tester.
4. *Test program effectiveness.* The performance of the individual test program for a UUT. This is a function of the capabilities of the test program generation software, the debugging tools and the amount of time and effort put into the programming task.

It should be clear from this that the performance of testers can vary considerably. This being the case, it is essential that the economic analysis includes the cost of the following test stages because the cost of these stages will be directly affected by the choice of equipment for the earlier stages.

Field-service costs

At some point in any test process the *following test stage* becomes the *customer*. The end user will have your product for far longer than you will. In that time the product will be exercised in ways that were not exercised in any of your test operations. As a result the user can detect defects that were undetectable by your testing due to limitations of fault coverage. In addition to this defects caused by uncertainties that you did not test for, and latent defects caused by the early life failure of components, will be discovered. Again the quantity of defects reaching this stage of the process will be directly affected by your choice of test strategy and the choice of testers to implement the strategy. This is also the most critical of all the stages. The cost of repair is very much higher, but significantly higher costs can occur if you lose the customer through having poor quality, or if a product liability suit is filed because your product caused damages in some manner.

This life cycle approach to economic analysis, being more accurate, will also make the justification easier in many cases. For example, if you try to justify the purchase of a tester for the incoming test of ASICs, based purely on the savings that will be made within the factory, you may have a problem. Including the potential savings in the field will make the justification easier.

4.6 The 'rule of tens'

Possibly the most frequently quoted concept in test economics is the 'rule of tens'. The rule states that the cost to detect and rectify a defect increases by a factor of ten as you progress through each major stage of the manufacturing and test process. The rule is usually shown graphically with a logarithmic cost scale as shown in Fig. 4.3. The test stages are normally defined as being incoming inspection, board test, system test and field service. In cash terms a component defect might cost one dollar at incoming inspection, ten dollars at board test, one hundred dollars at system test and one thousand dollars if it escapes to the field to be found by the customer.

This ten-to-one relationship was first recognized by Kemon Tashioglou and Jim Prestridge at the Teradyne Company in the early seventies. The concept is a very

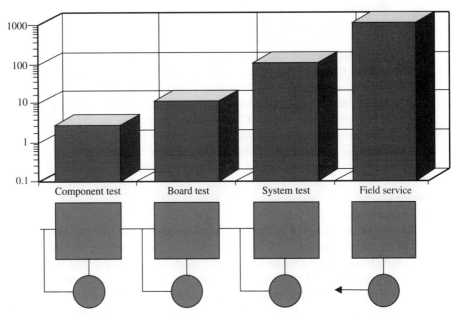

Figure 4.3 The 'rule of tens'

useful one to use when discussing test economics in a global manner. The idea that it becomes more expensive as you progress through the process can readily be understood by non-technical managers by using this concept. However, it is only a concept. It is not accurate enough to use for any analysis that may lead to a purchase decision. It was surprisingly accurate when first recognized, but this accuracy was rather short-lived. Apart from the fact that the four stages are not necessarily a good model of the real process, rapid progress in test technology changed things rather quickly.

When the relationship was first recognized the only in-circuit testers around were analog systems. Testing digital circuits was the big problem of the day and the solution offered by the ATE industry was the *functional tester*. At the time, however, functional testers had no diagnostics. Therefore, armed only with a simple logic probe, a logic diagram and a display screen full of ones and zeros, a skilled technician would *manually diagnose* defects. This would typically take 15 to 20 minutes per defect. Fortunately the average board only contained about 40 to 50 SSI (small scale integration) and MSI (medium scale integration) devices. It was not until late 1972 that such major innovations as computer guided probes were introduced. At the same time, component defect rates were much worse than they are today—around 1 per cent even for relatively simple ICs (integrated circuits). As we will see in a moment, the higher the defect rate, the cheaper it is to locate a defect at component test. These factors conspired to create the ten-to-one cost difference between incoming inspection and board test. With the advent of automatic diagnostics, better quality components and in-circuit testers, the ten-to-one relationship quickly changed.

More importantly, a simple concept such as the rule of tens cannot take account of the diagnostic capability of the various test stages. A component defect missed because there was no incoming test will not necessarily be detected at the board test stage. Depending on the nature of the defect, it might not get caught at the system test stage. If it now causes a failure in the field the cost of not testing at incoming inspection, according to the rule of tens, will be one thousand times the cost of testing at that stage. With today's levels of component quality and the better performance of the board testers the cost ratio is now in favour of the board tester by quite a margin (between 10 and 100 to 1).

A similar relationship to the testing rule of tens exists in the design area where design mistakes cost more to rectify as you progress through the design process.

Calculation methods

It is important to be consistent in terms of the method of calculating the costs at each test stage, but there are some good reasons why it should not necessarily be done in exactly the same manner for each of the stages. The way the calculation is made will depend to some extent on the purpose of the calculation. There is an analogy here with the use of loaded and unloaded labour rates when performing an economic analysis. If you want to know what an operation *costs* you for purposes of determining a *price* then you should load the labour costs with all of the applicable overheads. However, if you want to calculate the *savings* you will make if you can eliminate some of the labour costs you should not use a fully loaded labour rate. The reason is quite simple. Eliminating some direct labour costs will not usually save any of the overhead costs. The rent, the utilities, the supervisory staff, the management, the loans, the local taxes, etc., will all still be there. Including some part of these in a savings calculation will simply make that calculation invalid. Therefore costs are not always directly comparable to savings, and this can apply in the calculation of testing costs and savings for other reasons than the use of loaded or unloaded labour rates.

The cost of finding and rectifying a defect at any test stage is not simply the direct cost of diagnosis and repair (if appropriate) of the average defect. The cost must also include the cost of testing all of the defect-free units that had to be tested before finding a bad one. For example, if you test a batch of components that contain 1 per cent defectives, you will have to test one hundred of them to find one bad one. On average you will test 99 good units for every defective one. If it costs one cent to test one component, it will cost one dollar to find a defect. If the defect rate is 0.5 per cent, you will have to test two hundred devices to find one bad one, and the cost to find a defect goes up to two dollars. At a defect rate of 0.01 per cent (100 ppm) the cost per defect would be one hundred dollars, but if this type of defect cannot be found easily at other test stages, it might still be possible to justify incoming inspection even at this level of defect rate.

With today's relatively high levels of quality, both for the components and for the manufacturing process, we will spend a lot of time testing defect-free UUTs. Maybe this is what gave rise to the notion that testing adds no value but, as I explained in Chapter 1, the high cost of defects at the later stages makes this a false notion.

The one exception to the rule of including the cost of testing the good units is of course field service, where hopefully we mostly get called out to defective units and not the entire installed base of products. Another possible exception is the final test stage, and this will depend on your reasons for making the calculation. If you want to determine the total cost of the final system test stage then you should include the cost of testing the defective units and the defect-free units, the cost of diagnosis for the defective units and the cost of repairing the defective units. Here I include the cost of any re-test after repair with the testing costs. Alternatively, if you want to determine the savings you will make at the final system test by preventing some of the defects from reaching that stage, then you should not include the cost of testing but only the cost of diagnosis and repair. The reason is that the testing would probably have to take place regardless of how many defective systems there are, so you will not save the cost of this operation. Over the years I have seen many exaggerated claims about saving costs at the final system test that have resulted from the wrong calculation being made. If you take two extremes of test costs and diagnosis costs the potential for error becomes obvious. At one end of the scale you can have a situation where the diagnosis time for a defect will be many times greater than the time it takes to test a defect-free unit. This would occur if the test is fairly automatic as it would be if the test is essentially a software-intensive test routine, such as would be used for the final test of a PC. The diagnosis time could, however, take several hours or even a couple of days. Under these conditions, including the test costs when calculating any savings will only introduce a small error. At the other end of the scale is the situation where the final test procedure is mainly manual and involves many adjustments or calibrations. In such a case it is possible that the test time for a good unit might be long compared to the time taken to diagnose a fault and under these conditions there would be a large error introduced to the calculation. There is an example in Chapter 6 that discusses this in detail.

The effect of different fault types

As mentioned earlier, an obvious shortcoming of any simple concept such as the rule of tens is that it cannot consider the effect on costs of different types of defect. All tester types tend to have some strengths and some weaknesses when it comes to detecting defects spanning the entire fault spectrum. A functional board tester may be able to detect interactive problems involving two or more components whereas an in-circuit tester, with its usual 'one at a time' approach, can not. Similarly, an in-circuit tester has no problem measuring the value of a pull-up resistor, but a functional tester is unlikely to know if it is present or not—let alone whether it is the correct value.

A digital IC has three major failure modes:

1. It can fail *functionally*.
2. A *d.c. parameter* may be out of spec.
3. An *a.c. parameter* may be out of spec.

A low-priced IC tester will usually test for functional and d.c. parametric failures. A higher capability, higher priced, tester will also be able to perform a.c. parametric tests, but a board tester is unlikely to be able to detect anything other than functional failures. Even so-called dynamic functional testers or high-speed in-circuit testers have a very limited capability for detecting parametric failures. High-speed cycle-based (performance) board testers that work in a similar manner to a VLSI (very large scale integrated) component tester stand a better chance, but these are fairly expensive systems. Even at system test, an 'out of spec' device may well go undetected, depending on just how bad it is. Therefore a parametric defect that goes undetected at incoming inspection, because there was no incoming inspection, stands a good chance of causing a failure in the field when it degrades enough to affect the performance of the product. Most IC failure mechanisms do degrade with time and temperature, which is why burn-in techniques are used to accelerate the degradation to a point where it is measurable or the device fails completely. Even if the defect rates of components are very low, if the other test stages have little hope of detecting the type of defects that are likely to be present, it may still be possible to justify incoming test for certain types of component.

4.7   The extended rule of tens

As a concept for the general discussion of test economics with non-technical people, the rule of tens continues to be a useful tool. The increased importance of quality in today's market-place leads me to suggest an extension of the model beyond its traditional four stages. Modern quality philosophies require improved relationships with both customers and suppliers. By working closely with component suppliers so that defects are eliminated at source the cost may well be one-tenth that of finding them at incoming inspection. This is the rationale for vendor rating programs and 'co-destiny' programs, with the goal of being able to 'ship to stock'.

In just the same way, the equipment manufacturer must realize that if it costs him $1000 to fix a defect in the field it may well cost the customer ten to one hundred times that amount in lost production or in the higher cost of having to use some alternative approach. Ultimately, if you start to lose business because of quality problems caused by inadequate testing, the cost to you may well be one hundred to one thousand times the field-service cost of fixing a defect.

4.8   Statistics

Affectionately known to many as sadistics, the study of statistical and probability theories is a fascinating subject. At least that is what my maths teacher told me. Fortunately, for our interest in test economics, we only need to understand a few simple relationships. These are, however, quite fundamental to a good understanding of what goes on in the testing process. Most of what we need to know falls into the area of probability theory. Probabilities tell us how often an event will take place when considered over a long period of time or what proportion of a batch of items will have certain characteristics when a large number of these items are considered. Obviously, for test economics, we are going to be concerned mainly with such things as the probability that a component, a board, a sub-assembly or a complete system contains a defect.

This is all important because most of the time we will be predicting the future. Almost all economic analysis and financial justification is based upon guesswork. We may call it estimating, forecasting, predicting or something else to disguise the fact that it is guesswork, but that is what it is. Most business decisions are guesswork and the most successful managers are the ones that guess right more often than they guess wrong. Of course, we use as much real data as we can get to minimize the risk, but if we wait until all the data we need is on hand and fully verified then it is probably too late. The opportunity will have passed us by. We also need to use whatever techniques we have available to us to minimize the risk of a bad guess, and this is where statistics and probabilities come in. By using well-established techniques and relationships we minimize the risk of getting some of the basic assumptions wrong.

The need for large numbers

Probabilities do not work well for small quantities or for short periods of time. If you flip a coin the probability that it will come down heads is 0.5, but on two flips you will often get two tails or two heads. Flip the coin 1000 times and the result will be very close to the theoretical probability of 0.5.

We therefore have to be careful when applying probabilities to testing situations to involve quantities that are large enough to give a meaningful answer. Since we will normally be dealing with a one year volume of components or boards this is not usually a problem.

Probability situations

There are three common situations that arise when dealing with probabilities:

1. Mutually exclusive events. Mutually exclusive events take place when the occurrence of an event excludes or prevents the occurrence of all others.
2. Independent events. Events are independent if the occurrence of an event in no way influences the occurrence, or the non-occurrence, of other events.
3. Conditional events. Events are conditional when the occurrence, or non-occurrence, of an event is dependent on some other event having already taken place.

We do not really need to concern ourselves with conditional events so let us look at a few examples of the others to clarify the differences.

Mutually exclusive events

At a test stage the UUT may pass or it may fail the test. Either of these events will exclude the other so they are mutually exclusive. There are two possible outcomes.

Similarly, an analog test to pre-defined limits has three possible outcomes that are mutually exclusive. The test result may be above the high limit, below the low limit or it may be between the limits (within specification).

When mutually exclusive events are involved, a very simple rule exists that says that the probability that *any one* of the events will take place is determined by adding up the individual probabilities. This is typically written like this:

$$P(A \text{ or } B \text{ or } C) = P(A) + P(B) + P(C)$$

This is called the *special rule of addition*.

In the two simple examples above, the individual probabilities will add up to 1.0 because at least one of the events considered must occur. A probability of 1.0 denotes absolute certainty. If the probability is 0.7 that a board will pass the test at a specific board test stage, then the probability that it will fail must be 0.3 since one of these mutually exclusive events must occur.

Independent events

Component defects are a good example of independent events because, unlike apples, a bad IC in a batch cannot really affect the others. Most defects on boards can also be considered to be independent of each other but there can be exceptions to this. For instance, the presence of a short on a board that is then powered up on a functional tester may lead to the occurrence of another defect. If the short had been removed before the functional test took place, the second defect would not occur. This is an example of a conditional event. The second defect is conditional on the presence or the absence of the short. However, the added accuracy to be gained by trying to account for this kind of situation is rarely worth the extra effort. The easiest way to account for this effect if you are considering a functional test with no prior shorts test among your strategy options is to modify the expected fault spectrum. For example, if you expect that 40 per cent of defective boards will contain a short, you can make some estimate of what proportion of these would induce additional defects and then add these into your fault spectrum when analysing the cost of the functional test strategy. Naturally these defects would not be included in the fault list for any strategy that would involve the removal of the shorts before any power is applied to the board.

When two events are independent they can occur singly or together. The probability that they will both occur together is determined by multiplying their individual probabilities of occurrence together. This is usually written as

$$P(A \text{ and } B) = P(A) \times P(B)$$

This is called the *special rule of multiplication* and it holds true regardless of the number of possible events.

For example, if we know that the probability of getting a component-related defect, such as a faulty or a mis-oriented component, on a board is 0.7 and we also know that the probability of getting a short is 0.55, then we can determine the probability of getting both a component-related defect and a short on the same board by multiplying these probabilities together:

$$P(\text{component}) \times P(\text{short}) = P(\text{both})$$

$$0.70 \times 0.55 = 0.385$$

Therefore, in the long run, over some reasonable quantity of boards we could expect to see 38.5 per cent of the boards containing both a short and a component-related defect. This situation is illustrated graphically in Fig. 4.4.

This information is essential for correctly determining test times, costs and throughput figures for in-circuit testers or any form of tester where it is usual to stop the test if shorts are found. The normal practice would be to have the shorts removed at a repair station before returning the boards to the tester to complete the testing. This practice is recommended because the presence of the short can affect the ability of the tester to make other tests on the components. As indicated earlier, when power is applied other damage could occur, but even if no damage is caused the diagnostic performance of the tester will be degraded. For example, the short could be directly across a resistor. This would probably result in the diagnosis of two failures, a short

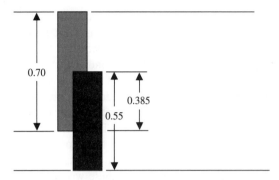

Figure 4.4   Independent events can occur singly or together

between two nodes and a resistor whose value is below the low limit. Two repair actions would be required, one of them being unnecessary.

The more important fact from an economic analysis point of view is that a board with a short and another type of defect will require one additional visit to the tester and to the repair station—in other words, one more pass around the diagnosis/repair loop. This has obvious implications for both cost and throughput calculations. A board with both kinds of defect will usually pass through the tester three times:

1. Once to find the short.
2. A second time to find the other defect.
3. And a third time to confirm that the repair action was performed correctly.

Assuming that we know the individual probabilities for a short and for other types of defect (0.55 and 0.7 in the example), we will also need to determine the probability that the board will fail for either reason. This will then also give us the probability that the board will be defect-free. This probability, usually expressed as a percentage, is of course the *yield* of good boards.

If we simply add the two probabilities we get

$$0.55 + 0.7 = 1.25$$

This of course is impossible. You cannot have a probability greater than 1.0 because this represents absolute certainty. The problem here is that the *special rule of addition* only applies to mutually exclusive events and not to independent events. A short and some other type of defect can occur independently or both together. They are not mutually exclusive. For independent events we have to use the *general rule of addition*, which is normally written as

$$P(A \text{ or } B) = P(A) + P(B) - P(A \text{ and } B)$$

For our board test example this formula says that the probability that a board will fail for any reason [$P(A \text{ or } B)$] is the probability that there will be a short plus the probability that there will be any other type of defect, minus the probability that both types of defect will occur together. This makes sense because if we do not subtract the probability of both occurring together, these defects will get counted twice.

For our example this becomes

$$P_{\text{short or other defect}} = 0.55 + 0.7 - (0.55 \times 0.7)$$
$$= 0.865$$

Now we can determine all of the relevant probabilities. Since the probability of a board failing at all is 0.865 then the probability that it will pass the test (the yield) is

$$P_{\text{pass}} = 1.0 - 0.865$$
$$= 0.135$$

Since 0.865 fail in total and 0.7 fail due to a fault other than a short, then $0.865 - 0.7$, or 0.165, must be the probability that the board will fail due to a short with no other type of defect present.

Similarly, since the probability of a short occurring is 0.55 then $0.865 - 0.55$, or 0.315, must be the probability that the board will fail due to a defect other than a short with no other shorts present on the board. We now have a complete picture of the various probabilities that will determine the flow of boards around the diagnosis/repair loop. This in turn will enable us to calculate costs and throughputs with the best possible accuracy:

1. 13.5 per cent of boards will pass. This is the yield.
2. 16.5 per cent of boards will contain shorts.
3. 31.5 per cent of boards will contain other defects.
4. 38.5 per cent of boards will contain both shorts and other defects.

This is illustrated graphically in Fig. 4.5. Hopefully your yield will not be as bad as this. The numbers here were chosen to illustrate what is happening in the test process and are not intended to represent a typical yield situation.

The special rule of multiplication revisited

Let us now take another look at the use of the special rule of multiplication. This you may recall is used to determine the probability of a number of independent events occurring together. Let us assume that you are experiencing component defect rates of 1 per cent. This is an unthinkable figure today but it keeps the maths simple. This would imply that the probability that an individual component is good is 0.99:

$$P_{\text{good}} = 0.99$$

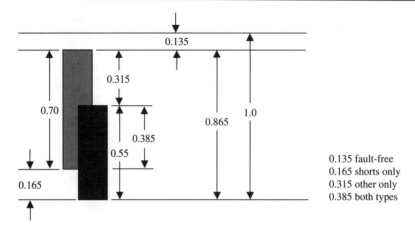

0.135 fault-free
0.165 shorts only
0.315 other only
0.385 both types

Figure 4.5   Graphical illustration of the example in the text

Therefore, if you build a printed circuit board containing 100 components then the probability that you will get *100 good components* will be 0.99 multiplied by itself 100 times. This is the same as raising 0.99 to the power of 100. If your calculator has a *Y to the X* key then you can quickly see that the probability is 0.366:

$$P_{100\ good} = (0.99)^{100}$$

Therefore 36.6 per cent of the boards will be free from component defects even though 100 components with a 1 per cent defect rate would imply that on average each board would contain one defect. On average they will, but in practice the defective components will not be distributed quite so evenly among the boards. It would be amazing if they were.

If we were to have an average of one fault per board, but 36.6 per cent of the boards are defect-free, then the 63.4 per cent of the boards that are defective must contain all of the defects. It follows then that the defective boards will each contain an average of 1.0/0.634, or 1.58, faults. Again the laws of distribution will be at work and since you cannot have 0.58 of a fault on a board (faults are only available in whole numbers), some faulty boards will contain one defect, some will contain two defects, some three, and so on. This is covered more thoroughly in the next section.

The special rule of multiplication applies to boards and sub-assemblies as well as to components. For example, if the probability that boards are defect-free is 0.98 and there are ten boards in a sub-assembly, then the probability that the sub-assembly is defect-free is 0.98 raised to the power of ten:

$$P_{10\ good} = (0.98)^{10}$$
$$= 0.817$$

If there are five similar sub-assemblies in the final product then the probability that it will be defect-free would be 0.817 raised to the power of 5:

$$P_{5\,good} = (0.817)^5$$
$$= 0.364$$

Amazing as it may seem, if you raise 0.98 to the power of 50 (the total number of boards in the system) you will get the same answer. Whichever way you make the calculation this is not a very good yield at final test so it would probably be a good idea to have some form of sub-assembly test stage.

If the various items being considered have different probabilities of being good, then it will be necessary to use the longer approach of multiplying the individual probabilities together rather than raising the average probability to the power of the number of items.

Distribution theory

We saw earlier that, even if there is an average of one fault per board (FPB), not all of the boards will be defective. This is because the faults will not be evenly distributed among the boards. For the 1 FPB case the percentage that will be defect-free (the yield of good boards) will be 36.6 per cent.

Note. Throughout the book I use the abbreviation FPB to denote the average number of faults per board across all boards. I use the abbreviation FPFB to denote the average number of faults per faulty board. For the 1 FPB case 36.6 per cent of the boards will be defect-free and the 63.4 per cent that contain defects will have an average of 1.58 FPFB.

In practice some production departments collect data on the total number of faults and the total number of boards produced in some time period. This equates to an FPB figure. Others count the faults and the number of boards that pass to the repair stations and so have an FPFB figure. It is worth checking on how your defect data is collected because using the wrong numbers will make a big difference to any detailed analysis. The relationship between FPB and FPFB is explained in Appendix D.

I also use the abbreviation DPU for defects per unit and DPDU for defects per defective unit. Both sets of terms (faults and defects) are in common use, which is why I have used both. Some companies differentiate between defects and faults because not all defects will result in the faulty behaviour of a product. In this book, however, I use the terms quite interchangeably.

In practice the statistical relationships between the number of defects and their distribution over the boards will only hold true if the defects occur randomly, which they should do if the manufacturing process is under control. If some part of the manufacturing process is not under control to the extent that it should be, then it is possible that some faults could occur to a pattern rather than randomly. As an example, if a wrong component is included in the assembly kits then all boards would contain this fault, and some proportion of them would also contain the randomly occurring defects. Under these conditions the fault distribution will not match the theory. However, as long as the process is under some reasonable degree of control, the occasional occurrence of some consistent defect should itself be random in nature. For our purposes in test economics we can assume that over some reasonable period of time or some reasonable volume of production, defects will occur randomly. This is a fair assumption since we will normally be concerning ourselves with annual production volumes for a five year period, five years being the most popular period over which to compute an ROI.

Earlier we determined the probability of a board being defect-free by multiplying together the individual probabilities of the components being defect-free. Unfortunately there can be other types of defect, each with varying probabilities of occurrence, so this approach could become quite cumbersome. What we need is some simple way of relating the average faults per board (FPB) to the expected yield. This need is satisfied by this simple formula:

$$Y = e^{-FPB}$$

where e is the exponential constant 2.718. A scientific calculator with an *e to the x* key makes this calculation very easy. If you perform this calculation for the 1 FPB case, you may notice that this method gives a slightly different answer to that obtained by the special rule of multiplication:

$$0.99^{100} = 0.366, \qquad e^{-1} = 0.368$$

This is because e^{-FPB} gives us a Poisson distribution and the special rule of multiplication gives us a binomial distribution. However, the differences are small and either method may be used.

Appendix E contains tables of both binomial and Poisson distributions. These can be very useful for determining defect distributions for a variety of situations. For example, the table of cumulative Poisson probabilities can tell us the probability of r or more random defects occurring when the average number of defects is m. With such a table we can determine the probability of getting 0 or more defects, 1 or more defects, 2 or more defects and so on. Obviously there will always be 0 or more defects so the probability is always 1.0 for an r of 0.

For our 1 FPB example m is 1.0 so if we look down this column we can see that 0.6321 is the probability of getting 1 or more faults. Subtracting this from 1.0 and rounding to three figures gives us our yield of 0.368, the same figure we obtained by raising e to the minus 1.0. The next line down tells us that the probability of getting 2 or more defects is 0.2642. For 3 or more defects it is 0.0803 and so on.

Since these are cumulative probabilities, if we subtract each of them in a column from the one above it, we will get the absolute probabilities for boards with one fault, two faults, three faults and so on. If we do this for the 1 FPB case, round to two figures and convert to percentages, we will see that

37 per cent of boards should be defect-free
37 per cent will contain 1 fault
18 per cent will contain 2 faults
 6 per cent will contain 3 faults and
 2 per cent will contain 4 or more faults

The Poisson tables can be used to determine the distribution of faults for any number of average faults per board. This information is essential for a good understanding of what is likely to be happening within the various test stages.

For higher levels of yield some simplification is possible depending on the level of accuracy required. For an average of 0.1 FPB the defect distribution would be as follows:

90.48 per cent of boards will be defect-free
 9.05 per cent will contain 1 fault
 0.45 per cent will contain 2 faults
 0.02 per cent will contain 3 or more faults

Less than 0.5 per cent of the boards will contain more than one defect, so if you are calculating some overall figures for costs, throughput or capacity of a test stage you may well be able to ignore the distribution effects and simply assume that each board contains one defect. The errors will be greatest when calculating the workload at the repair stations since about 5 per cent of the defective boards will contain multiple faults.

Another simplification that can be made for yields above 90 per cent, which also relates to this distribution effect, is in the relationship between the average number of faults and yield. For an average of one fault per board the yield is about 37 per cent ($e^{-1.0} \times 100$) rather than the zero yield that an average of 1 FPB would imply. When the average faults per board is 0.1 the yield is 90.48 per cent ($e^{-0.1} \times 100$), which is only 5.3 per cent different to simply taking $(1 - FPB)$ as being the yield. Obviously this error reduces at higher yields. With an average of 0.05 FPB the error is only 1.26 per cent.

4.9 Yield

Moving on now from the basic statistical relationships that we need to be aware of, we can take a more detailed look at the issue of yield. As indicated above, yield is the probability that the items being considered are defect-free. More often than not it is expressed as a percentage, and this is simply obtained by multiplying the probability by 100.

The yield you get from a process is a measure of the performance of that process; yield quite simply means money. If you improve the yield of any process it will cost less, the capacity will increase and you will ship better quality products.

We can use some of the techniques outlined in the statistics section to determine the expected yield of a single process step or a complete set of process steps. For example, if we know that the probability that components will be good is 0.995 and that there will be 150 of them on a board, then we know that 0.995 raised to the power of 150 will give us the probability that the board will be free from component defects. This is 0.47. If we also consider that the bare board that these components will be assembled on to has a P_{good} of 0.90 then the yield will now be

$$0.47 \times 0.90 = 0.42$$

If we also know that the probability that we will assemble the board correctly is 0.90 and that we will solder it without defects is 0.95, then our overall yield going into the first test stage will be

$$0.42 \times 0.90 \times 0.95 = 0.36$$

Therefore each step in the manufacturing process has an error rate associated with it. The defects introduced at each stage are cumulative and we can estimate the overall yield of a process by using the special rule of multiplication. This example is shown graphically in Fig. 4.6. This shows what I call a yield progression. As the product progresses through the process, the yield will change. For a manufacturing process step the yield will become smaller. Following a test step, with repair, the yield will increase. The yield will also decrease following any stressing step such as a burn-in or a vibration screen. For any test stage where repair is not possible, as with

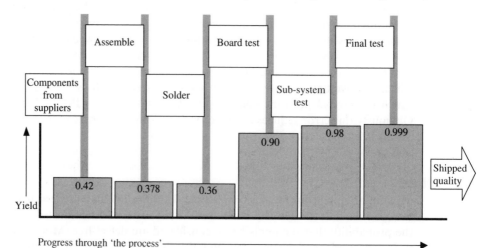

Figure 4.6 The 'yield progression'

semiconductor component manufacturing, the yield will also increase because defective items will be discarded. However, there will be fewer products as a result of this scrapping of defectives. Component manufacturers often refer to this as 'shrinking' because this is lost product that cannot be recovered. It is easy to see why yield has always been such a critical issue in the semiconductor industry. With so many process steps involved, a small improvement at each will make a big difference to the capacity and the profitability of a production line.

Equipment manufacturers are in a somewhat better position because their products are repairable. This may be the reason why they are less paranoid about yield, but, as will become clear from some examples, they ought to be just as concerned about yield as their semiconductor counterparts are. Since they can repair or re-work their products, equipment manufacturers can in theory ship 100 per cent of the products that they set out to make. In the yield progression any test or inspection stage should result in an increase in yield as defects are detected and repaired, with no loss of production volumes. The in-circuit test stage will detect the manufacturing defects and some of the component defects. Sub-system test will find additional defects, as will the final system test stage. In practice, of course, none of these will detect 100 per cent of all the defects present and some will always slip through to be shipped to the customer.

Therefore the effectiveness of each tester, and that of the overall test strategy, will have a major impact on the yield progression, the cost of achieving any improvement and the number of defects shipped to the field. This then brings us to the subject of fault coverage.

4.10   Fault coverage

Fault coverage is a measure of the ability of a tester to detect defects. Fault coverage can be one of the major differences between competing test systems and will be central to any economic justification for a high-performance tester versus a lower priced model with less performance. Fault coverage affects yield in two ways:

1. It influences the yield 'seen' by the tester.
2. It determines the yield coming out of the tester.

To understand what I mean by this it is necessary to consider that any process stage has an input yield and an output yield. For a manufacturing stage the output yield will be lower than the input yield since the stage will have the possibility of inducing defects. For a test stage the output yield should normally be higher than the input yield since any detected defects will be either repaired or discarded. However, the output yield from a test stage will not be 100 per cent because of the fault coverage limitations. In terms of faults, there will be some faults present on the UUTs entering the test stage that will determine the input yield. There will also be the faults detected by the test stage, which will then be repaired. The difference between these will be the faults that escape detection, and these will determine the output yield from the test stage.

4.11 Apparent yield

The yield that is 'seen' by the test stage is a function of the number of faults that are detected by it rather than the number of faults that are actually present. As the fault coverage drops the yield seen by the test stage will increase. This should be quite obvious because if the fault coverage is so bad that no defects are detected, the yield will appear to be 100 per cent. For this reason I refer to the yield seen by the tester as the *apparent yield*. Figure 4.7 shows the three yields associated with a test stage.

You may recall that the relationship between the number of faults and yield can be expressed by the formula

$$Y = e^{-FPB}$$

This formula relates the actual yield to the actual number of faults present and so this is the formula for the input yield to the stage. The apparent yield (Y_a) at the tester is a function of the number of *detected faults per board* (DFPB) and so can be expressed as

$$Y_a = e^{-DFPB}$$

Alternatively, since the number of detected faults is a function of the number of actual faults and the fault coverage of the test stage, we can also express Y_a as follows:

$$Y_a = e^{-(FPB)FC}$$

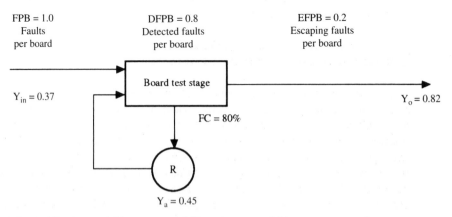

Figure 4.7   Input yield, apparent yield and output yield

where FC is the fault coverage expressed as a factor, as opposed to the usual way of expressing it as a percentage.

The output yield (Y_o) from the stage is a function of the number of *escaping faults* (EFPB) and can be expressed as

$$Y_o = e^{-EFPB}$$

Output yield can also be expressed as

$$Y_o = e^{-(FPB-DFPB)} \quad \text{or} \quad Y_o = e^{-[FPB-FPB(FC)]}$$

As an example, if we have 1 FPB coming into this test stage and the fault coverage is 80 per cent, then we will have 0.8 DFPB and 0.2 EFPB. This will give us an input yield of 0.37 or 37 per cent, an apparent yield at the tester of 0.45 or 45 per cent and an output yield from this stage of 0.82 or 82 per cent. The difference between the input yield and the apparent yield is one reason why some companies think that their manufacturing yields are higher than they really are. There are, however, other factors that may influence this view which relate to the way in which data is collected from the production process.

At any point in time there will be batches of newly manufactured boards and batches of repaired boards entering a test stage. If the data collection system being used differentiates between these batches then the yield measured at the tester for the new boards should be close to the apparent yield that could be calculated from a knowledge of the FPB and the fault coverage of the test stage. If the data collection system does not differentiate between these batches of new and repaired boards, then the yield measured from the data would be very much higher than any calculated value because the high yield of the batches of repaired boards will skew the results.

Using the example of 1 FPB and a fault coverage of 80 per cent we determined that the apparent yield would be 45 per cent. Therefore 55 per cent of the boards will fail (be detected) and go off to be repaired. Assuming for simplicity that the repair action is performed correctly and that none of the boards with 2 or more defects require a second pass around the diagnosis/repair loop, then the repaired boards will all pass the test when they are re-tested. If the data collection system differentiates between these two batches of boards entering the test stage, then we will see apparent yields of 45 per cent for the newly manufactured boards and 100 per cent for the repaired boards.

4.12 Perceived yield

If the data collection system does not differentiate between these batches then for every 100 boards produced the test stage will see 155 tests, of which 55 fail and 100 pass. The yield based on these numbers will be 100/155 or 0.65 (65 per cent). This is shown in Fig. 4.8. To avoid confusion with apparent yield, which simply accounts for the effect of fault coverage, we call this the *perceived yield* (Y_p). In this example,

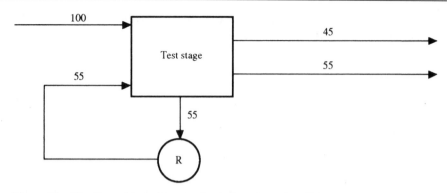

Figure 4.8 The flow of boards around a test and repair cell

as in practice, this erroneous measure of yield caused by deficiencies in the method of data collection is usually well above the actual yield of 37 per cent and the true apparent yield of 45 per cent.

It is easy to see from this example why the yield calculated from a knowledge of the number of faults can vary quite widely from the yield measured from production data. Even if the measured number of actual faults is used to calculate yield there may well be discrepancies depending on how the passes and fails at the test stages are counted. Automatic quality data collection systems do not allow for this incorrect counting of data. As a result, almost every installation of such systems quickly shows the yields to be lower than the customer thought they were. We need to be aware of these sources of error if we are to develop reasonably accurate economic justifications. Many justifications will be based upon improving the yield or on lowering the cost of achieving acceptable yields. To attempt to make any cost calculations without a knowledge of all of the 'different' yields involved would be a bit risky.

For the example above we made the simplifying assumption that all defective boards would be repaired correctly and completely with one visit to the repair station. However, for a variety of reasons a 'functional' test system can usually only find one fault at a time, so if there is more than one defect present multiple passes around the diagnosis/repair loop will be needed. In the section on statistics we also saw that, contrary to popular belief, an 'in-circuit' tester will not necessarily find all defects in one operation. It will find all shorts in one test and all component-related faults in one run of the test program, but not a short and a component-related defect together. It is possible on most in-circuit testers to override the 'stop-on-shorts' mode but it is not a good idea to do so. It is also good practice to 'stop-on-bus-faults' because such a defect may also affect the validity of some of the following tests.

Let us now look again at the example we used earlier to illustrate the use of simple probability rules to determine the proportion (probability) of boards with both a short and some other type of defect. This time we will analyse the flow of boards around the diagnosis/repair loop assuming a 100 per cent fault coverage for the tester.

We will then repeat this analysis with a less than perfect fault coverage to see the effect that this has on the operation. The diagram in Fig. 4.9 shows what happens when the fault coverage is 100 per cent and the repairs are performed correctly. This diagram makes it fairly obvious how the shorts and the other defects are distributed over the boards. The diagram in Fig. 4.10 shows how the boards pass around the diagnosis/repair loop. Initially 1000 boards enter the tester. Of these (the yield) 135 pass the test and migrate on to the next process step. Then 865 fail the test and move to the repair station. Of these 865, some 550 will have failed the shorts test part of

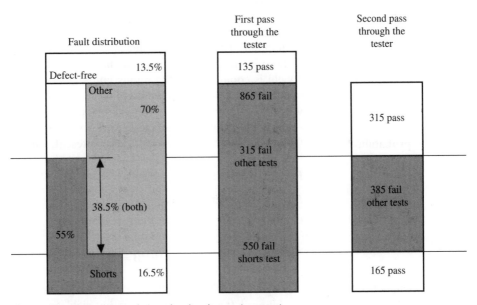

Figure 4.9 Defect detection at in-circuit test (see text)

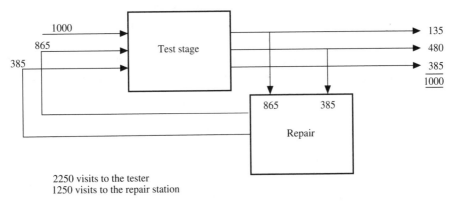

2250 visits to the tester
1250 visits to the repair station

Figure 4.10 Board flow diagram for the example in the text

the in-circuit program and the other 315 will have passed through the shorts test but failed in the component testing portion of the program. After repair the 865 boards return to the tester for re-test. Now the boards that originally only had one defect type will pass the test. These will be the 165 that only had shorts and the 315 that only had other defect types present. These 480 boards will now migrate on to the next process step. The 385 boards that had both types of defect present will now pass through the shorts test but will fail the remainder of the test program. These now return to the repair station, are repaired and then return to the tester where they should now pass the entire test program and migrate on. In total there have been 2250 visits to the tester and 1250 visits to the repair station in order to get the 1000 boards through to the next step.

Now let us see what happens if the fault coverage is less than 100 per cent. Before we do this, however, there is another little formula that we need. So far we have covered the fact that yield is equal to e to the minus faults per board, but now we need to calculate how many faults per board there are when we know the yield. This is necessary because we need to determine the detected faults per board by multiplying the actual faults per board by the fault coverage factor. So far all we know is the probability of there being shorts or other types of defect present. Thus

$$P_{\text{shorts}} = 0.55; \text{ therefore } Y_{\text{shorts}} = 1 - 0.55 = 0.45$$

$$P_{\text{other}} = 0.70; \text{ therefore } Y_{\text{other}} = 1 - 0.7 \quad = 0.3$$

Now

$$Y = e^{-\text{FPB}} \quad \text{and} \quad \text{FPB} = \text{LN}.Y$$

where LN is the natural logarithm. This actually produces a negative answer for FPB. The negative sign can be ignored. It results from the fact that yield is e to the *minus* FPB.

From the above we can calculate that

$$\text{Shorts per board} = 0.80$$

$$\text{Other defects per board} = 1.2$$

Let us now suppose that the fault coverage of the tester is 95 per cent for shorts and 85 per cent for all other defects. This will result in detected faults per board figures of

$$\text{Detected shorts per board} = 0.76$$

$$\text{Detected 'other' faults per board} = 1.02$$

The apparent yield for each defect type will therefore be

$$Y_{\text{a shorts}} = e^{-0.76} = 0.468$$

$$Y_{\text{a other}} = e^{-1.02} = 0.361$$

Therefore the probability of detecting a short will be

$$P_{short} = 1 - 0.468 = 0.532$$

and the probability of detecting other defects will be

$$P_{other} = 1 - 0.361 = 0.639$$

The probability of getting both types of fault on the same board is obtained by multiplying P_{short} by P_{other}:

$$P_{both} = 0.532 \times 0.639$$

$$= 0.340$$

The overall apparent yield can be obtained from e to the minus 1.78 (the total number of detected faults):

$$Y_a = e^{-1.78} = 0.169$$

From this we can see that the overall probability of new batches of boards failing the test will be 0.831. We can then determine the probability of boards failing with shorts only and with other defects only in the same way that we did earlier:

$$P_{shorts\ only} = P_{fail} - P_{other}$$

$$= 0.831 - 0.639$$

$$= 0.192$$

$$P_{other\ only} = P_{fail} - P_{short}$$

$$= 0.831 - 0.532$$

$$= 0.299$$

From all of this we can now construct the diagrams that show the distribution of the faults across the boards, and the flow diagram showing boards passing around the diagnosis/repair loop. These are shown in Fig. 4.11. Now that the fault coverage is less than perfect we see that the tester is doing less work. The batch of 1000 boards now result in only 2171 visits to the tester as opposed to 2250. This is simply because some defective boards are not being detected and so do not require any re-test. Similarly, there are now only 1171 visits to the repair station. On the first trip through the tester 831 boards fail, 532 as a result of shorts and 299 due to other defects. Thus 169 boards pass the test and migrate on to the next stage in the process. However, we know that 34 of these must contain defects because the actual yield is only 13.5 per cent as opposed to the apparent yield of 16.9 per cent. The 831 boards are repaired

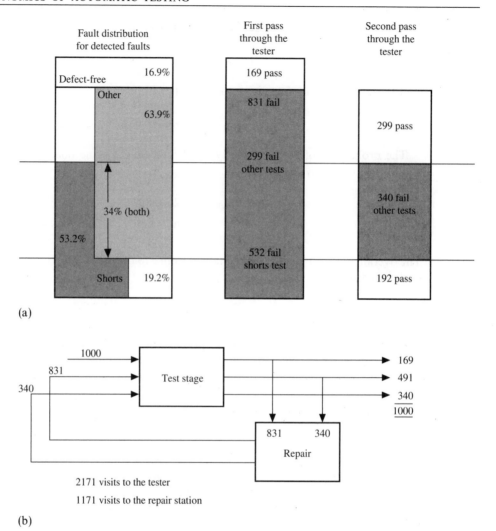

Figure 4.11   Defect distribution and flow diagram when the fault coverage of the tester is allowed for

and returned to the tester. Now, on this second trip 340 fail, these being the boards that originally had both shorts and other defects present. These boards pass around the diagnosis/repair loop a second time.

Incidentally, if the data collection system is unable to differentiate between new batches of boards and batches arriving at the tester from the repair station, then the perceived yield would be 46 per cent. This is well out of line with the actual yield, but in a higher yield situation the differences between the actual yield, the apparent yield and the perceived yield will not be quite so obvious. This is all the more reason then to be sure that the data is valid.

4.13 The effects of incorrect repair action

So far we have assumed that the repair action will be performed correctly and completely for all defects on each visit to the repair station. In practice this rarely happens for a number of reasons:

1. The diagnosis from the tester may be incorrect or ambiguous.
2. The diagnostic message may be interpreted incorrectly.
3. The repair action may cause another defect.
4. The tester may have 'detected' a non-existent defect (see Fig. 4.2).

Whatever the reason there will usually be more repair actions than the actual number of faults originally present on the boards would require. To allow for this we need to apply a correction factor to any calculations involving diagnosis or repair. If we do not do this then any calculations of cost, throughput and capacity will be in error. This correction factor is usually referred to as the *loop-number multiplier*. There are several ways that this factor can be determined, but the simple thing to do is to estimate a percentage of incorrect or unnecessary repair actions. This can be based upon past experience, if the data exists, or a best guess at what the figure might be. Including an estimate of the loop-number multiplier is still preferable to ignoring it. Leaving it out would be equivalent to a wrong guess on the low side.

Since there will normally be the same number of diagnosis actions and repair actions, the correction factor will be applied to both. However, it will not impact the number of 'passing' tests. In any model of the test process it is important to note that a board, or anything else for that matter, will normally only get *one* passing test at each test stage. It will either pass on its first visit because no faults are detected or it will eventually pass after one or more trips around the diagnosis/repair loop when all faults have been successfully repaired. This may sound pretty obvious but I have seen numerous formulae over the years that have effectively modelled two passing tests for boards with defects. This erroneous approach is, I assume, based upon the idea that boards are tested, repaired and then re-tested. However, the first test would be a 'diagnosis' and not a 'passing test'; otherwise the repair would not have been required. Since a diagnosis will often take longer to perform than a passing test, it is important to differentiate between the two if accuracy is to be preserved. Since the loop-number multiplier will be applied to diagnoses and not to passing tests, any errors caused by not applying this differentiation would be further increased.

Let us now return to the previous example and apply the correction factor to see the effect that this incorrect repair action will have on the test process. To simplify the analysis we will assume that any incorrect or unnecessary repair action will be performed correctly on its second visit to the repair station. This will not always be

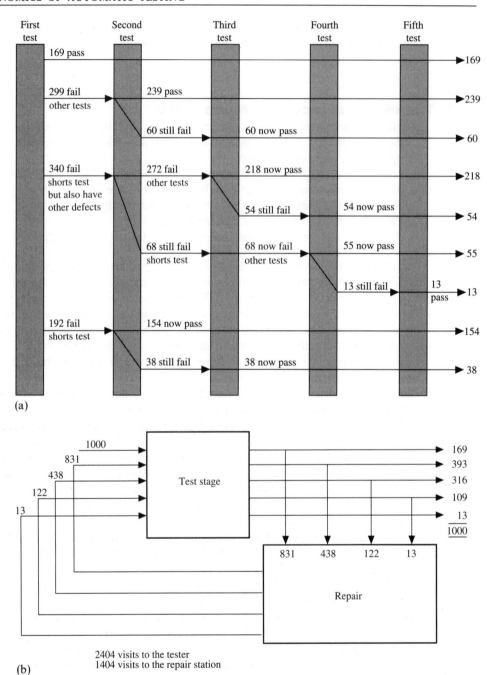

(a)

(b)

2404 visits to the tester
1404 visits to the repair station

Figure 4.12 The example from Fig. 4.11 with the effects of the diagnosis/repair 'loop-number multiplier' added. This correction factor accounts for the unnecessary diagnosis and repair actions that take place

the case. It will require a system that logs previous unsuccessful repairs to provide a good probability of this happening. It will not happen by chance.

The diagram in Fig. 4.12 should help to clarify what is happening here. The loop-number multiplier has been set at 1.2, which implies that 20 per cent more diagnosis and repair actions than are theoretically required will take place. A further simplification is that the same correction factor of 1.2 will apply to both shorts and other types of defect. In practice, for best accuracy, any model should make it possible to set the loop-number to a different value for each class of defects since shorts are more likely to be repaired correctly.

We can see from the diagram that of the original 831 failing boards, 166 (20 per cent) are not correctly repaired. Of these 38 are boards with shorts only, 60 are boards with other defects only and 68 are boards containing both types of fault. The incorrectly repaired (or added) shorts on these 68 boards will continue to mask the other defects on these boards. Therefore on the second visit to the tester only 272 of the 340 boards with both types of fault can now be tested for the 'other' category and 54 of these (20 per cent) will be incorrectly repaired. The 60 boards with 'other' defects are now correctly repaired on their second visit to the repair station, as will be the 38 boards with shorts. On their third visit to the tester all but 13 of the 68 boards with remaining 'other' defects will now be correctly repaired, as will be the 54 that were incorrectly repaired on the last visit to the repair station. Finally, the last remaining 13 boards will now be diagnosed and repaired correctly.

The flow diagram shows the flow of boards around the test and repair stage. The totals have now risen to 2404 visits to the tester, from 2171, and 1404 visits to the repair station, from 1171. This will reduce the *capacity* of the test system by approximately 10 per cent. This is only a first-order approximation based on the number of visits. A more accurate assessment would require the actual test times, handling times and diagnosis times to be used. The repair station would lose about 20 per cent of its capacity. It is interesting to note that some boards are now passing around the diagnosis/repair loop four times even though we simplified the model by assuming that all boards will migrate to the next process step after a maximum of two visits to the repair station. In practice this is not necessarily the case and some problem boards may pass around the loop many times.

The flow diagrams we have been using to illustrate these different test situations provide a useful way of checking the accuracy of any mathematical models of test stages within an overall test strategy.

4.14 Checking your data

At times it will be necessary to use historical defect and yield data in the preparation of a justification for a new test strategy or a new tester. If the yield calculated from the defect numbers does not correlate with the measured yield it may be necessary to determine why so that the correct information can be used in the justification. As explained earlier, the most common reason for errors is that the data collection system does not differentiate between new batches of boards and repaired batches.

This gives rise to a perceived yield that is based on the total number of test actions and the total number of passing tests:

$$Y_p = \frac{\text{number of passing tests}}{\text{total number of test actions}}$$

This would be multiplied by 100 if you want a percentage.

If you examine the flow diagrams for the various examples we have looked at you will see that the total number of test actions is equal to the number of passing tests added to the number of repair actions (visits to the repair station). We can therefore rewrite the above formula as

$$Y_p = \frac{\text{number of passing tests}}{\text{passes plus repair actions}}$$

If we normalize the numbers to probabilities, or the proportions of the batch that pass and that visit the repair station, we can rewrite the above formula simply as

$$Y_p = \frac{1}{1 + \text{proportion visiting repair station}}$$

For an in-circuit tester we determined that two repair actions will be needed for boards that contain both shorts and other types of defect, so the total number of repair actions is the number failing the first test plus the number failing the second test. The proportion failing the first test will be $1 - Y_a$ and the proportion failing the second test will be the proportion that have both types of defect present, P_{both}. $1 - Y_a$ can also be written as P_{fail}, the overall probability of a board failing, so the average number of repair actions per board will be

$$\text{Repair actions per board} = P_{fail} + P_{both}$$

To account for any unnecessary repair actions we simply need to multiply this by the loop-number multiplier (LP). Perceived yield then becomes

$$Y_p = \frac{1}{1 + (P_{fail} + P_{both})\text{LP}}$$

As a check on this, if we insert the numbers from the last example we will see that

$$Y_p = \frac{\text{number of passes}}{\text{number of tests}}$$

$$= \frac{1000}{2403}$$

$$= 0.416$$

Using the generalized formula,

$$Y_p = \frac{1}{1 + (0.830 + 0.342)1.2}$$

$$= 0.416$$

The defect data should enable us to determine P_{short} and P_{other}. From these we can determine P_{both} and Y_a. If the yield data does not agree with Y_a we can use the above formula to determine Y_p. If this agrees reasonably with the yield data we have found the cause of the lack of correlation and we can correct the figures. If Y_p does not agree with the yield data then we will have to look elsewhere for the source of the error. Hopefully the methods covered here will be of some help in finding the problem.

4.15   Perceived yield for functional test

The formula we derived above covers the in-circuit test case. For a functional test the formula for Y_p is a little simpler. Since functional testers generally find only one fault at a time, regardless of the nature of the fault, the number of repair actions is simply the number of detected faults per board (DFPB). This figure has then to be multiplied by the loop-number multiplier to allow for any unnecessary repair actions. Thus

$$Y_p = \frac{1}{1 + (\text{DFPB})\text{LP}}$$

4.16   Using yield progression to study the process

As stated in the introduction chapter, the analysis of test economics will normally be a three-stage process. The first step will be to determine which test strategy is ideal for the situation under review, the second step will be to determine which specific equipment will maximize the effectiveness of the chosen strategy and the third step is to prepare the financial justification. For the most thorough analysis of the problem the first two steps will require the development of a mathematical model that accurately represents the process. In most cases the model can be created using a spreadsheet program running on a personal computer. The model would be used at stage 1 to determine which of the alternative test strategies is the best fit. In this mode the various test stages modelled would be given generic performance capabilities. For stage 2 the performance of the tests would be modelled in more detail so that alternative systems, possibly with widely different prices, can be compared. These two stages could be analysed with a single model that can accept different degrees of detail, or it may be easier to use two separate models. Either way it is important that the models accurately reflect what happens in the process because some very big decisions will be based upon the results.

One approach to the modelling of the overall process at the macro level is to base the model on the yield progression. In this way it is possible to track where the defects are sourced, where they are detected and which of them 'escape' detection. For this overall view it is usually sufficient to model the test performance in terms of the simplified fault spectrum where the defects are split into three main categories: design induced defects, supplier induced defects and manufacturing process induced defects. Such a simple model can be useful for the initial investigation stages when a number of 'what-if' simulations can be performed very quickly without the need to supply masses of input data or to analyse masses of results. Once the most promising alternatives have been determined, a more detailed model should be used for the final decisions.

4.17 Capacity calculations

Capacity calculations are required to see if the test strategy under consideration can cope with the workload and how many test and repair stations will be needed. It will also be necessary to predict when any spare capacity at each stage of the test process will be used up so that appropriate plans can be made to increase capacity. This will usually involve adding an additional test station or increasing the time available with overtime or second shift operation. Some discussions on test economics tend to equate capacity with throughput. They are directly related but they are not usually equivalent. Throughput, usually measured in units per hour, will be a function of the test and diagnosis times, the yield and the amount of unnecessary diagnosis that takes place. Capacity will be a function of the throughput and the amount of time that is available for testing. This in turn will be a function of the total available time at each test stage and the time lost as a result of other factors. The lost time will typically come from test program and fixture debugging on the testers, downtime, routine maintenance, program changeovers, housekeeping operations on the tester and 'human' functions such as tea breaks and key discussions about last night's football match.

When will additional capacity be required?

Having determined the amount of spare capacity at a particular stage it is often useful to know when this capacity will run out. This is the same problem as a compound interest calculation in that we have a present value (the current workload), a future value (the maximum capacity), an interest rate (the rate at which the workload is increasing) and a time period (the number of years before the present value equals the future value at the current interest rate). The formula for compound interest is

$$FV = PV(1 + i)^n$$

where

$$FV = \text{future value}$$

$$PV = \text{present value}$$

$$i = \text{interest rate (as a factor)}$$

$$n = \text{number of periods (usually years)}$$

Solving for n we get

$$n = \frac{\log(FV/PV)}{\log(1+i)}$$

This is all right as long as the maximum capacity (future value) is fixed. However, in most cases this will be changing. It may be reducing if the factors that make up the 'lost time' are increasing, such as when the programming load is increasing due to more programs or to increased complexity. Alternatively, the maximum capacity may be increasing if the factors making up the 'lost time' are diminishing.

Capacity calculations when estimated maximum capacity is not constant

If the estimated maximum capacity on a test system is likely to change, the formula for determining when additional capacity will be required become slightly more complex. For the case where the capacity is likely to fall, perhaps due to an increase in programming workload, the situation shown in Fig. 4.13 will occur, where

$$MC = \text{present estimated maximum capacity}$$

$$FV = \text{capacity at the time additional capacity will be required}$$

$$PV = \text{present volume}$$

$$i_1 = \text{factor (percentage) decline in capacity}$$

$$i_2 = \text{growth rate of production volume}$$

$$n = \text{time (number of periods) when } MC = PV$$

$$MC(1+i_1)^{-n} = FV$$

and

$$PV(1+i_2)^n = FV$$

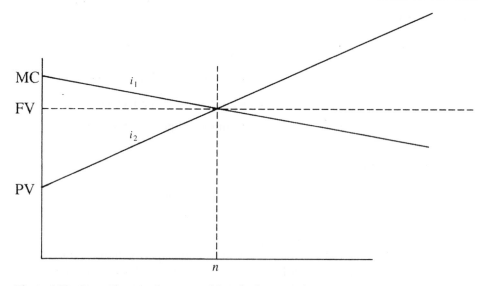

Figure 4.13   Increasing requirements with reducing capacity

Therefore

$$MC\left(\frac{1}{1+i_1}\right)^n = PV(1+i_2)^n$$

$$\frac{MC}{PV} = \frac{(1+i_2)^n}{[1/(1+i_1)]^n}$$

$$\log\left(\frac{MC}{PV}\right) = \log\left[\frac{1+i_2}{[1/(1+i_1)]}\right] \times n$$

and

$$n = \frac{\log(MC/PV)}{\log[(1+i_2)(1+i_1)]}$$

For the case where the maximum capacity is likely to increase, due to declining programming efforts or to the introduction of something that will take some of the workload off a tester, the situation shown in Fig. 4.14 will occur, where i_1 is now the rate of growth of system capacity. Therefore

$$MC(1+i_1)^n = FV$$

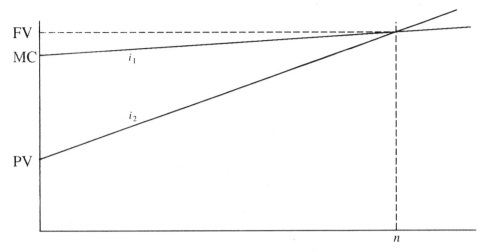

Figure 4.14 Increasing requirements with increasing capacity

and

$$PV(1+i_2)^n = FV$$

Therefore

$$\frac{MC}{PV} = \frac{(1+i_2)^n}{(1+i_1)^n}$$

and

$$n = \frac{\log(MC/PV)}{\log[(1+i_2)/(1+i_1)]}$$

4.18 Testability

The 'testability' of a chip, a board or a system is an indicator of the ease with which tests that will detect all defects can be generated. The complexity of developing test programs is such that the UUT will usually be broken down into smaller blocks (partitioned) and then a set of tests developed for each of the blocks. This usually means that the test signals or patterns will have to work their way through other blocks before reaching the block that is currently being tested. This in turn means that we need to be able to control these other blocks in such a way that the desired test patterns actually reach the inputs of the block under test. However, this is only half of the battle. We then need to make sure that all of the responses of the block under test can be observed by the tester so that any incorrect responses caused by a fault condition can be detected. This is shown diagrammatically in Fig. 4.15. The primary inputs to the UUT are those that the tester has direct control over and

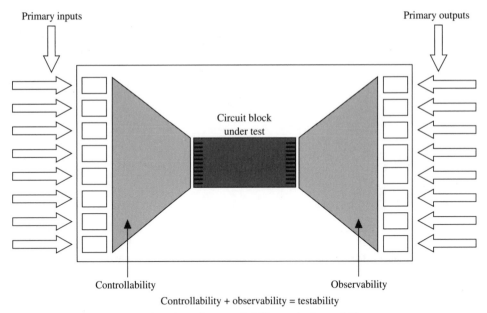

Figure 4.15 Testability—a function of controllability and observability

the primary outputs are those that the tester can observe directly. In other words, these are the 'driver/sensor' pins of the tester. Digital functional board testers that drive and sense from the edge connector tend to suffer most from testability problems and additional test points are often required to gain observability. ASIC designs can also have poor testability, but here there is no possibility of adding test points. In-circuit board testers suffer less from poor testability because they can usually access all of the nodes of the board and therefore bypass the controllability and observability problems. Boundary scan techniques also seek to provide full access to all nodes and so overcome controllability and observability problems. Other techniques of building testability into designs are also used and major organizations such as the IEEE are constantly looking at new methods. Testability problems are not exclusive to digital circuits. Many analog designs can also be very difficult to test.

Closely related to 'testability', which is primarily the problem of generating the right electrical tests, is the issue of accessibility. This is the problem of the physical access needed to apply the tests. Surface mount technology with its small pin spacing can sometimes lead to problems concerned with getting the test probes on to all of the nodes of a board. Good physical board layout rules are needed to ensure that the right tests can be applied to the right places.

Testability has nothing to do with the testers. It is purely a function of the design. Some tester types may cope with testability problems better than others, but poor testability is still poor testability.

5. Design and test economics

One of the major changes over the past ten years or so has been the gradual coming together of the design and test operations in terms of better communications, more empathy for each other's problems and closer working relationships. There is also quite a lot of automatic transfer and translation of the physical and behavioural elements of a design which enable parts of the test generation process to run more smoothly. The elimination of a lot of duplicate effort not only saves time but also reduces the chances of making mistakes. This has been a very necessary trend but one that, unfortunately, does not always find the same rate of acceptance in all organizations. This is nothing new. All new ideas and new technologies tend to be accepted at different rates by different organizations. There will always be the early adopters, the fast followers, the main group and the late adopters. Unfortunately some of the early adopters of the 'integration of design and test' did not do the right thing in the right manner and this has resulted in some scepticism about the effectiveness of the idea because of their reports that the results did not justify all of the work that went into the program.

5.1 The 'over the wall' problem

The biggest obstacle to making design and test integration work effectively is not a technical one but an organizational one. The traditional separation of the design engineering and the test engineering functions into different parts of the organization, with separate budgets and separate reporting structures, is at the heart of the problems that have been encountered. The two functions are frequently in different buildings and often at completely different locations. In some cases they may even be in different countries. This all tends to make communication very difficult, even when there is every good intent to work together as a team. Over the years this has led to a high degree of sub-optimization, as discussed in Chapter 1. It led to a situation generally referred to as the 'over the wall' problem. The engineering department would complete a new design and then throw it over the wall to test engineering who would then be expected to produce high fault coverage test programs in a short time with little or no communication with the designer. Test engineering were not entirely blameless in all of this as they would frequently solve a testing problem but fail to communicate the problem or the solution to engineering. There was only very minimal cooperation between the two departments.

The more reasonable managers involved in all of this began to realize that this was a silly way to go about things. They should be combining their talents to fight the competition not each other. They realized that survival required team-work in order to meet their cost reduction goals along with the need for better quality and reduced time to market. The technology issue was important also. The complexity of new designs was such that it was quite easy to design something that could not be tested adequately—at least not with the test strategy and the testers that were in use at the time. Designers were frequently unaware of the limitations of testers or the problems of test generation.

Therefore educational programs were implemented to cross-train design engineers and test engineers about each other's problems. The more enlightened companies realized that team-work was the only way to make the integration work. Unfortunately some companies tried to force the issue with less than ideal results and it was this that led to some bad publicity about the effectiveness of design and test integration.

5.2 Design ... for ... to ... and ... test (the right philosophy)

It is possible to describe several different approaches or philosophies to implementing design and test integration using the above sub-heading. *Design for test* describes an approach where the test engineering group take something of a leading role and define (or even dictate) a set of testability rules that have to be followed closely by the design engineering group so that new designs will be testable using the chosen test strategy. This is not to be confused with 'design for test' or 'design for testability' (DFT) features that are built into the design to improve testability, and which are usually a good thing.

Design to test can describe an approach where the design engineering group take the lead and provide testability features and possibly some portions of the test program. This is then passed on to the test engineering group who have to make the most of it and try to get an acceptable level of fault coverage from something that they have not been involved with from the beginning.

Although both of these approaches will improve the situation relative to the 'over the wall' approach they are still not ideal. Both are sequential in nature. Design engineering do their thing and then pass the ball to test engineering to do their thing. This results in time to market delays that may jeopardize the success of the product. Perhaps more importantly, these slightly dictatorial methods can result in a degree of ill feeling between the two groups that is counter-productive to the overall objectives. The really successful implementations have employed a *design and test* approach where each group has equal status. They work together as a team for the overall good of the project and therefore of the company. For example, the merits of the various DFT options that could be utilized in the design are discussed up front rather than being presented to test engineering as part of the design release.

The design and test approach is essential if concurrent engineering is to work effectively, and this is really essential if time to market goals are to be met. There is still a long way to go. Design and test need to move even closer together than they are today. Eventually they will share a truly common database and a common set of tools. It is possible, of course, that the two activities may eventually be combined, but this will depend on the nature of any technological advances, particularly in the area of automated test pattern generation (ATPG). At the present time there is still a need for people with rather different viewpoints about testing. This has been made very clear in the ASIC design and test arena. The test patterns that a design engineer produces to prove that the design works can be very different to the test patterns needed by the test engineer to detect and diagnose a defect.

5.3 What are the economic issues?

There are several quite different situations which require an economic approach to the analysis of the alternatives. The first involves the justification of the purchase of the electronic design automation (EDA) tools that are required to get the design job done. EDA or CAE (computer aided engineering) equipment is mandatory for the design of ASICs. Simulation and logic synthesis techniques are essential to this activity because there is really no viable alternative. You cannot use traditional breadboarding techniques to verify the design of such complex devices. However, the use of similar approaches to the design of the rest of the circuitry on the PCB is a different matter. Here the design and the design verification more often make use of the more traditional techniques of building prototypes and testing with benchtop instruments rather than computers and simulation. It will therefore be more difficult to convince management to invest in the tools that are required to use the more modern methods. There are very good reasons to apply the techniques of chip design to the other areas of a PCB. Various research has shown that between 30 and 50 per cent of all ASIC designs do not work correctly with the board prototype when the two are first brought together, even though both check out correctly in isolation. It is easy to see why this will happen. You are bringing together the results of the work of different people that have been designed using different techniques and verified using very different techniques. When you consider the complexity involved it is surprising that so many of them do work together. However, when they do not there are usually only two alternatives: redesign the ASIC or redesign the board. Both are expensive activities in terms of cash and of time to market. The solution to this problem is to use the same techniques for the design of both parts. There have been many technological improvements to make this feasible but it needs investment in the right tools and that will require a financial justification.

The second major area where an economic analysis approach is needed involves the decisions about including DFT features in a new design. Nothing is free. Including DFT in an ASIC costs additional design effort, it may increase the size of the chip and it may impose a performance reduction. If the product will not run as fast you may have to reduce the selling price. On the positive side you should get a more testable product, improve your time to market and improve the field maintainability of the product. An economic analysis is the only way to sort out these kinds of trade-off.

The third major area involves economic analysis for the justification of the tools needed to integrate design and test. These are the specialized packages needed to take the data from the design and convert it into the form needed by the test generation system. Some of these tools are provided by the CAE vendors and some by the ATE vendors, but there are also a number of companies who specialize in the supply of these links between the CAE and ATE equipment. These tools cost money to purchase and they cost money and time to use. These costs have to be compared to the cost benefits or the value that comes from using the tools, so yet again we have a need for an economic analysis approach to decide on a technical solution.

5.4   PCB simulation

As indicated earlier, the use of two different design and verification methodologies for different parts of the same printed circuit board cause problems when the two are brought together. Nobody designs an ASIC without simulating it, but very few board designs pass through any form of simulation for the circuitry that surrounds the ASIC. The two elements are also usually designed by different individuals or teams of individuals. In fact the modern PCB is so complex that it usually does require a team of designers with specialists in the various disciplines that are involved. The job is so complex that the simulation of a board is usually referred to as 'system level simulation'. Various reports indicate that as many as 50 per cent of ASICs do not function correctly first time at the system level. Even if this number is exaggerated there is obviously plenty of room for improvements and the cost of ASIC development in both cash and time means that a viable solution should be easily justified.

As with any economic analysis, the first task is to identify the costs and the savings. System level simulation tends to have different hardware and software requirements to ASIC simulation. Most ASIC designs can be simulated on engineering workstations with only the very complex devices requiring more powerful computers. However, simulating an entire board, or even just the part of it that is surrounding the ASICs, is a much bigger task. To avoid having the whole operation grind to a halt, with a major impact on the time to market goals, it is necessary to use some different approaches. ASICs are almost always simulated at the gate level so that the designer can examine functionality and timing in great detail. In any event it is necessary to design the device using the library of basic elements provided by the ASIC vendors who will eventually manufacture the devices. The more gates there are in the design the longer it takes to simulate it so more computing power is required. There are hardware accelerators available to speed up this process but this is not the only problem. Many of the devices surrounding the ASICs on a board will be complex components for which you cannot obtain gate level models so other modelling techniques have to be used. *Behavioural* models effectively describe the operation of the device and these are widely used in system level simulation. Other techniques include *register transfer logic* (RTL) models and *functional* models. In many cases these software techniques alone cannot model the complete system, mainly because of the lack of accurate information about all of the devices. The way around this problem is to use a *hardware modeller*. The hardware modeller interfaces to the simulator and provides it with the necessary data from a real device that is plugged into the hardware modeller.

The investment needed to make use of system level simulation will be the cost of buying a powerful computer, a suitable simulator with a model library and a hardware modeller. The benefits to be derived from such a system will span all areas of the product life cycle and most of the benefits will accrue outside of the engineering department. Within engineering most of the cost savings will come from shortening the design cycle, from reduced ASIC development costs, from reduced ASIC prototype costs, from reduced board prototyping costs and from fewer board prototype cycles. The direct cost savings in engineering are likely to be small relative to the benefits of improved time to market and quality. The results of the McKinsey study referred to in Chapter 3 indicated that the profitability of a product is relatively insensitive to the development cost. However, this example was for a product with a long life expectancy (five years) relative to many electronic products. As the product lifetime reduces the sensitivity to development costs will increase but so will the importance of time to market (TTM). In most cases the benefits of a faster TTM will overshadow any direct cost savings in engineering, but these savings should not be ignored. To begin with, they will be more believable than the larger but potentially difficult to predict TTM savings. The techniques outlined in Chapter 3 can be used to determine the TTM cost benefits. Alternatively, it may be possible to include a simplified calculation of these in a spreadsheet model that looks at the overall effects of system level simulation.

5.5 Quality benefits

The quality benefits that can be derived from system level simulation could in some cases overshadow the TTM benefits, but unless careful records have been kept about testing costs in the factory and the field this may be difficult to prove. The simplified defect occurrence model described in Chapter 1 splits the source of defects into three major areas: design induced defects, supplier induced defects and manufacturing induced defects. Of these the design induced defects are without doubt the most difficult and the most expensive to find. They are also the most likely defect type to escape to the field to be detected by the ultimate tester—your customer. There is a subtle difference between design induced defects and the other two source categories that is not always appreciated. *A design error will be present on all boards.* Supplier induced defects and manufacturing induced defects are random in nature, but a design error will be built into every board you make. It will not cause all of the boards to fail, but it will be there like some hidden time bomb waiting to go off. The nature of design defects is that they result in marginal or variable performance. They frequently take the form of complex interactions that manifest themselves as timing or speed-related problems or as intermittencies. They usually only account for between 2 and 5 per cent of the overall fault spectrum but they cannot be detected by any but the most elaborate and expensive types of board tester—the so-called 'performance' tester that can cost several million dollars. These testers are architected along similar lines to a VLSI component tester with cycle-based testing and accurate control of timing. The more commonly used in-circuit testers, functional testers and combinational testers are unlikely to find any but the most glaring kind of design mistakes.

Most commonly the next test stage would be a sub-system test, also referred to as a hot mock-up or a functional verification test. This is usually a functional test of a sub-assembly, running at full operational speed. As a result some of the design induced defects will be detected here but the diagnosis can often be a real headache. It is often a very lengthy process involving a highly skilled engineer using some elaborate test equipment. Although the test equipment costs will be far less than the cost of the board tester at the previous test stage, the fact that much more time is spent in test and diagnosis means that you frequently need a lot of these test stations so the capital cost can be quite high.

Some of the design induced defects that escape detection at the functional verification test will get caught at the final system test stage, but because of their nature and the relatively poor fault coverage of the final test stage some of these defects will escape detection and pass on to the customer. The fault coverage of the final test stage will vary enormously depending on the product and the test methods employed, but in many cases it is not very high. There is a good analogy here to the test vectors developed by the ASIC designer to prove that the design works. If you run these vectors through a fault simulation, the fault coverage obtained is often quite low. The vectors are good enough to prove that the device functions the way that it should, but they are not necessarily good enough to detect faulty behaviour. The same is often true of the final test stage. It is designed primarily to prove that the system functions correctly. It is not usually designed to propagate obscure defects to the outside world.

Another factor in all of this that has to be borne in mind is that it is the nature of semiconductors that have any inherent problems to become worse with time and temperature. The marginal nature of design induced defects means that many of these may not be detectable until a certain degree of degradation has taken place. This virtually guarantees that the failure will occur in the field and not in the factory.

System level simulation will eliminate many of the design induced defects because the simulator, or its timing analyser, will flag most of the errors and marginal behaviour. The designer is also able to play 'what if' games and experiment with different circuit implementations. Minimum, typical, maximum and worse case combinations of timing specifications can all be tried out to make sure that the design is really solid. At the same time the testability improves because the designer is able to spot problem areas and do something about it. The end result of all of this is that you have a better quality product with fewer of the most problematic of faults. The testability improvements also result in higher fault coverage at the board test stage for some of the non-design-related defects. System level simulation can reduce the design induced defects down to about 10 to 25 per cent of their previous levels depending on the complexity of the board and how bad they were to begin with.

PCB SIMULATION ECONOMIC ANALYSIS

	YEAR 1	YEAR 2	YEAR 3	YEAR 4	YEAR 5
Number of board designs per year	10	10	11	12	14
Number of prototype cycles per design	3	3	3	3	3
Average cost of a prototype cycle	11000	12000	12500	13000	14000
Number of ASIC designs per year	10	12	14	16	18
Average number of gates per ASIC	12000	13000	13000	15000	15000
Number of ASIC re-designs required	5	6	7	8	9
NRE costs per gate	2	2	2	2	2
Test program development costs per board type	25000	26000	27000	28000	30000

INPUT PARAMETERS WHEN PCB SIMULATION IS USED

	YEAR 1	YEAR 2	YEAR 3	YEAR 4	YEAR 5
Number of prototype cycles per design	2	2	2	2	2
Number of ASIC re-designs required	2	2	2	2	2
Time To Market improvement (weeks)	6	6	7	7	8
Additional sales from early time to market	600000	600000	700000	700000	800000
Gross profit margin (%)	25	25	25	25	25
Test program cost reduction (%)	40	40	45	45	50
Simulation costs per design	15000	16000	17000	18000	19000
Modelling costs per design	12000	12000	12000	12000	12000
Training costs (Annual)	12000	12000	12000	12000	12000
Hardware maintenance costs (Annual)	12000	12000	12000	12000	12000
Software maintenance costs (Annual)	12000	12000	12000	12000	12000

MANUFACTURING COSTS AND SAVINGS

	YEAR 1	YEAR 2	YEAR 3	YEAR 4	YEAR 5
Annual production volume for new board types	40000	42000	44000	47000	50000
Type of board tester. Process (0) Performance (1)	1	1	1	1	1
First pass yield at board tester (%)	80	80	80	80	80
Fault coverage of Process Tester for design faults (%)	10	10	10	10	10
Fault coverage of Perf. tester for design faults (%)	90	90	90	90	90
Design related faults (as % of fault spectrum)	10	10	10	10	10
Reduction of design related faults with sim. (%)	80	80	80	80	80
Fault coverage of system test stage	80	80	80	80	80
Average cost to test one board (Proc. test)	4	4	4	4	4
Average cost to diagnose one fault (Proc. test)	3	3	3	3	3
Average cost to test one board (Perf. test)	4.43	4.43	4.43	4.43	4.43
Average cost to diagnose one fault (Perf. test)	11	11	11	11	11
Average cost to diagnose one fault (at Sys. test)	100	100	100	100	100
Average rework cost for a defective board	10	10	10	10	10
Average cost of a field service repair	1500	1500	1500	1500	1500
PCB development costs (prototyping only)	330000	360000	412500	468000	588000
PCB development costs (simulation & proto.)	370000	400000	462000	528000	658000
ASIC NRE costs (prototyping only)	360000	468000	546000	720000	810000
ASIC NRE costs (simulation & proto.)	288000	364000	416000	540000	600000
Savings in board test program development	100000	104000	133650	151200	210000
Savings in board test/diagnostics/rework	163427	171598	179769	192026	204283
Savings in system test/diagnostics	6284	6598	6912	7383	7855
Savings in field service	21422	22493	23564	25171	26777
Increased profit from earier TTM	150000	150000	175000	175000	200000
TOTAL SAVINGS	167132	202689	244395	274780	318915

RETURN ON INVESTMENT AND PAYBACK ANALYSIS

Life of Project (Yrs.)	5
Tax rate (%)	50
Hurdle rate (%)	15
Depreciation	20
Incremental investment (Hardware)	160000
Incremental investment (Software)	40000

PAYBACK ANALYSIS		YEARS
Simple payback		1.162
Payback after tax/depr.		1.795
Discounted payback		1.795

Year	1	2	3	4	5
Total savings	167132	202689	244395	274780	318915
Tax on savings	83566	101344	122198	137390	159458
Depreciation	40000	40000	40000	40000	40000
Tax saved by depreciation	20000	20000	20000	20000	20000
Net cash flow	103566	121344	142198	157390	179458
Present value factor	1.000	1.000	1.000	1.000	1.000
Discounted cash flow	103566	121344	142198	157390	179458

NET PRESENT VALUE	503956	PROFITABILITY INDEX	3.52

Figure 5.1 An example of a spreadsheet model for the economic analysis of board level simulation

This can result in some big cost savings even though this class of defect may only have been in the region of 2 to 5 per cent of all defect types.

Figure 5.1 shows an example of a spreadsheet program that was developed to look at these cost benefits. This should provide a starting point to develop your own model that fits your particular situation. In case you are not familiar with the terminology, ASIC NRE costs are the non-recurring engineering costs associated with the design process. These costs do not include the manufacturing cost of production quantities. If it is possible that problems with an ASIC might only be discovered after production quantities have been produced then the better design process could also save this cost. NRE costs typically vary between 2 and 5 dollars per gate depending on the actual technology used. Most of the other entries in the spreadsheet should be fairly self-explanatory. This particular model also enables the comparison of two different board test strategies defined as *process test* and *performance test*. The process tester is essentially something that is primarily aimed at finding manufacturing process defects. This description is usually applied to an in-circuit tester, but many functional and combinational testers can only really detect manufacturing induced defects and some of the more basic supplier induced (component) problems. The 10 per cent fault coverage assigned to this tester in the example might be overgenerous. The performance tester would be one of the cycle-based (multi-strategy) testers referred to earlier, which have very large numbers on their price tags. These are capable of detecting some types of design defect because of their timing control and accuracy.

One potential area of cost saving that is not included in this example is that of engineering change orders (ECOs). The use of system level simulation should also reduce the need for this activity.

5.6   The economics of DFT

Testability has been an issue in design and test ever since I first became involved in ATE in the late sixties. It was probably a big issue before that also but it was around this time that complex digital products started to emerge. The complexity of the testing problems was nowhere near the situation we have today but the defect rates of the devices were considerably higher. Even so, the tendency to make complex products by using large quantities of relatively simple boards made the board test problem more manageable. Commercial board testers (all functional testers at that time) did not acquire any form of automated diagnostics until 1973 with the introduction of 'guided probes' by Teradyne and GenRad. Prior to this breakthrough the diagnosis could take a skilled technician between 20 and 30 minutes per fault, and there were frequently several faults on each board. The testers were not very powerful by today's standards. They were typically based upon 12 bit computers that were initially limited to 4 kilobytes of core memory. Yes that really is kilobytes, and the memory really was magnetic beads threaded on to thin wires. Because of these limitations, poor design practice, from a testability point of view, could cause serious problems. Unused reset pins tied to ground or V_{cc}, long counter chains that had no reset and all the other little features that designers love to use caused major headaches. The sceptics might say that nothing has really changed and until the mid eighties they would probably be right. With the exception of a few brave attempts to improve testability, which usually had to be forced upon the designers, there was no effective effort until this time. A group formed in Europe under the name of the Joint Test Action Group (JTAG) met to try to address what they (correctly) felt to be a major testing problem. In-circuit testing had become the predominant method for testing complex PCBs. The divide and conquer approach to testing, the relatively easy programming and the high throughput of these systems had made them very popular. However, there were problems looming for this type of tester that related mainly to its method of interfacing to the board. In-circuit testers require electrical connection to each electrical 'node' of the board and this is achieved with a 'bed of nails' fixture using spring-loaded test probes. The move towards manufacturing printed boards using a 'surface mount' technology rather than the traditional 'through hole' technology meant that this access would become increasingly difficult to achieve. Apart from the fact that not all of the electrical nodes would appear on the 'bottom' side of the board there were the 'pin grid array' components with contacts beneath them, which might not be accessible from either side of the board. In addition, surface mount technology (SMT) boards are often populated with components on both sides of the board, and this can complicate the fixturing substantially. With these and other problems in mind the JTAG committee set about trying to develop a solution. It is now widely known that their solution was to recommend a 'boundary scan' test technique which would provide electrical access to every node on the board using just four test lines. Better yet, the approach enables access to all points on each electrical node and this permits full connectivity tests to be performed for shorts and opens. In fact at the present time this has become the main application for boundary scan technique. In parallel with the JTAG work the IEEE in the United States were also working on the issue of testability, and to cut a lengthy story short they eventually became involved with the JTAG work and adopted the same approach. The IEEE standard P1149.1 is the relevant standard that defines the boundary scan architecture.

Although boundary scan has become the flavour of the month it is not the only way to build testability into a board design. The big achievement of the JTAG, however, was to convince component manufacturers to build the boundary scan cells and the four additional test pins into their ICs (integrated circuits). It is the only technique that has this distinction. ASIC designers have even more choice and flexibility since they are not restricted to what is commercially available. There are many different implementations of scan-based DFT, each with their pros and cons, their advocates and their critics. There is also a commercial technique developed by CrossCheck Technology that is licensed to several ASIC vendors. This system does not require any modification to the functional logic of the component, as scan methods do, so there is very little performance degradation. It uses an array of transistors to provide observability into the ASIC. The approach to the economic analysis of DFT versus no DFT will, however, be much the same regardless of which of the available techniques is used. They all have costs and benefits (value) associated with their use. If the value exceeds the cost then it is worth doing. If not then it is a waste of time.

As with the analysis of the economic merits of system level simulation, most of the costs are incurred in the engineering department and most of the value comes from downstream savings. This may make it difficult to get the concept accepted unless people are prepared to look at the bigger picture and avoid sub-optimization. Many engineers still talk about the 'overhead' of DFT—the extra work, the lost performance, the extra real estate on chips and boards, etc. This is clearly not the way to look at it. I make no apologies for repeating Armand Feigenbaum's definition of product and service quality: 'The total composite product and service characteristics of marketing, engineering, manufacture, and maintenance, through which the product and service in use will meet the expectations of the customer.' We must also constantly bear in mind that 'the market' is simply a collection of customers, and it is these customers who dictate the four market forces of cost, quality, time and technology. Building DFT into a product design is simply one of many options available to get the job done correctly. It is not an overhead if it creates value.

DFT benefits the design engineer as well as the test engineer. This fact needs to be communicated to more and more designers before the use of DFT can become more widespread. The value of DFT increases exponentially with complexity. For an ASIC it starts to become really effective for devices of around 7000 to 10 000 gates. At this level of complexity, the time it takes the design engineer to develop a set of test vectors with a high enough fault coverage becomes a significant proportion of the total development time. The days are gone when the engineer could simply send a design to the ASIC vendor with a limited set of test vectors that would simply prove that the ASIC did what it should do under fault-free conditions. This led to so many disputes that the ASIC vendors started to insist on 95 per cent fault coverage test vectors. At the present time (1992) the majority of ASIC designs are at this level of complexity. The use of DFT techniques will save time at this level of complexity, typically cutting the test vector generation time by 30 to 70 per cent, depending mainly on the experience of the designer with DFT. Since manual test development could easily take 50 per cent of the total design time this can represent a significant time saving. This will need to be balanced against the added design time, the extra cost of the ASICs and any performance degradation. In most cases there will be a net gain in terms of the direct engineering costs, since the extra design cost will be less than the savings in test development. The extra cost of the ASIC comes mainly from the increased area of the die and the decreased wafer yield that will usually result from this increased size. This cost will also be a function of the production volume for the ASIC. The cost of any performance degradation will come from the fact that the product may have to be sold at a lower price to reflect the lower performance. This is therefore also volume dependent. On the value side, the additional revenues generated by an earlier time to market and the lower cost of quality that results from a more thorough test will need to offset any price reductions.

6. Component test economics

One of the biggest changes that has taken place since the first edition of this book was published has been in the typical defect rates of electronic components. The opening paragraph of this chapter in the first edition included the statement: 'for fairly low failure rates of 1 per cent or less'. By today's standards such a defect rate would be regarded as exceptionally high. For the user of components, as opposed to the manufacturer, the testing of components has turned full circle in the past ten to twelve years. In the late seventies equipment manufacturers mistrusted the quality of components to such a degree that fairly high levels of incoming inspection, burn-in and the use of outside test laboratories were commonplace. Today the real and the perceived quality levels of components has led to a widespread belief that there are no longer any bad components. Incoming inspection, except for the more complex and custom devices, has almost become a thing of the past.

There is a danger here. A belief that there are no more bad components to worry about can lead to the selection of an inadequate test strategy. The usual consequence of this will be excessive field service problems, a reputation for poor quality and lost business. There has always been a tendency for component defect rates to be underestimated, especially by those companies that did not perform any component testing. The main reason for this, I believe, is that many test strategies have very poor fault coverage for component defects so the test results simply confirm the belief that the defect rate is low. The problem is that there is rarely any good analysis of field failures so it is difficult to make any meaningful correlations. Components replaced on field return boards are assumed to be the result of abuse or an early life failure. In many cases these defects could have been prevented from reaching the customer by doing the right kind of testing.

Another possible reason for assuming that component defect rates are better than they really are is the vast amount of 'quality' promotion that takes place. Almost every trade journal that you could pick up over the past ten years will have had a 'quality success story' or an advertisement featuring quality contained within its pages. It is true of course that tremendous improvements have been made, but they have not been made at the same rate for all classes of component. Probably the biggest improvements have been achieved in the high-volume areas such as MSI and RAM (random access memory) devices. In fact the rate of improvement probably correlates well with the volume produced. This makes sense, of course, because there is a much better chance of getting a high-volume process under tight control. At the other end of the scale come the custom and semi-custom devices that are produced in relatively small quantities and have much higher defect rates. Typically these are used in smaller quantities on boards so the higher defect rate tends to balance out at the board level. Even so, they will still contribute towards field failures unless the test strategy is designed to detect them somewhere in the factory or at the vendor's location. There is also a potential problem of poor fault coverage for these complex devices. If the test program for an ASIC is developed manually it may well have a lower fault coverage than for a device with better testability because it has built-in DFT circuitry. This enables a more automated generation of the test vectors, which provide a better, and measured, degree of fault cover. Many companies do not use fault simulation because of the time and the computing power that it takes. This is another candidate for using economic analysis techniques to determine whether it should be done or not. It is important, however, to consider the full life cycle costs when performing such an analysis. Even if a fault simulation is performed it will usually be based only on 'stuck-at' fault models, and even if you could achieve a 100 per cent coverage there are failure mechanisms that will still not be detected. In order to minimize testing costs some ASIC foundries only test the devices at 1 MHz and some restrict test vector application to 4k blocks to fit in with their testers. Linear devices and mixed-signal devices also still suffer from higher defect rates than are typical for volume digital devices.

Having said all of this, however, one thing remains abundantly clear. The 'rule of tens' went out of the window some time ago when it comes to comparing the cost to find defects at incoming inspection and board test. The board test stage has now become the cheapest place to find defects by a wide margin, but only for those defects that are detectable by the board tester that you decide to use. You still cannot escape the fact that a good component tester can perform a better test on a component than any other form of test stage.

The 'rule of tens' used to be frequently misinterpreted in the following manner: 'If I do not test at incoming inspection I will find the defects at board test but it will cost me more money.' The reality is that most board testers cannot detect the kind of defects that even a relatively simple component tester can, so most defects would pass through the board tester to the more costly test stages that follow it. Even here the fault coverage may not be adequate to detect a high percentage, so many will escape to the field—the most expensive test stage of all. The possible exception to this situation is the 'performance' or 'multi-strategy' board tester. These are architected along similar lines to a VLSI component tester and do have some chance of detecting component and even design induced defects. However, these are extremely expensive systems that are also complex to program, and they still do not perform like a true VLSI tester. Even so, if you have the kind of problems that these testers are designed to solve they can still be cost-justified. One problem to be aware of, however, is that a number of ATE vendors have begun to use the term 'performance' to describe some of their better products, even though they have no capability to test the performance of the unit under test. A true performance tester has to be 'cycle-based' and have a high degree of flexibility and control over the timing of the application and observation of the test vectors in order to detect performance defects. A tester that simply applies test vectors at high speed with the same inter-test time cannot detect such problems.

The key thing to consider about potential component defects is how the different types of failure mechanisms might be detected.

For active devices there are three major failure modes to consider in terms of their effect on a design and the ability to detect problems with the test equipment available or under consideration:

1. *Functional failures*: failure of the basic functionality of a device. For a digital device this is most likely to be a failure to perform to its specified truth table, or possibly the failure of one section of a multi-function device.

   From a testing point of view it is necessary to distinguish between *static functional* performance or *dynamic functional* performance, the difference being 'Will it function at its rated speed (or frequency)?' In general, the lower priced testers will not operate at the full operational speed of many device types. The relevance of a dynamic functional tester (FNT) will be dependent on the device itself and the design requirements of the product in which it is to be used. For example, it may be adequate to test transistor-transistor logic (TTL), small scale integration (SSI) and medium scale integration (MSI) devices at lower than rated speed, even for a design where they will be operating close to maximum-rated speed. However, you would probably not want to take such a risk with a dynamic random access memory (RAM) component. As the functional complexity of digital devices increases, so does the need to use one of the more expensive testers. The simpler low-cost testers just cannot cope with the long and complex test sequences required to test the functionality of large scale integration (LSI) and very large scale integration (VLSI) devices, even if the issue of testing speed is unimportant.

2. *D.C. parametric failures*: failures of such things as voltage and current driving capability, or input loading of any device driving the device under test, etc. The lower priced component testers available usually perform very little or no parametric testing; medium price-range testers usually do quite a good job; and the high-performance systems usually offer more accuracy and resolution. D.C. parametric failures can be important to detect at an early stage since they often result in marginal performance of a product that is difficult and time consuming to diagnose. In digital devices, parametric failures often result in a loss of noise immunity that is a particularly difficult problem to locate.

3. *A.C. parametric failures*: timing-related faults that include parameters such as rise-times, fall-times, propagation delay, bandwidth, speed of operation, etc. Only the more expensive test systems will be capable of performing good a.c. parametric tests. In making decisions about the cost/performance trade-offs in a lower priced system the capability for a.c. parametric tests is normally the first thing a test equipment designer will leave out. This is reasonable since a.c. failures usually form the smallest percentage of all failure modes, the possible exception being for memory devices. However, these devices usually require specially designed memory testers to cope with their special testing needs.

   The decision to purchase a $500 000 high-performance digital IC tester rather than a $60 000 medium-performance benchtop system will be largely determined by the complexity of functional test patterns required and the need to perform a.c. parametric tests.

6.1 Do users receive bad components?

The short answer to this question is obviously 'yes', but they do receive far fewer than they used to. Many manufacturers talk about 100 per cent testing but sometimes this is only performed at the wafer stage of the process. This is done for the benefit of the manufacturer rather than the user. Packaging the chip is one of the most expensive parts of the process and they do not want to add value to a bad die.

When sample testing is used there will inevitably be some small number of defective devices escaping into the factory. Another source of defects is the handling of the components. Some of the component manufacturers claim that their shipped quality is so high that the potential for damage caused by handling them for an incoming test is greater. This could well be true for those devices that have very low defect rates but it is also a marvellous excuse if someone finds that the shipped quality is not what it should have been.

6.2 Can incoming inspection still be justified?

The simple, non-committal, answer to this question is: 'It all depends.' It will depend on:

1. The actual performance of your suppliers and the volume of components consumed.
2. The nature of your products.
3. The expectations of your customers and the performance of your competitors.
4. Your own quality goals.
5. The cost of not doing it.
6. etc.

The following simple example shows a method of getting a quick idea about what can or can not be justified. This approach is not accurate enough to form a base for any major decisions. A more complete modelling of the test strategy is needed for that. Also, the approach of including the amortized cost of the testers should never be used if you plan to calculate any form of ROI (return on investment), because this would result in the cost of the equipment being counted twice. See the financial appraisal chapter (Chapter 11) for more details. In addition this type of analysis does not include the effects of taxation and depreciation so it should be regarded as a preliminary method only.

6.3 Example

The production situation
1. 40 000 boards per year with an average of 150 ICs per board.
2. Component defect rates average 100 ppm (0.01 per cent).

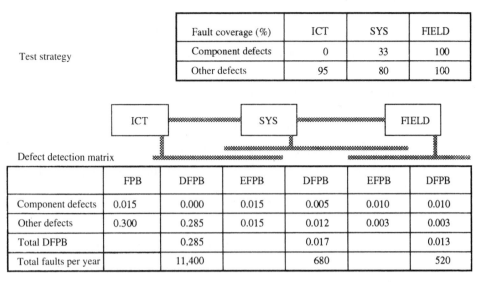

Test strategy

Fault coverage (%)	ICT	SYS	FIELD
Component defects	0	33	100
Other defects	95	80	100

Defect detection matrix

	FPB	DFPB	EFPB	DFPB	EFPB	DFPB
Component defects	0.015	0.000	0.015	0.005	0.010	0.010
Other defects	0.300	0.285	0.015	0.012	0.003	0.003
Total DFPB		0.285		0.017		0.013
Total faults per year		11,400		680		520

Figure 6.1   The defect detection matrix assuming no incoming inspection

3. A 100 ppm defect rate is equivalent to a probability that each device is good of 0.9999. Therefore the probability that each set of 150 devices needed for a board contains no defective devices is 0.9999^{150} or 0.985.
4. Therefore each board will, on average, contain 0.015 defective devices.
5. The actual yield out of manufacturing is 73 per cent and this equates to an average of 0.315 faults per board. There are therefore 0.300 other defects present on each board.
6. The company use a very good in-circuit tester which has a 95 per cent fault coverage for these other defects but no fault coverage for the component defects.
7. The final test can detect one-third of the component defects and the other two-thirds will escape to the field.

Figure 6.1 shows what I call the 'defect detection matrix'. This indicates which of the defects are detected at each test stage and which of them escape to the next test stage. This simple matrix, with varying degrees of detail, forms the core of any test strategy model.

Component testing costs

8. The component tester under consideration can test 700 components per hour, with an automatic handler.
9. Operator costs are $15 per hour.
10. The tester and handler will cost $500 000 and be depreciated over a five year period. Amortized tester costs will therefore be $100 000 per year.

11. Annual component volume is 6 000 000 per year (40 000 × 150).
12. Tester capacity is 700 × 7 hours × 5 days × 50 weeks or 1 225 000 per shift per year. Therefore two testers will be required operated over 2 to 3 shifts.

Annual costs will therefore be

$$\text{Amortized tester cost} = 2 \times \$100\,000 = \$200\,000$$

$$\text{Labour cost} = 6\,000\,000/700 \times \$15 \text{ per h} = \$128\,571$$

$$\text{Total cost} = \$328\,571 \text{ per year}$$

Cost per defect will be

$$6\,000\,000 \text{ devices at } 100 \text{ ppm} = 600 \text{ defects}$$

$$\text{Cost per defect} = 328\,571/600$$

$$= \$547 \text{ per defect}$$

Board test costs

13. Test time (including handling) = 1 minute per board.
14. 0.285 detected faults. Therefore the yield at board test will be $e^{-0.285}$ or 0.75 (75 per cent). Therefore 25 per cent of the boards will fail at the board test stage. The average board test time will therefore be 1.25 minutes due to the re-test of the 25 per cent of boards requiring repair. It is assumed for simplicity that all repairs are performed correctly first time.
15. Throughput = 60/1.25 or 48 boards per hour.
16. Total time required to test the annual production volume is therefore 40 000/58 or 833 hours. Therefore only one board tester will be required since there are 1750 hours available per shift (7 × 5 × 50).
17. The board tester under consideration costs $500 000. Therefore the amortized cost will be $100 000 per year.

Annual board test costs will be

$$\text{Amortized cost of the tester} = \$100\,000$$

$$\text{Labour costs are } 833 \text{ hours} \times \$15 = \$12\,495$$

$$\text{Repair cost is } 11\,400 \text{ defects} \times 10 \text{ min}/60 \times \$15 = \$28\,500 \text{ per year}$$

$$\text{Total cost} = \$140\,995 \text{ per year}$$

$$\text{Cost per defect} = 140\,995/11\,400 = \$12.37$$

System testing costs

Considering only the cost of diagnosis and repair at the system test stage the costs will be determined as follows:

18. Capital cost of each final system test set-up is $50 000 (also depreciated over five years straight line).
19. Average diagnosis time is 1 hour per defect.
20. Labour cost is $30 per hour.

$$\text{System cost per defect is } [10\,000 \times (680/1750)]/680 = \$5.71$$

$$\text{Labour cost per defect} = \$30.00$$

$$\text{Repair cost per defect is } (10\,\text{min}/60) \times \$15 = \$2.50$$

$$\text{Total diagnosis and repair cost at system test (per defect)} = \$38.21$$

Total system testing costs

If we calculate the total system testing costs, as opposed to only the cost of diagnosis and repair, the costs will be determined as follows:

21. With 40 000 boards per year and 8 boards in each system there will be 5000 systems to test.
22. With 0.017 faults per board detected at system test the board yield here will be $e^{-0.017}$ or 0.983. With 8 boards in each system the system yield will be 0.983^8 or 0.873. Therefore 12.7 per cent of the systems will fail at final test.

The total time required for system test and diagnosis will be $(5000 \times 4\,\text{h}) + 680\,\text{h}$ for a total of 20 680 hours. With 1750 hours available per shift 20 680/1750 or 11.82 shift years are required to get the work done. If two shifts are operated then six final system test set-ups will be needed. Therefore the annual amortization cost of the equipment will be

$$(6 \times 50\,000)/5 = \$60\,000 \text{ per year}$$

The total test, diagnosis and repair (TDR) cost at system test will therefore be

$$\text{Equipment cost} = \$60\,000$$

$$\text{Labour cost} (20\,680 \times 30) = \$620\,400$$

$$\text{Repair cost} (680 \times 10/60)15 = \$1700$$

$$\text{Total cost} = \$682\,100$$

If we calculate the cost per defect from this figure we get

$$\$682\,100/680 = \$1\,003 \text{ per defect}$$

This is a very different figure from that obtained when we only included the cost of the diagnosis and repair. This difference highlights a very important fact. *For a high yield situation it will usually cost more to test the defect-free units than it costs to diagnose and repair the defects.* This is very obvious for any incoming inspection tests where yields should be well above 99 per cent. However, even at a sub-system test stage or a final system test stage this situation can also arise. This does not necessarily mean that you should consider eliminating the test stage, because it may still cost a lot more at the later test stages. This has frequently led to some gross overstatements of the potential cost savings.

I have seen numerous articles and conference papers over the years that have made two specific mistakes that have tended to inflate the savings. The first mistake is to use loaded labour rates. Using labour rates fully loaded with all of the overhead costs will overstate the savings because the overhead will not usually be saved. There can be some exceptions to this but in general changing your test strategy or one of your test tactics will not result in savings in the rent, the utilities, the management or any other overhead items. The second mistake is the one highlighted in this example. If you calculate that the cost per defect at final test is $1003 and you then assume that every defect you prevent from reaching that stage will save you $1003, then you will be in for a surprise. The real savings per defect will usually be the $38.21 we calculated from the diagnosis and repair costs alone. Why should we not include the cost of testing the defect-free systems in the same way that we count the cost of testing the good devices at incoming inspection? The reason is quite simple. The incoming inspection test is usually optional but the final system test is usually not. Even with a high yield at this stage, how many companies would be prepared to risk shipping products without a final test? If we are going to perform the test anyway, we can not save any of the testing costs. We can only save the diagnosis and repair costs.

There are, of course, situations quite different from the one described above where the diagnosis time at system test can be very long relative to the test time for a good unit. A complex system that requires little in the way of calibration and adjustments and has much of the final test procedures loaded as software rather than requiring many manual measurements may well have a short test time but a diagnosis time of many hours or even a few days. Under these conditions the diagnosis costs may well dwarf the cost of testing the defect-free units, especially if the yield is relatively poor at final test. The laws of probability being what they are, it will be the more complex products that are likely to have the poorer yields. In this situation it becomes mandatory to maximize the yield out of manufacturing and the fault coverage of the tests prior to the final system test. There are now products available that use heuristics and artificial intelligence techniques to speed up and automate some of the system level diagnostics. These could be cost effective in many cases; however, they do not usually have any impact on the fault coverage of the test stage.

If we now return to the example we can work out the cost in the field for those component defects that escape detection in the factory. There will of course be other defect types escaping, but this example is specifically looking at the impact of component testing and so these have been ignored for simplicity.

Field-service costs

The usual way to calculate field-service costs on a 'per call' basis is to take the total cost of running the field-service operation, including personnel costs, spares inventories, travel and so on, and then dividing this cost by the number of calls made in a year. For this example we have the following:

23. Total cost of the field-service operation = $10M.
24. Number of service calls per year is 5000. Therefore the average cost per call is $2000.

Since 520 defects escape to the field each year the cost will be $1 040 000. The other field calls will be to early life failures, products shipped in the previous year, false alarms, routine maintenance, etc.

We now have a complete picture of the costs for the strategy shown in Fig. 6.1:

$$Board\ test\ costs\,(TDR) = \$\ \ 140\,995$$

$$System\ testing\ costs = \$\ \ 682\,100$$

$$Field\text{-}service\ costs = \$1\,040\,000$$

$$Total\ cost = \$1\,863\,095$$

Adding incoming inspection to the strategy

If we test all of the components we will have the added cost of incoming inspection, which we worked out earlier to be $328 571. On the positive side we should see a decrease in costs at the system test and field-service stages. There will be no change in the board test costs because we assumed for simplicity that the board tester has no fault coverage for component defects. Figure 6.2 shows the defect detection matrix assuming that 90 per cent of the defective components have been detected at incoming inspection and replaced with good devices.

System testing costs (TDR) with incoming inspection

Testing costs will remain the same at $600 000 but diagnosis and repair costs will fall as follows, where there are 0.0125 defects per board and 40 000 boards there

Test strategy

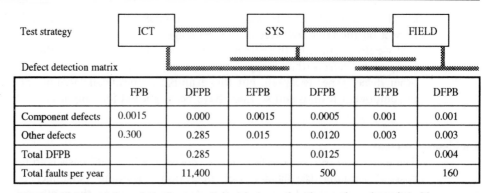

Defect detection matrix

	FPB	DFPB	EFPB	DFPB	EFPB	DFPB
Component defects	0.0015	0.000	0.0015	0.0005	0.001	0.001
Other defects	0.300	0.285	0.015	0.0120	0.003	0.003
Total DFPB		0.285		0.0125		0.004
Total faults per year		11,400		500		160

Figure 6.2   The defect detection matrix with incoming inspection detecting 90 per cent of component defects

will be 500 defects detected at system test:

$$\text{Diagnosis cost is } 500 \times 1\,\text{h} \times \$30 = \$15\,000$$

$$\text{Repair cost is } 500 \times (10/60) \times \$15 = \$1250$$

Equipment costs will be the same since there is no substantial reduction in the workload. Therefore the total costs become

$$\text{Equipment costs} = \ \ \$60\,000$$

$$\text{Testing costs} = \$600\,000$$

$$\text{Diagnosis costs} = \ \ \$15\,000$$

$$\text{Repair costs} = \ \ \ \$1\,250$$

$$\text{Total costs} = \$676\,250$$

Field-service costs with incoming inspection

The total number of defects per board escaping to the field has dropped to 0.004 FPB. Multiplying this by the 40 000 boards result in 160 defects escaping to the field for a cost of $320 000.

The total costs with incoming inspection therefore become

$$\text{Component testing costs} = \ \ \$328\,571$$

$$\text{Board testing costs} = \ \ \$140\,995$$

$$\text{System testing costs} = \ \ \$676\,250$$

$$\text{Field-service costs} = \ \ \$320\,000$$

$$\text{Total costs} = \$1\,465\,816$$

For this example the strategy that includes incoming inspection results in an annual cost difference of $397 279. Maybe incoming inspection is still a viable tactic after all. The results of this simple approach would certainly indicate that a more accurate analysis might be worth while. There are two important things we have to be aware of that emanate from this example. Almost all of the savings occur in the field so we have to be sure that we account for the escaping defects as accurately as we can and we have to calculate the cost benefit of reducing the field failures as accurately as we can.

6.4 Predicting escapes to the field

The mathematics involved in determining the number of defects that will escape to the field are fairly simple. The example 'defect detection matrix' in Fig. 6.1 shows this quite clearly. The real problems, for the purposes of any economic analysis, fall into three main areas. The first involves the determination of the fault coverage that the latter (usually manual) test stages will have; the second is to determine how many of the escaping defects will cause a system failure; and the third is the determination of the cost of such failures and therefore the calculation of the savings that can be made if the number of defects reaching the field can be reduced.

Most test generation systems used for component and board testing will provide some measure of the fault coverage that can be expected. This will not be an absolute measure because at best it can only relate to the defects that have been considered when making the estimate. Other defects that could occur may or may not be detected, we really do not know, and this is usually the case with any sub-system test stage or the final system test. Short of resorting to physical fault insertion there is little that can be done to measure fault coverage. In any case the type of defects that can be 'simulated' with physical fault insertion are the very defects that would almost certainly have been detected by an earlier automatic board test. If no such test takes place there may be some value in performing physical fault inspection at the sub-system test stage, but this can be very time consuming, potentially dangerous to the unit under test and limited in terms of the type of defects that can be inserted. Even if this is done, the problem of estimating the fault coverage at the final system test still remains.

Opinions about the degree of fault coverage that is achieved at these latter stages of test tend to polarize into opposing camps. Some people believe that the coverage will be very high since these tests will be run at system operating speed, under typical operating conditions, with a thorough exercise of the product. Others believe that the coverage is actually quite poor because it is not possible to really exercise a complex system to such a degree that all fault mechanisms will be detected. So which group is right? Well, they are probably both right because the real situation will be heavily dependent on the nature of the product being tested. It may well be possible to perform a test on a sub-assembly, or even the final product, that will exercise it so thoroughly that most defects that might be present would result in some form of malfunction that could lead to the diagnosis of the defect. However, as the complexity of the product increases it becomes less and less likely that this can be achieved. We only have to consider the difficulty of achieving a high fault coverage at the component test or board test stages, where we have a high degree of automation in the test generation process, to understand how difficult it must be to achieve a high fault coverage at a system test stage where the test procedure is developed manually. It may be useful here to remind ourselves about the definition of testability. For an area of a complex circuit to be testable we have to be able to apply appropriate test patterns to its inputs and measure the response of the circuit to those patterns at its outputs. This is simple enough but the problems occur because there is usually other circuitry sitting between the inputs and outputs of the block we are trying to test and the primary inputs and outputs that we can access with the tester. Now we have the problem of controlling this additional circuitry in such a way that we can get our required test patterns to the piece we are trying to test, and at the same time we must set up the circuitry on the output side so that we can observe the response of the piece we are testing to those input patterns. *Controllability* and *observability* are what *testability* is all about (see Fig. 6.3). The essence of the problem is to find a set of test vectors that will cause a different response in the presence of a defect and at the same time make sure that this different response is made visible to the outside world, i.e. the tester. Only when both of these requirements have been met will the defect be detected. Consider how difficult this must be for a multi-board sub-assembly or a complete multi-board product.

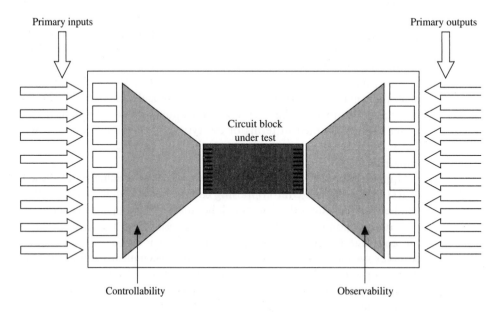

Figure 6.3 The testability diagram

Another important consideration here is the fact that most test procedures for these test stages are developed to prove that the product performs correctly. The test procedures are rarely developed to detect specific defect mechanisms. There is a good analogy here to the development of the test program for an ASIC. It is a well-established fact that the test vectors developed by an engineer to prove that the ASIC design works will usually have a poor fault coverage and would not be good enough for the testing of the devices. As indicated in the chapter on design and test economics (Chapter 5) there used to be many problems caused by these test patterns being supplied to the ASIC vendors with a design. Many disputes took place with each party blaming the other for the quality problems. Eventually the ASIC vendors started to insist on 95 per cent or better fault coverage for the test vectors to overcome these problems. I believe that in many cases the system test and sub-system test situation is the same. A test procedure that is developed to prove that the product works will not necessarily have the ability to detect a high percentage of the defects. Essentially the controllability and the observability may not be present at the same time.

It may be possible over time to develop a knowledge of the typical coverage for a sub-system test from a knowledge of the nature of the defects that are being found at the final system test stage. Any defect that should have been detectable at the sub-system test is obviously an escape from that stage. This requires quite a lot of work so relatively few companies do it. A similar approach could be used to estimate the coverage of the final system test stage, but this involves even more work obtaining defect data from field failures. Often the best approach may be to estimate a range of values for these fault coverages to see the impact they will have on the life cycle cost of the product. This can then be compared with actual levels of field failures after making some allowance for genuine early life failures and latent defects as opposed to the defects that were present when the product was shipped.

Whatever method you use to estimate the fault coverage of these latter test stages it is important to remember the concept about estimating that I covered in Chapter 1 (Sec. 1.12 on forecasting, estimating and guesswork). Any reasonably intelligent guess will be far more accurate than ignoring the issue because you do not have accurate data.

6.5 Will all defects result in a field failure?

The answer is 'probably not'. The argument that the final system test will not catch all defects because you cannot fully exercise the product in an economic manner also applies to the customer's use of the product. The more complex the product is and the greater the number of ways in which it can be used, the more likely it is that some defects may never be detected. This will be true even if there is no redundancy or fault tolerance designed into the product. It is quite feasible that a defect, present from the day the equipment was built, will only cause a malfunction if the customer uses the product in a particular manner with some particular combination of settings. If this combination was not tested for in the factory and the customer never uses it in this way, the defect will not be a problem. One way to account for this would be to apply a fault coverage factor to the defects that you calculate might escape to the field. However, is this the right approach? If you subscribe to the modern quality ethic and the pursuit of maximum customer satisfaction, can you really trust to luck that some of the field escapes will not cause a problem? It is probably better to err on the pessimistic side and assume that all escaping defects will cause a problem at some time. Some of these problems may occur outside of the warranty period so there may not be a direct cost involved; in fact you will be getting paid to fix a problem that has been there from day one, but it will dent your quality reputation.

6.6 Determining field-service savings

Before we can compare the life cycle cost of alternative testing strategies we need to develop a knowledge of the cost of field service. Knowing the costs we can then determine the savings that we would make if the quality of the shipped product is improved. There is a similarity here with the potential for miscalculating the final system testing costs, which was highlighted in the earlier example. What we calculate to be the average cost of each field-service call is not necessarily what we will save if we reduce the number of calls. In fact the overheads in some types of service organizations are so high that the cost per call may go up if we reduce the number of calls by improving quality.

The usual approach taken is to determine the overall annual cost of running the field-service operation and then dividing this figure by the total number of failures in a year. This typically gives a cost per failure of several hundred to several thousand dollars for most large industrial electronic products. For the military market this might be as high as 10 to 20 thousand dollars depending on the equipment and the logistics involved. This method gives a realistic, and a dramatic, picture of what your current warranty failures cost. However, the incremental cost of one more service call, or the savings made if you make one less call, would only be the direct cost of the visit. Even the cost of the engineer would not be saved because you probably pay them monthly rather than on an as-required basis. The counter argument to this is that each call made should bear its proportion of the overheads so the 'per call' cost if you make 2001 calls in a year is not very different from the cost of making 2000 calls. However, we still can not use these 'per call' costs to determine the savings. The situation here is the same as in manufacturing. If you use the fully loaded labour rates to determine savings in a testing situation then these savings will be grossly overstated because you will not save much, if anything, of the overheads. The big difference in field service, however, is that there is a lot of potential to reduce these overheads substantially in a relatively short time. This is because the size of the service organization, the spares inventory required and many of the other overhead elements are a function of the number of failures each year. If you can reduce the number of failures then many of the overhead elements can, over time, be reduced by a similar amount. The following example shows a method for making a reasonable estimate of the potential field-service savings for an increase in the shipped quality of the product.

Example of field-service cost analysis

This example uses the following assumptions:

1. To simplify the calculation there is no growth in the volume of units shipped each year.
2. The warranty period is one year.
3. A new product with a higher level of shipped quality (half the number of field defects) replaces the current product.
4. With a constant production rate, a constant defect rate and a one year warranty, half the calls made in a year will be to products shipped the previous year and half will be to products shipped in the current year that has just ended.
5. The overhead is set at three times the direct costs. It is assumed that there will be no reduction in the overhead during the first year of production at the higher' quality level. In the second and third years of production at the higher quality level the overhead is set at three times the direct costs for the previous year. In effect the overhead reduction lags the reduction in direct call costs, as it would in practice.

Table 6.1 shows the result of this simple analysis. As can be seen, the cost per call actually rises for a couple of years because the overhead is greater than three times the current year's direct costs. The cost per call then settles back to its original value. The more important figure is the 'savings per call saved' which is derived from the total savings and the reduction in the number of calls. This is a more realistic measure of the savings than simply taking the overall cost per call from the base year. Since most justifications will be based upon a five year period it should be possible to use this approach to determine the field savings for each year being considered. In practice, of course, the mix of products and the more gradual improvement of the quality will usually result in a less abrupt change.

Table 6.1   Field-service savings example

	Year 1	Year 2	Year 3	Year 4
Calls to last years shipments	2500	2500	1250	1250
Calls to current years shipments	2500	1250	1250	1250
Direct costs	$2.5M	$1.875M	$1.25M	$1.25M
Overheads	$7.5M	$7.5M	$5.625M	$3.75M
Total cost	$10.0M	$9.375M	$6.875M	$5.0M
Cost/call	$2000	$2500	$2750	$2000
Number of calls saved	0	1250	2500	2500
Savings	$0	$1.625M	$3.125M	$5.0M
Savings per call saved	$0	$1300	$1250	$2000

The example presented earlier and this discussion of field costs have been presented here in the chapter about component testing because the bulk of the cost savings that might be achieved by component testing will accrue in the later test stages and in the field. Other actions to reduce the number of defects will also generate savings at these stages and the comments made in this section will be applicable to these also.

6.7 Component failures and board failures

Statistically the probability of building a board free from defective component is given by

$$Pn = (P)^n$$

where Pn is the probability of selecting n good parts from a batch whose probability of being good is P. For example, if the average defect rate of the components is 0.01 per cent then P will equal 0.999 (99.9 per cent sure that the batch is defect-free). If our boards contain 200 components then the probability that a board will contain no defective components will be

$$P_{200} = (0.999)^{200}$$

Using a scientific calculator to compute Y^x gives us 0.819. Therefore we can expect that approximately 82 per cent of the boards will be free from defective components and that the remaining 18 per cent of the boards will contain all of the defective devices.

In practice different types of component will have different defect rates and a more accurate approach to calculating the board defects may be required. The more accurate approach is to calculate the probability of each component type causing a defect at the board level and then to multiply these resultant probabilities together. This is the way that test strategy analysis models should be constructed. The EVALUATE model described in Chapter 8 uses this approach to preserve accuracy. Table 6.2 shows the relationships between parts per million (ppm) defect rates, percentage defect rates (parts per hundred) and the probability of no defect occurring. Table 6.3 shows the calculation of the individual probabilities of no defects for a board containing several device types. Multiplying these defects together gives the overall probability that the

Table 6.2 Defect rate conversions to probabilities of being good (defect-free)

Defect rate	Percentage	Probability of being good
1 ppm	0.0001%	0.999999
Six-sigma (3.4 ppm)	0.00034%	0.9999966
10 ppm	0.001%	0.99999
100 ppm	0.01%	0.9999
1000 ppm	0.1%	0.999
10 000 ppm	1.0%	0.99

Table 6.3 How the probability of being defect-free falls as the number of opportunities for a defect increases

	Quantity	Defect rate, ppm	Probability	Cumulative probability
Resistors	80	100	0.9999	0.992
Capacitors	60	200	0.9998	0.996
Digital (simple)	150	100	0.9999	0.985
Digital (complex)	20	300	0.9997	0.994
ASICs	4	1000	0.999	0.996
PLDs	8	3000	0.997	0.976

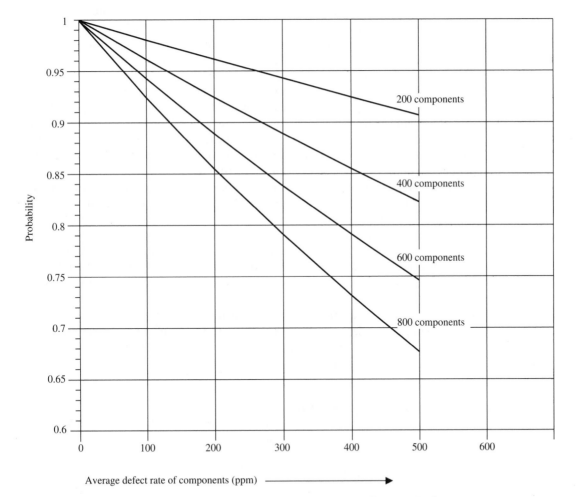

Average defect rate of components (ppm) ⟶

Figure 6.4 The probability that the board is free from defective components versus component defect rates, for various board sizes

boards will be free from defective components:

$$0.992 \times 0.996 \times 0.985 \times 0.994 \times 0.996 \times 0.976 = 0.940$$

Therefore 94 per cent of boards should be free from defective components and 6 per cent will contain defective devices. Figure 6.4 shows the probability that a board will be free from component defects for different component defect rates and board complexities.

6.8   Sample testing and the new quality ethic

Over the years there has been widespread use of sample testing both at the supplier's site and in incoming inspection departments. The most common system used for electronic components has been the acceptance quality level (AQL) sampling plans specified by many military and standards organizations. In the first edition of this book I criticized the AQL system because it is misleading and, as a result, widely misunderstood. The reasons for this are explained in the next section of this chapter. I was pleased to see that the system also came under attack by the leading quality experts during the quality revolution of the eighties. In essence the AQL system represents an agreement between a supplier and a consumer of components that some small quantity of defectives in each batch purchased is acceptable. The quality gurus argue that 'no defect is acceptable' and that the system merely gives the component manufacturer a 'licence to manufacture defective devices'.

The concept of acceptable defects is completely contrary to modern quality thinking. This is very neatly explained in a now famous story about a large American company which, before the quality revolution in the West, decided to test the performance of a Japanese component manufacturer. They placed an order for 10 000 components specifying that they would accept three defective devices in the batch of 10 000. This was a quality level that was almost unheard of at the time in the United States. The Japanese supplier's command of the English language was not perfect. They supplied the 10 000 components along with three defective ones neatly packaged in a separate bag. In an accompanying letter they expressed an interest in knowing why the customer wanted these defective devices which they had had to manufacture specially.

Even though the system has now fallen into disrepute it is still widely used so the section describing it has been retained for completeness.

6.9   How effective is sample testing?

In AQL testing, a sample size is defined based upon the batch size. Table 6.4 shows the sample size code letters for three different 'inspection levels'.

Note: tables are based upon MIL-STD-105.

Having selected a sample size code letter, reference is then made to another table (Table 6.5) to find the actual sample size that is a function of the required AQL.

Table 6.4 Sample size code letters

Batch size	Inspection level		
	I	II	III
151 to 280	E	G	H
281 to 500	F	H	J
501 to 1 200	G	J	K
1 201 to 3 200	H	K	L
3 201 to 10 000	J	L	M
10 001 to 35 000	K	M	N

Table 6.5 Single sampling plans for normal inspection (level II) (Ac = accept; Re = reject)

Sample size code letter	Sample size	AQL							
		0.25		0.40		0.65		1.0	
		Ac	Re	Ac	Re	Ac	Re	Ac	Re
G	32	0	1	0	1	0	1	0	1
H	50	0	1	0	1	0	1	0	2
J	80	0	1	0	1	1	2	2	3
K	125	0	1	1	2	2	3	3	4
L	200	1	2	2	3	3	4	5	6
M	315	2	3	3	4	5	6	7	8

Table 6.6 Tabulated values for operating characteristic curve for 0.65 AQL sampling plan ($n = 125$, $c = 2$). P_a = probability that the batch will be accepted based on the sample test; P = percentage of defects in the batch

P_a	P
99.0	0.349
95.0	0.654
90.0	0.882
75.0	1.32
50.0	2.14
25.0	3.14
10.0	4.26
5.0	5.04
1.0	6.73

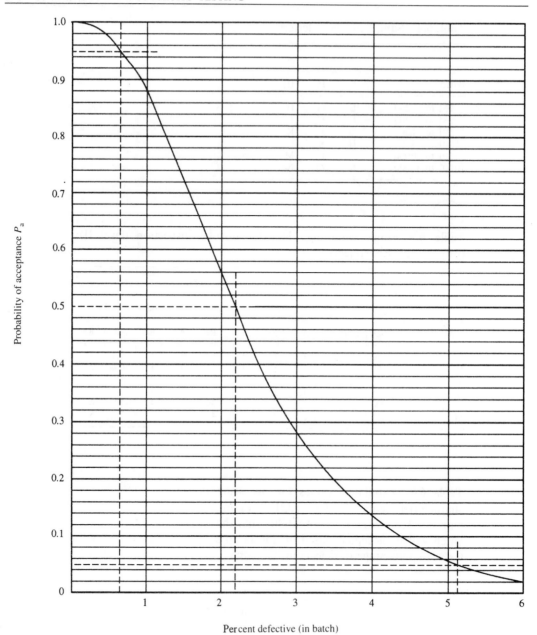

Figure 6.5 The 0.65 per cent AQL operating characteristic curve ($n = 125$, $c = 2$)

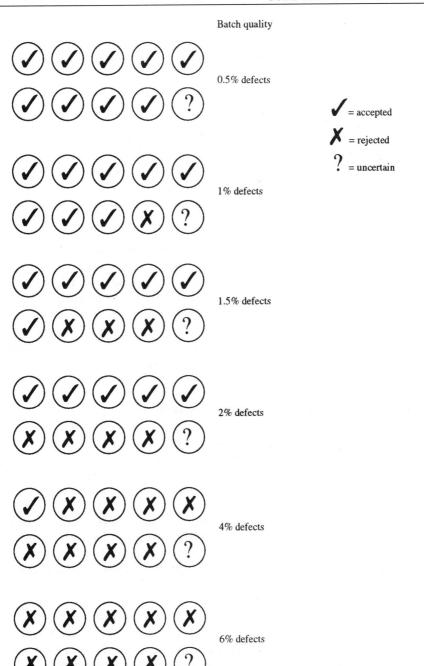

Figure 6.6 The effect of 0.65 per cent AQL sample plan on batches of consistent quality ($n = 125$, $c = 2$)

Table 6.5 defines the number of defects in the sample that would cause the entire batch to be rejected. For example, if testing a batch of 4000 components to level II inspection, the code letter is L. For an AQL of 0.65 per cent, the sample size would be 200 and the batch of 4000 would be accepted if there were three or fewer failures in the sample or rejected if there were four or more failures. It should be noted that in this example the recommended sample size is only 5 per cent of the batch. The probability of the sample being unrepresentative of the batch is, therefore, quite high. It is also worth noting that sample testing does not seek to eliminate defective devices, but to indicate that there are probably only a few.

For any given AQL an 'operating characteristic (OC) curve' can be plotted, which defines the probability of a batch of components being accepted for different percentages of defects. Table 6.6 lists the relevant values for an AQL of 0.65 per cent, that is a commonly used AQL.

Note: n = sample size, c = critical (accept) number of defects.

This operating characteristic is shown in graphical form in Fig. 6.5. The whole concept of using a sampling approach is based upon an effective agreement between the supplier and the consumer of the components that a few defective components is acceptable. By agreeing to this concept they each accept a certain amount of risk.

The effect of performing this particular sample test on batches of consistent quality is shown diagrammatically for several quality levels in Fig. 6.6.

6.10   Supplier's risk and consumer's risk

1. *Supplier's risk.* The supplier runs a risk that a good batch will be rejected if the supplier or the consumer performs a sample test on a batch and the sample selected is not representative of the batch.
2. *Consumer's risk.* The consumer runs a risk of accepting a bad batch if the sample selected is unrepresentative of the whole batch.

In the above statements 'good' and 'bad' mean 'less than' or 'more than' the agreed-to percentage of defects. It is the OC curve that defines these risks. The usual points selected are the 95 per cent probability points. The interpretation of the OC curve is as follows: for the 0.65 per cent AQL curve shown the supplier is 95 per cent sure that if the batch contains 0.65 per cent or fewer defective components, then it will be accepted based upon the selection criteria of the sampling scheme. In practical terms this means that, 95 per cent of the time, if the sample size required is 200, there will be three or less defective components in the sample (see Table 6.5).

The consumer, on the other hand, is 95 per cent sure that he or she will not accept a batch containing more than 5.04 per cent defective components.

We can now see why AQL can be misleading and misrepresentative of the quality of a batch of components, because it refers to the supplier's risk rather than the consumer's risk. Inspection of the OC curve shows that a batch containing 2.14 per cent defective components would be accepted 50 per cent of the time. This might be a more meaningful number to quote than the 0.65 per cent AQL.

The problem becomes clear if you calculate the accept/reject numbers as a percentage of the sample size. Referring to Table 6.5, we see that for a sample size of 125 and AQL of 0.65 per cent the accept/reject numbers are two and three respectively. Therefore two failures in the sample would be acceptable and the whole batch would be accepted. However, two in 125 is 1.6 per cent, not 0.65 per cent.

With the AQL procedure we are assuming that a sample of 5 per cent of the batch will be representative of the whole batch. We then say we shall accept the entire batch even if the sample contains 1.6 per cent defectives, and we call the whole thing an *acceptable* or *acceptance* quality level of 0.65 per cent.

The whole concept becomes even more absurd if you calculate what the failure rate needs to be before the batch would be rejected. Three (the reject number) in 125 is 2.4 per cent and so our sample would have to contain 2.4 per cent or more failures before the batch would be rejected.

The average of 1.6 and 2.4 per cent is 2.0 per cent, which approximates to the 2.14 per cent at the 0.5 probability point of the OC curve shown earlier. This 0.5 probability point is sometimes referred to as the indifference quality level (IQL). The 95 per cent point is the acceptable quality level (AQL) (the acceptance quality level more correctly refers to the overall process rather than to just this point on the curve). The 5 per cent probability point (consumer's risk) is called the rejectable quality level (RQL), or the limiting quality (LQ), or the lot tolerance per cent defective (LTPD).

The effect of reject numbers expressed as a percentage being considerably different from AQL is worse when the sample size is small and the AQL is low. This is because a difference of one must exist between the accept and reject numbers, and this difference is a large amount when the sample size or AQL is small. For example, the accept/reject criteria for an AQL of 0.1 per cent and a sample size of 125 is 0 and 1. It stays this way for AQLs up to 0.25 per cent since you cannot have a reject number of less than 1. For 0.1 per cent AQL this means the failure rate in the sample for rejection has to be 0.8 per cent—almost an order of magnitude worse than the AQL figure. For a sample size of 800 the reject number is three, implying a failure rate in the sample of 0.38 per cent—still four times the AQL. From the MIL standard or BS 9000 tables, a sample size of 800 would normally imply a batch size of 150 001 to 500 000 components. Not too many users buy components in these kinds of batch size.

6.11 Average outgoing quality level (AOQL)

A testing set-up can be shown as follows:

$$\text{Incoming quality} \rightarrow \text{test procedure} \rightarrow \text{outgoing quality}$$

If the incoming quality of components is very bad, whole batches will be rejected and replaced or 100 per cent tested to select the good components. On the other hand, if the incoming quality is very good, the outgoing quality will also be very good. At each extreme of incoming quality the outgoing quality will be very good. At some point between the extremes, the outgoing quality will reach a peak. This point is known as the maximum average outgoing quality level (AOQL) or sometimes as the average outgoing quality limit. If you use AQL procedures in your incoming inspection of components what AOQL can you expect to get? This is important since the AOQL determines the maximum average level of bad components reaching your board assembly area.

Table 6.7 AOQL values for a 0.65 per cent AQL sampling plan

P_a	P (for 0.65% AQL)	AOQL
0.99	0.349	0.346
0.95	0.654	0.62
0.90	0.882	0.79
0.75	1.32	0.99
0.50	2.14	1.07
0.05	3.14	0.78
0.10	4.26	0.42
0.05	5.04	0.252
0.01	6.73	0.067

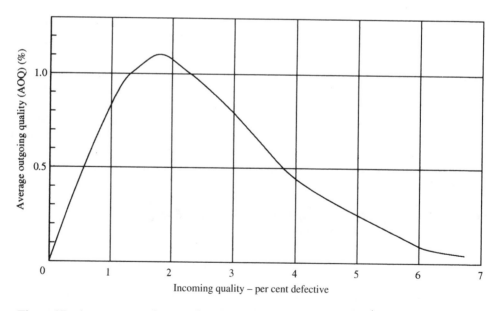

Figure 6.7 Average outgoing quality limit (AOQL) for a specific 0.65 per cent AQL sampling plan ($n=125$, $c=2$)

Table 6.8 AOQL factors for normal inspection levels

Code letter	Sample size	AQL				
		0.25	0.40	0.65	1.0	1.5
G	32		1.2			2.6
H	50	0.74			1.7	2.7
J	80			1.1	1.7	2.4
K	125		0.67	1.1	1.6	2.5
L	200	0.42	0.69	0.97	1.6	2.2
M	315	0.44	0.62	1.0	1.4	2.1

AOQL can be calculated by multiplying the probability of accepting a batch by the percentage defects in that batch, as shown in Table 6.7 for an AQL of 0.65 per cent. A plot of this is shown in Fig. 6.7, which shows the maximum AOQL as being 1.1 per cent.

Table 6.8 (from US-MIL-STD 105) shows AOQL factors for different AQLs and sample sizes. In practical terms the maximum average quality over a long period of time that you can expect to get if you buy components to 0.65 per cent AQL, or if you test yourself using this criteria, is in the region of 1 to 1.1 per cent. In terms of its effect on board failures this could be significant. However, bear in mind that this is an average. AQL sample testing will not prevent the occasional 2, 3 or even 5 per cent batch getting through, and this could be potentially disastrous, especially if you are working close to maximum in terms of testing capacity at later testing stages.

6.12 Other sampling plans

Most of the discussion of sampling plans has been centred on AQL sampling plans since these are the most commonly used, possibly because they favour the supplier. There is increasing pressure on suppliers, particularly in the United States, to use sampling plans that set the consumer's risk at some agreed level rather than the supplier's risk. Naturally, any plan will have both a supplier's risk and a consumer's risk associated with it, but the shape of the OC curve can vary widely depending on the sample size and the acceptance number, as shown in Fig. 6.8. It will be seen from this that different sampling plans for a given AQL can give different levels of consumer risk.

6.13 Lot tolerance per cent defective (LTPD)

LTPD is the level of consumer risk on the OC curve. Sampling plans exist for specified levels of LTPD rather than AQL. Different LTPD plans would produce OC curves all hinging around the chosen LTPD, as shown in Fig. 6.9.

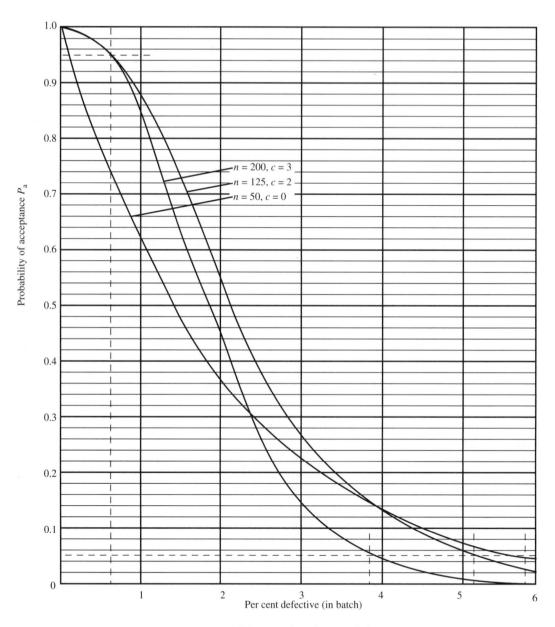

Figure 6.8 The 0.65 per cent AQL operating characteristic curves

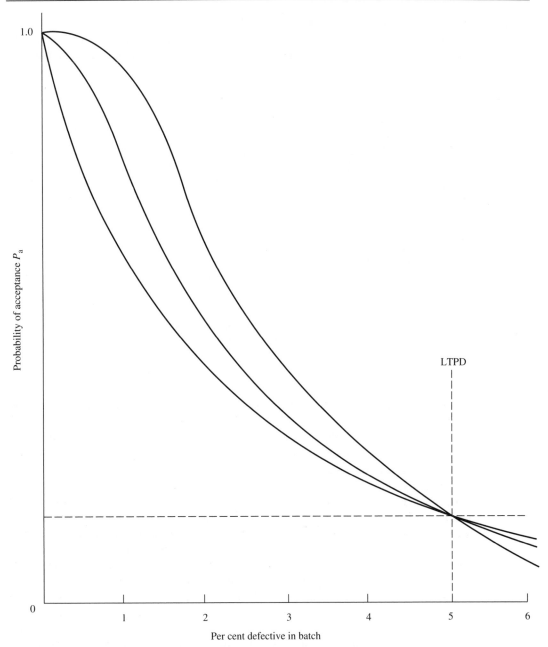

Figure 6.9 OC curves for LTPD sampling plans

6.14   Average outgoing quality level (AOQL)

AOQL was defined earlier as the average outgoing quality level, assuming that rejected batches are fully tested. This implies that if the incoming quality is very bad (all batches fail) then the outgoing quality will be very good since 100 per cent inspection takes place. Somewhere between the two extremes of very good and very bad incoming quality the AOQL reaches a peak—the maximum AOQL. Sampling plans also exist for specified levels of AOQL. For more information on this subject see *Juran's Quality Control Handbook*, Sec. 25 (1988).

As I indicated earlier, the modern view of quality and the low level of defect rates that are being achieved put the use of the AQL system into serious doubt for anything other than destructive testing where a 100 per cent test would be a little inappropriate. However, because it is still used, and still widely misunderstood, I decided not to leave it out of this new edition.

6.15   Component testing alternatives

There are three basic alternatives to look at when considering whether or not to perform any component testing. These are:

1. Do nothing—rely on the supplier's test procedures.
2. Perform a sample test—basically repeating the supplier's test procedure as a cross-check.
3. Perform a 100 per cent test—testing every component to some degree.

The first of these options is probably the most common choice today, with the possible exception of ASIC devices and PLDs. ASIC prototypes have to be 'characterized' before the production volumes are produced and this is a task for the designer. It is possible that the same test equipment required for this characterization could also be used for incoming tests on the production devices.

Obviously there are sub-options that are possible as to the extent of testing performed. For instance, on a digital IC this could be a simple functional test, or include d.c. parametric tests and even a.c. parametric tests. The greater the degree of testing the greater (usually) the cost of the equipment needed to perform the tests. The benefit is, of course, a more thoroughly tested device, but can the cost of testing be justified? The answer to this question will depend on a number of factors. One of these, as stated at the beginning of this chapter, will be the failure modes that are important. This will affect the cost of a tester to find the important failure modes and the cost of finding those faults at later stages of test if no incoming inspection is performed.

Another factor will be the design of the product in which a given component is to be used. Many designs may be tolerant of fairly wide variations of certain parameters whereas other designs will make use of the maximum performance of a good device.

Unfortunately, there is very little published information or statistics on failure modes and what percentage of total failures can be located with a simple functional test, what percentage would require d.c. parametric testing and what percentage would require a.c. testing. It is, however, possible to generalize a little on how these ratios shift with IC complexity. The chart in Fig. 6.10 shows how the predominant failure mode shifts from parametric failures with SSI devices to predominantly functional failures with LSI devices. A.C. parametric failures tend to remain fairly constant as a percentage, although their importance will tend to increase with complexity, especially for memory devices. This graph is very much a generalization. The actual percentages will be different for different device types and device

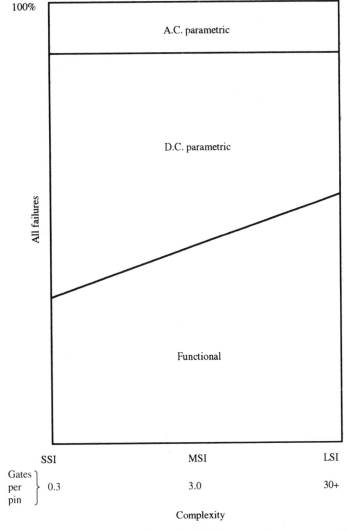

Figure 6.10 Approximate change in IC failure modes versus complexity

technologies, but since most companies use a wide selection of components it is a not too unreasonable generalization to make. It is fairly easy to visualize why the graph should have this form. In general, the functional complexity increases much more rapidly than does the pin count. Since parametric measurement can only be done on the external pins it tends to make sense that the ratio of functional to parametric failures should shift in this way. Even if no change takes place, the validity of the main argument still holds, i.e. that board testers will not detect parametric

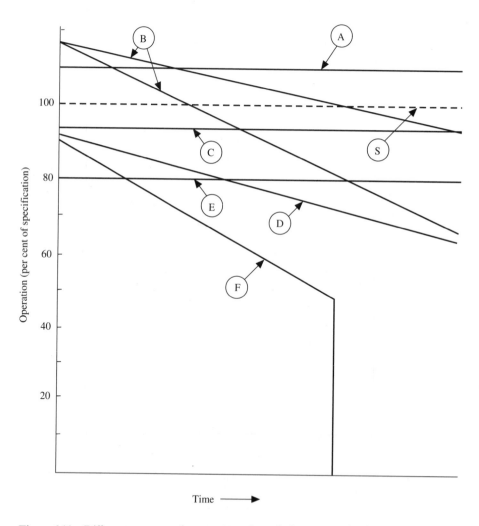

Figure 6.11 Different patterns of component degradation versus time

problems and system testing will not find all of them, so some will escape to the field. The reason why some of these failures will not be detected during production testing is that many of the problems will not be bad enough at that time to cause a system failure.

For a component to be reliable it must be both operational and stable. Unfortunately, many of the failure mechanisms in ICs are due to electrochemical defects and are a process of continuous degradation with time. The rate of degradation is a function of the extent of the problem, temperature, current density and bias conditions. These are some of the failure mechanisms that result in 'early-life failures'. The whole philosophy of stress testing such as burn-in of components is based upon the ability to accelerate the degradation process and so eliminate 'weak' devices. Figure 6.11 shows a number of components and their possible behaviour versus time:

S = the specification of the device
A = a good reliable device—it is well inside (above) its specification and it is stable with time
B = two devices initially good but degrading with time at different rates; it would be necessary to use stress testing to weed out these unreliable devices
C = a device below specification but stable; depending on the design this might not cause any long-term problems
D = a device below specification and degrading
E = a device well below specification but stable; it would probably cause problems
F = a device below specification, degrading and eventually failing catastrophically

A 100 per cent component test will weed out failure types C, D, E and F. It could be argued that failure type E is very unlikely to occur in practice since, if it is this bad to begin with, it must have something wrong with it that is also likely to degrade with time.

6.16 Comparing the costs of no testing, sample testing and 100 per cent testing

Given the following:

B = typical batch size
S = sample size (usually derived from published tables)
P = probability of accepting a batch of components
A = AOQL (again obtainable from AQL tables) expressed as a factor
C = cost to test one component
K = average cost to locate, repair, replace and re-test when a faulty component is found at a later stage of test than component test

Then the cost of our first option of 'no testing' is simply

$$BAK$$

The cost of our second option of sample testing becomes

$$(SC) + [(B-S)AK] + [(B-S)(1-P)C]$$

where the first term is simply the cost of performing the sample test. The second term is the cost of components that escape to be found at a later stage and the third term is the cost of a 100 per cent test on any batch that fails. This term would be removed if failed batches are returned to the supplier for replacement.

Our third option of 100 per cent testing is simply

$$BC$$

The expression for the cost of sample testing can be simplified if we assume that:

· 1. The sample size is much smaller than the batch size. This is valid since AQL sample sizes are typically 5 per cent of the batch.
2. The probability of accepting the batches is high. This is also valid if the same procedures that the supplier used are used by the user (P should be 0.95).
3. The cost to test one component is relatively low.

Given these assumptions, the cost of sample testing can be simplified as follows:

$$(SC) + [(B-S)AK] + [(B-S)(1-P)C]$$

Small relative to B
Small since $1-P$ is small and so is C
Small since S and C are small

leaving us with BAK, which is the same as the cost of no testing. This being so there is no economic advantage to performing a sample test if the supplier has done a similar test.

Example

The following example is based on an AQL of 0.65 per cent which would be totally unacceptable today as a quality level, even for the most complex of devices. The purpose of the example is to show the futility, from an economic point of view, of sample testing so the actual quality level chosen is of no consequence. Many published sample tables only go down to 0.01 per cent AQL and even this level requires batch sizes of over 500 000.

XYZ Electronics decide to calculate the cost of the three alternatives for the following data and a total of 600 000 components per year:

$B = 3000$ (typical batch size)
$S = 125$ (sample size of 0.65 per cent AQL)
$P = 0.95$ probability of acceptance
$A = 0.011$ AOQL for an AQL of 0.65 (from published tables)
$C = \$0.121$ cost to test one component
$K = \$143.60$ average cost of finding a defective component at later stages of test

Calculation of C. 600 000 components per year tested at a rate of 400 per hour require 1500 hours of labour at $15 per hour. This part of the cost is therefore $22 500. The tester will cost $200 000 and be depreciated over four years for an annual amortized cost of $50 000 per year. (Remember that this simple approach does not include any tax allowances for the capital equipment depreciation. As a result the operating costs will be overstated.) Thus

$$C = (\$22\,500 + \$50\,000)/600\,000$$

$$= \$0.121$$

Calculation of K. This is more tricky. There are three places other than at incoming inspection where a faulty component may be detected:

1. During a board test.
2. During a system test.
3. In the field after shipment of the product.

Let

b = average cost to diagnose, repair and re-test a fault at board test
s = average cost to diagnose, repair and re-test at system test
f = average cost of a warranty repair
P_b = probability of finding faulty component at board test
P_s = probability of finding faulty component at system test
P_f = probability of faulty component escaping to the field and causing a failure during the warranty period

Then

$$K = (P_b b) + (P_s s) + (P_f f)$$

If

$$P_b = 0.3, \qquad P_s = 0.6, \qquad P_f = 0.1$$
$$b = \$12, \qquad s = \$100, \qquad f = \$800$$

then K will equal $(0.3 \times 12) + (0.6 \times 100) + (0.1 \times 800) = \$143.6$

The cost of 'no testing'

$$BAK = 3000 \times 0.011 \times 143.6$$
$$= \$4738.80$$

The cost of 'sample testing'

$$(SC) + [(B-S)AK] + [(B-S)(1-P)C]$$
$$= (125 \times 0.121) + (2875 \times 0.011 \times 143.6) + (2875 \times 0.05 \times 0.121)$$
$$= \$4573.87$$

This confirms the previous simplification of the 'sample test' formula since there is only about a 3.5 per cent difference between the cost of 'no testing' and the cost of 'sample testing'.

The cost of 100 per cent testing

$$BC = 3000 \times 0.121$$
$$= \$363.00$$

These costs are for each batch of 3000 components. Since the total usage of these particular devices is 600 000 per year then the total costs will be

$$\text{No testing} = \$4738.80 \times 600\,000/3000$$
$$= \$947\,760$$
$$\text{Sample testing} = \$4573.87 \times 600\,000/3000$$
$$= \$914\,774$$
$$\text{100 per cent testing} = \$363.00 \times 600\,000/3000$$
$$= \$72\,600$$

The 100 per cent testing option could therefore save up to \$875 000 per year compared to the other alternatives. This is for a very high defect rate compared to what is usual today, but the vast different in costs suggests that the 100 per cent testing option might be viable for much lower defect rates.

6.17 Break-even analysis

It would be useful to know the defect rate at which 100 per cent testing becomes more cost effective than 'no testing' or 'sample testing'. This break-even failure rate is simply

$$C/K \times 100 \text{ per cent}$$

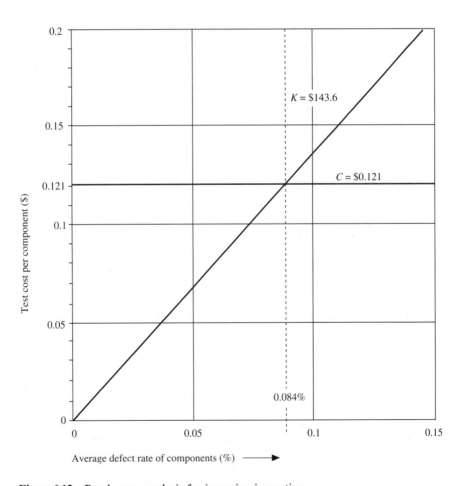

Figure 6.12 Break-even analysis for incoming inspection

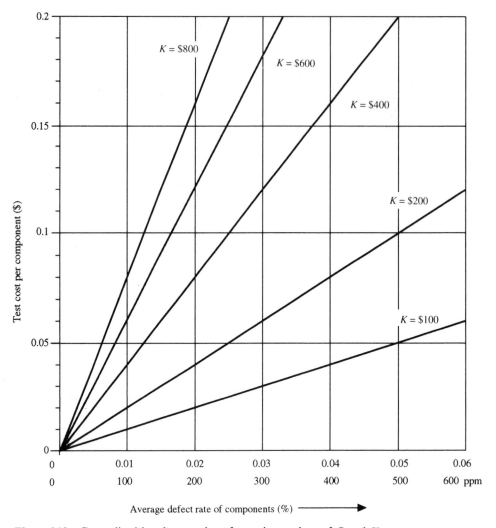

Figure 6.13 Generalized break-even chart for various values of C and K

For the example above this becomes

$$(0.121/143.6) \times 100 \text{ per cent} = 0.084 \text{ per cent}$$

Therefore, for defect rates above 0.084 per cent (840 ppm), 100 per cent testing is cheaper than 'no testing' or 'sample testing' for these levels of C and K. Obviously, as C decreases and K increases, the break-even defect rate will go down. Figure 6.12 shows the break-even chart for the example and Fig. 6.13 shows a generalized break-even chart for a range of values of C and K.

6.18 Increasing capacity with an auto handler

In the example above XYZ Electronics had decided it could meet its current testing needs with one digital IC tester used with manual insertion of components to the test head. Its relatively conservative estimate, based upon one shift, seven hours per day and twenty days per month, gave it capacity for 672 000 component tests per year. When this capacity needs to be increased there are a number of alternatives available:

1. Buy a second tester.
2. Add an auto handler.
3. Revert to sample testing of the better quality devices.
4. Add a second shift.

As we have seen, there is a high degree of risk attached to the third alternative so we shall compare the first two. Throughput using an auto handler can be calculated as being the reciprocal of the sum of test time and index time of the handler:

$$r = \frac{1}{I + t}$$

where

$$r = \text{throughput rate}$$

$$I = \text{index time of handler}$$

$$t = \text{test time}$$

Therefore, with an index time of 400 ms and a test time of 300 ms the throughput would be 1.429 components per second or 5144 components per hour. This would then be reduced by any loading and unloading time if this interrupts the testing.

Example: comparison between adding a second tester or an auto handler

Cost of the tester $= \$200\,000$

Cost of the handler $= \$50\,000$

Labour cost of the operator $= \$15$ per hour

Manual test rate $= 400$ components per hour per system

Auto handler test rate $= 4000$ per hour

New capacity requirement $= 1\,000\,000$ components per year

Depreciation period $= 4$ years

Cost to test manually with two testers

$$\text{Equipment armortization (2 testers)} = (200\,000 \times 2)/4$$

$$= \$100\,000 \text{ per year}$$

$$\text{Labour cost will be } (1\,000\,000/400) \times 15 = \$37\,500$$

Note that I have used the test rate for one system. With two systems the rate will be 800 per hour but two operators will be needed so the answer will still be $37 500.
As the total annual cost is $137 500 the cost per component will be

$$137\,000/1\,000\,000 = \$0.138 \text{ per component}$$

Cost to test with one tester and a handler. The equipment cost is now $250 000 so the annual cost will be $62 500. The labour cost will now be $(1\,000\,000/4000) \times 15 = \$3750$. As the total cost is now $66 250 the cost per component will be

$$66\,250/1\,000\,000 = \$0.066 \text{ per component}$$

In this simple example the amortized cost of the hardware has been 'allocated' fully to the job of testing the components. This is the right way to do it if the tester will not be used for anything else. If, however, the tester is also used for other things, such as characterization of ASIC prototypes, etc., then some of the equipment cost should be allocated to this activity. Be careful not to fall into the trap of calculating an hourly cost for the amortization since this will not be valid unless the equipment is used every available hour. If this were done for the tester plus handler part of the example we would get the following cost per component:

$$\$62\,500 \text{ amortization}/1680 \text{ hours per year} = \$37.20 \text{ per hour}$$

Adding the $15 per hour labour cost to this gives a total hourly cost of $52.70. Since we need 250 hours to test 1 000 000 components the cost per component becomes

$$(52.70 \times 250)/1\,000\,000 = \$0.013 \text{ per component}$$

One point three cents as opposed to six point six cents is a big difference. The problem is that not all of the equipment costs has been allocated to the job. If the tester and the handler are going to be sitting idle for much of the time then this idle time is still costing money and has to be accounted for in this simple type of analysis. Bear in mind, however, that this type of analysis is very much a 'first cut' approach. This technique should not be used for justification purposes. The detailed reasons for this are given in Chapter 11.

6.19 Stress testing of components

Introduction

So far all the discussions regarding component failures and the economics of testing components have hinged around the fact that the instant you receive a batch from your supplier some of them will be faulty. If these faulty devices are not detected at an early (incoming inspection) stage they will pass through to other levels of testing and you will effectively be adding value to bad parts.

There are other reasons why components may fail while in your plant or out in the field. These fall into two major areas:

1. Problems caused by handling of the devices during inspection, testing, stocking, withdrawal from stores, kit formation for assembly, assembly, heat shock from soldering, electrical stress when testing the board (due to another fault condition), etc.
2. Early-life failures—failures during normal operation of the device (i.e. within its specification) due to some defect that, because it gradually gets worse, did not show up when the device was tested at some earlier time.

This section deals with the second of these two areas, which is often also referred to as 'infant mortality'.

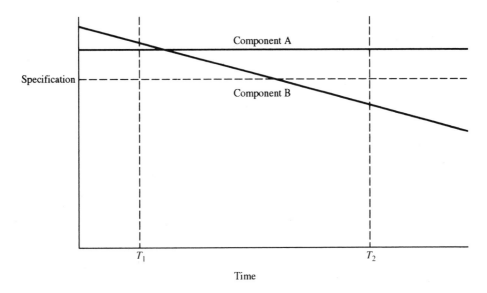

Figure 6.14 Reliable and unreliable devices

Component reliability and 'infant mortality'

Unfortunately, the testing of a device at incoming inspection tells us nothing about the device's reliability. To be reliable a device must be operational (meet its specification) and also be stable with time.

The simple diagram in Fig. 6.14 contrasts a reliable device with an unreliable device. Only by testing the devices at two different times can we get any measure of reliability. The problem is that the time it takes for an unstable device to go 'out of specification' may take several seconds or several months, or even years. For most active electronic components, such as integrated circuits, the probability of failure follows the fairly well-known *bathtub* curve shown in Fig. 6.15. For such devices, the infant mortality period is typically in the range of 1000 to 6000 hours depending on the nature of the device and its operating conditions. The 'lifetime' period might be many years, as will be the 'end of life' period when failure rates begin to increase.

Obviously, it is not practical to test a device and then re-test it after several thousand hours to determine its reliability. Additionally, many failure modes in components are only 'triggered' when the device is powered up and biased in certain ways, so just putting components into the stores for a few months would not work anyway. Fortunately, it is possible to accelerate many of the failure mechanisms by stressing the devices in certain ways. A program of testing, stressing and re-testing is usually referred to as a 'screening program'.

For semiconductor devices there are two elements to stability:

1. Mechanical stability.
2. Electrochemical stability.

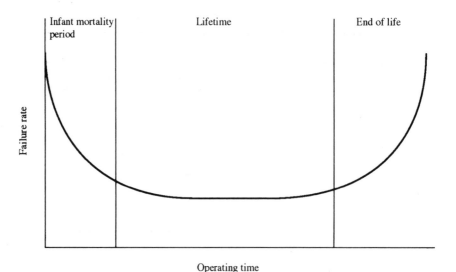

Figure 6.15   The 'bathtub' curve

Mechanical stability problems can be accelerated in an unstable device by shock, vibration, thermal shock, etc. Electrochemical stability can be accelerated by stabilization bake, thermal cycling and burn-in. Of these, burn-in is the more common and probably the more cost-effective approach for weeding out electrical stability problems. A rough rule of thumb for burn-in is that for every 10°C rise in temperature the rate of degradation doubles. Therefore

$$\text{Accelerated time} = \frac{\text{normal infant mortality time}}{2^n}$$

where n = number of 10°C increments.

As an example, if a device type has a normal infant mortality period of 5000 hours when its junction temperature is 50°C then a burn-in at a 120°C junction temperature would reduce this to

$$\frac{5000}{2^7} = \frac{5000}{128}$$

$$= 39 \text{ hours}$$

Therefore, a 39 hour burn-in should weed out most infant mortality failures. A more typical burn-in time is 168 hours or 7 days. In practice, burn-in is performed in chambers with components mounted on panels, with power and biasing applied. As a result of the bias, there will be a rise in junction temperature anyway, so the ambient temperature of the chamber has to be set such that the manufacturer's specifications for the device are not exceeded. Most manufacturers specify a maximum ambient temperature of 125°C for biased conditions and 150°C for storage of military-grade ICs. Therefore, the chamber temperature is set such that with power applied to the device its maximum junction temperature is 150°C.

Economics of burn-in

Unstable components that are likely to fail during the infant mortality period will either fail at some testing stage in the factory, during installation (if a large system) or during the warranty period (assuming that this is the usual six months to one year period). For high-reliability applications, such as military or aerospace, the normal level of potential failure rates is unacceptable and so vendors will either buy 'HI-REL' components, screen them, or have them screened by an independent laboratory. For applications where the maximum reliability is not essential the decision becomes an economic one.

Figure 6.16 shows how the probability of devices failing decreases as time goes on. The assumption here is that given a very large quantity of devices the probability is 1.0 (100 per cent) that at least one of them will fail when power is first applied to it. This implies that some failures at incoming inspection are really infant mortality

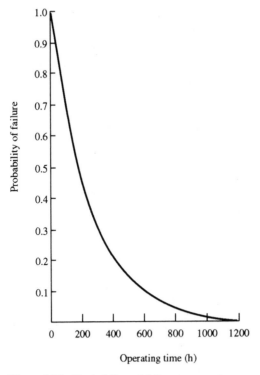

Figure 6.16 Probability of failure versus time

Table 6.9 An example of test times for various test stages

	Duration
Component test	100 ms–20 s
Board test (power applied)	100 ms–10 min
Sub-assembly test	30–120 min
System test	1–10 h
Soak test	24–120 h

failures. This is a valid assumption since if the supplier only conducted a sample test then the majority (possibly 95 per cent) of the devices have not been powered up since the wafer stage of production. This type of failure is normally a total failure rather than some parameter being below specification. Some devices will take longer to fail and so will be picked up at later stages. Table 6.9 gives typical times for testing at various stages for a relatively small product.

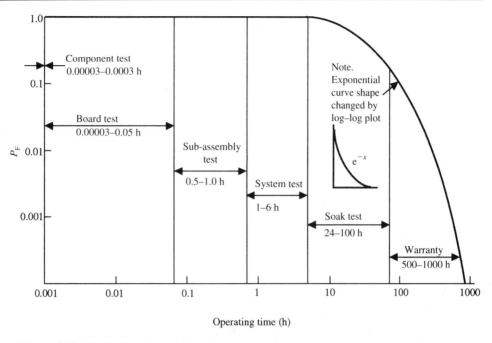

Figure 6.17 Typical 'power up' times for various test stages relative to the exponential portion of the bathtub curve

Figure 6.17 shows how these times might relate to the infant mortality period. This curve is still exponential in form but the shape is changed by the log–log plot. It highlights possibly the major disadvantage of ATE—it is too fast to see many of the infant mortality problems.

Just as with determining the cost effectiveness of incoming inspection, it will be necessary now to determine the cost of not finding these infant mortality problems. This can be done by making some estimates of what failures will be detected at each stage, and the cost at each stage. If we assume 100 per cent incoming inspection (which will have eliminated some failures) then the first infant mortality problems will show up at board test. At this point, there is a high probability of failure but a relatively short time for it to happen. System test is usually several hours and so is more likely to show up a problem, as is the field installation. Finally, some failure will occur during the warranty period, which will be several months long, but at a lower probability of failure. The failures at each stage will be relative to the area under the curve. It is possible to estimate this using the exponential probability distribution (see pp. 22–23 and 33–34 of Juran and Gryna, 1988). This technique can be used to determine the probability of the time between successive failures, in a complex system, exceeding any given figure, given in the mean time between failures (MTBF) and the fact that early-life failures occur in an exponential manner.

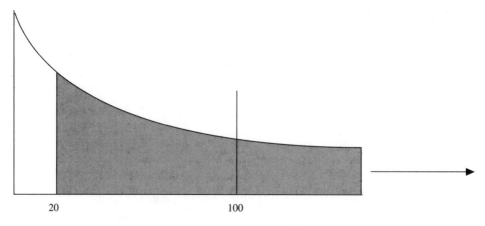

Figure 6.18 The area under the exponential curve from the 20 hour point

Example

Records taken over a period of time show that failures occur on a piece of equipment exponentially versus time and that the MTBF is 100 hours. What is the probability of successive failures exceeding 20 hours? What is needed is the area under the curve from the 20 hours point on out (Fig. 6.18). The formula is

$$e^{-x} \qquad \text{where } x = \frac{N}{\text{MTBF}}$$

and

$$N = \text{number of hours}$$

and

$$e = 2.718$$

For the example

$$x = \frac{20}{100}$$

$$e^{-0.2} = 0.8187$$

which is the probability of successive failures exceeding 20 hours.

If the total operational time in the factory is known, we can determine the probability of a failure in the field for a given duration of time by adopting this technique.

Example: potential cost of early-life failures

Expected operational time during warranty period = 1000 hours

Expected MTBF during this period = 200 hours

Total operational time in the factory including a soak test is 40 hours. The area under the curve from 0 to 1040 is what interests us (Fig. 6.19). We now need to find

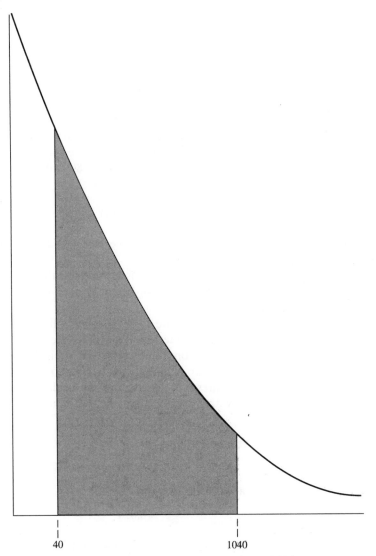

Figure 6.19 The area under the curve from 40 to 1040 hours

what proportion of this is the 40 to 1040 part since this will give us the proportion of failures during warranty:

$$\text{Area for 40 hours} = e^{-40/200}$$

$$= 0.8187$$

$$\text{Area for 1040 hours} = e - 1040/200$$

$$= 0.0055$$

Therefore,

$$\text{Area from 40 to 1040} = 0.8132$$

Since the area from 1040 on out is only 0.55 per cent of the total, we can ignore it. Therefore, the proportion of failures from 40 to 1040 hours will be 81 per cent and so 19 per cent of the failures will occur in the factory.

Since board testing takes only a few minutes, most of the factory failures will occur at system test. However, many of these can be diagnosed by re-testing failed boards on an ATE system. If the internal cost of a failure was, say, $80 and the average field cost $600 then the cost of these early-life failures would be

$$(80 \times 0.19) + (600 \times 0.81) = \$501.20 \text{ per failure}$$

If the components needed for the equipment can be screened for less than this cost then the screening would be cost effective.

This example is intended to illustrate an analysis technique—the data used may not be realistic in a practical case. Service records would need to be analysed in order to arrive at real numbers.

Intuitively, the longer the MTBF the more likely that early-life failures will be picked up in the field rather than in the factory. Also, the better the design from the point of view of not placing stress on components (i.e. designing well within operational limits) the longer it will take for unreliable devices to fail—thus almost guaranteeing that the early-life failures will show up in the field. As proof of this, if we repeat the example above with an MTBF of, say, 400 hours, we should see the difference:

$$\text{Area for 40 hours} = e^{-40/400}$$

$$= 0.905$$

$$\text{Area for 1040 hours} = e^{-1040/400}$$

$$= 0.074$$

The proportion of failures up to 1040 hours is $1 - 0.074 = 0.926$. In the first example we could ignore the proportion beyond 1040 hours as insignificant.

Here we cannot, so we must multiply our answers by 1/0.926 to get to 100 per cent within the 1040 hour period:

$$0.905 \times 1/0.926 = 0.977$$

Therefore, 97.7 per cent of the faults occurring within the 1040 hours will occur after 40 hours—during warranty.

This rather theoretical approach is intended to emphasize the problem. Fortunately, it is not quite this bad in practice since many of the failure mechanisms will cause a device to have a very short lifetime. As a result, the number of failures during the first few hours of operation will be higher than the exponential theory suggests.

It would appear then that the traditional bathtub curve, with its exponential 'early-life failures' portion, is not strictly applicable to semiconductors. For the distribution of time between failures to be exponential, the failure rate has to be constant (see pp. 23–82 of Juran and Gryna, 1988). Practical records and analysis show that the failure rate for complex semiconductors is not constant but starts at a level and decays with time. A paper presented by Belat and Montague of Honeywell at the 1979 IEEE Annual Reliability and Maintainability Symposium, based largely on an extensive database of real operation of devices (over 1.9×10^9 device hours), supports this and shows that the failure rate falls by a factor of 10 during the 100 to 1000 hours of operating period and by almost a further factor of 10 during the period of 1000 to 10 000 hours. It then starts to stabilize.

Therefore if the theory is not very accurate what kind of benefit can be gained in practice from burn-in? It is generally accepted, as a result of practical tests, that screening procedures will provide a minimum improvement in reliability of 10 to 1. This will result in a dramatic reduction in field-service costs for equipment that needs this kind of support. For military and aerospace applications, the benefits are even more dramatic. A secondary benefit is that of reducing production testing costs. Tests have shown that approximately 90 per cent of the failures accelerated by a burn-in program would be detected prior to shipping the product to a customer, even if the burn-in was not performed. This is considerably better than the 10 to 20 per cent that the exponential theory would suggest. Unfortunately, since component and board test times are very short, most of these failures would occur at the system testing stage, where the cost of diagnosing a fault is at its highest level. Burn-in followed by component test ensures that these failures are detected in the cheapest way.

Example: savings in warranty costs due to burn-in

XYZ Electronics estimates from research into the experience of other manufacturers that a burn-in program could reduce its warranty failures caused by semiconductors by a factor of 5 to 10. They expect to ship 2000 units of a complex new product in the first 12 months and expect to see half of these suffer from an early life failure

(MTBF = 24 months). Their average cost of a field-service call is $800 and they calculate the potential savings as follows:

Warranty costs without screening 1000 failures × $800 = $800 000

Warranty failures with burn-in (at 5:1 reduction) 200 failures × $800 = $160 000

Therefore the annual savings will be $640 000. If the cost of the screening operation is less than this annual saving it can be justified. This, however, is another situation where the difficult to quantify benefits of an improved quality reputation need to be considered. It is also a situation where a careful analysis of the real savings in field service have to be estimated as accurately as possible.

Burn-in at what stage of production?

Burn-in can be performed at:

1. Component level.
2. Board level.
3. Sub-assembly level.
4. System level.

Which one makes most sense? In most cases—but not all—it will make most sense to do burn-in at the component level, for the following reasons:

1. Components can be stressed to their limits, thus ensuring shorter burn-in times and maximum acceleration of degradation. Burn-in of a board or an assembly will be limited by the lowest common denominator as far as temperature and other stresses are concerned. This will mean that higher temperature devices will not be stressed enough.
2. ICs need specific bias conditions to trigger certain failure modes. It will be impossible to achieve this for all components at the board or assembly level.
3. Burn-in chambers need to be larger, as do power supplies, etc., for board or module burn-in.
4. Work-in-process costs are lower for components sitting idle in chambers than for boards or assemblies.

On the other hand, the burning-in of boards or assemblies has the advantage of weeding out some kinds of fault that burn-in of components cannot find. Probably the two major fault mechanisms detectable in this way are cold-solder (or dry-solder) joints and contact problems due to warping and expansion effects.

Soak test

The burn-in of complete systems is more often a 'soak test' than a burn-in since elevated ambient temperatures are rarely used other than during the testing of prototypes, etc. For physically large pieces of equipment space limitations usually prevent true burn-in.

6.20 Summary

There has been a tremendous degree of improvement in the defect rates of components shipped to the user over the past 12 years. However, the greatest improvement has been in the high-volume merchant devices. Over the same period of time there has also been a significant increase in the use of custom and semi-custom devices and these tend to have much higher defect rates than the high-volume parts. At the same time these devices are often only the subject of a rudimentary test at the board test stage so that defect detection is often only possible at the latter stages of test or in the field. Depending on the circumstances it may still be desirable to perform incoming inspection, at least on these devices, especially since some test capability is required in any event for the characterization of the prototypes.

Over the same period of time the cost of field repairs has also increased, as has the customer's expectations about quality and reliability. These factors need to be considered carefully before dismissing incoming inspection out of hand.

Component testers usually perform a more thorough test of the component than other test stages can. Many defects detectable by component testers cannot be detected by board testers.

Sample testing is not cost effective in the long term but may prevent really bad batches reaching your assembly area and causing testing bottlenecks. The results of AQL sample test procedures are often misunderstood and misleading.

For products that will require field service, with its associated high costs, 100 per cent incoming inspection is much easier to justify since it will detect problems that other production test stages will not see. It will also prevent below-specification devices, which may be degrading, from reaching the field.

The break-even failure rate at which the cost of 100 per cent testing equals the cost of sample testing or 'no testing' can be well below 0.1 per cent under certain conditions.

Burn-in procedures can reduce production testing costs by catching failures that would normally occur at system test. The reliability of devices and equipment is typically 10 or more times better following burn-in.

Reference

Juran, Joseph M. and Frank M. Gryna (1988) *Juran's Quality Control Handbook*, 4th edn, McGraw-Hill, New York.

7. Board test economics

7.1 Overview

The automatic printed circuit board test system has been at the heart of most production test strategies since the early seventies. As boards have become progressively more complex the option of manual testing, even for relatively low volume operations, has become less and less viable. For most companies the choice is not whether to use an automatic board tester but which one to use. The available options are extremely wide in terms of the test methods employed, the price and the performance. For anyone involved in making their first purchase decision it can be a bewildering and nerve-racking task. So much depends on making the right choice. The first section of this chapter provides some background about the development of board testers from the mid sixties up until the end of 1992. Then the next section looks briefly at each of the basic types of tester with some indication of their advantages and disadvantages. This is followed by a discussion about some of the key issues that affect the selection and the economic performance of board testers. A more detailed look at the economic analysis needed to compare alternative testers is covered in Chapter 8 along with a discussion of how to go about developing an economic analysis model.

Selecting the most appropriate board tester has not only become more difficult to do but it has also become more important to do it correctly. Section 1.9 in Chapter 1 addresses what I believe to be a misunderstanding of the quality experts who dictate that 'test and inspection should be eliminated because they add no value'. What the quality experts are saying is that any *unnecessary* test and inspection should be eliminated, that we should work to reduce what is left and that we should get away from the approach of testing quality into a product. The protestations of these experts have been taken rather too literally at times. They would be the first to agree that testing is much more acceptable than shipping defective products. Testing adds value so long as the cost of testing is less than the cost of not testing and this cost of not testing has to be looked at over the life cycle of the products being manufactured. The quality experts tell us we should 'Do it right the first time—every time.' The parallel concept in testing is that 'We should do the right testing right the first time—every time.' To be competitive we have to consistently reach the four main objectives of cost, quality, time and technology. In terms of the role that the overall test strategy plays in meeting these objectives the board tester will have the biggest influence for two important reasons:

1. The board test stage is now the cheapest place in the testing process to detect and repair defects.
2. The escape rate established at the board test stage determines an escape rate to the field that can only be improved at great expense.

The old 'rule of tens' stated that it becomes ten times more expensive to detect and rectify a defect as you progress through the test process. This process was usually shown as having four basic stages: incoming inspection, board test, final system test and field-service test. Today the average defect rate seen at incoming inspection is very much lower than it was in the early seventies when the rule of tens was first formulated. This means that the cost to find and rectify a defect at incoming inspection is now much higher than it was and very much higher than the cost at board test. There is a more detailed discussion of this concept in Chapter 4.

What this simplified view of test costs tends to overlook is that the incoming test, because it is a different kind of test, can detect defects that even the most elaborate board tester will not see. Parametric defects in ICs is probably the best example. Nevertheless, for defects that are detectable at the board test stage this is now the best place to find them in terms of meeting the four main objectives. The obvious conclusion is that we should work to maximize the fault coverage of the board test stage while at the same time minimizing the time taken to reach this high coverage level. In practice this requires a testable design, design and test integration, and an effective test program generation system.

The second major reason for the board tester's influence on meeting the objectives, as mentioned above, is that it 'sets' the escape rate to the field. Consider the following. Once the overall test strategy has been agreed upon and competing board testers are being evaluated, it is unlikely that the choice of tester will cause any change to the test plans of the downstream test stages such as the sub-system test and final system test. The escapes to the field will therefore be a function of the combined fault coverage of the test stages following the board test stage and the number of defects escaping from the board test stage. Any potential variations in the fault coverage at the board test stage will therefore directly affect the escape rate to the field. Here I am only considering defects present when the board is first tested as opposed to any early-life failures that crop up later. A more detailed discussion of this appears in Chapter 2 with a numerical example shown in Fig. 2.7.

For these two main reasons the decisions taken about board testing are critical to the success of the overall strategy that is established to meet your cost, quality, time and technology objectives. As emphasized in Chapter 1, you have to get the strategy right *and* you have to get the tactics right.

7.2 Background

The increasing availability of digital ICs in the mid sixties led to a testing problem that was mainly addressed with 'in-house' designed and built test rigs. The ICs at that time were very simple and relatively small quantities were used on each board type. As a result the testers were often fairly simple affairs using toggle switches to input test patterns and lamps to indicate the logic states of the output pins. The 'test program' consisted of a sequence of input patterns and the expected responses were written down on a testing specification. More sophisticated systems used punched cards or punched paper tape to input the test patterns. Then the first commercial testers appeared. The more popular units used a 'Gray Code' generator to apply all possible combinations of logic states to the inputs of the UUT and a 'known good' reference board simultaneously. The outputs of each board were fed to a series of comparators driving indicators to see whether the two boards responded in the same manner. The Gray Code generator was used in preference to a simple binary counter because it only produces one transition at each test step at the input to the UUT, thus avoiding any 'race' or 'hazard' conditions. Some of the testers using 'Gray Code' inputs simply compared the number of logic transitions on each output pin throughout the test sequence, rather than comparing the logic state at each test step.

These testers were fine for testing combinational circuitry but were much less effective for sequential circuits. A much higher degree of flexibility in defining the input patterns or the 'input vectors' was needed so in 1969 the first commercial computer-based digital board tester appeared. This was produced by the General Radio Company who were later to be re-named GenRad. This was quickly followed by systems from Digital General, Computer Automation and later still by Teradyne. Around the same time in the United Kingdom a tester was launched by Membrain (now Schlumberger), although this was not computer based. By 1971 the commercial automatic board test business had developed from a small start into a small industry.

All of these testers were 'functional testers' which set out to determine that the UUT functioned correctly in its entirety. The rate of pattern application was very slow relative to the speed at which the UUT would be running in the final product, so these were called 'static' functional testers. Input signals were applied to the edge connector on the board and the output responses were also measured there. As a result, functional testers are sometimes referred to as 'edge connector testers'. As complexity grew it sometimes became necessary to add extra test points to the board, either as part of its physical design or by using clips attached to the ICs.

At first these testers had no automatic diagnostics; if a defect was detected they would simply indicate which pins had the wrong logic states and the number of the failing test. The software was usually set up to stop testing and to turn off the power, as soon as a defect was detected. The technician would then use a logic probe, an oscilloscope and the logic diagram, to manually diagnose the cause of the failure. In the early seventies this process typically took from 15 to 20 minutes per fault. Defect detection was strictly on a 'one fault at a time' basis due to the 'stop on fail' approach used, so defects would be repaired one at a time and the boards returned for re-test and possibly the detection of a second defect. Yields were typically in the 30 to 50 per cent range so a good proportion of the boards would contain more than one defect. The need to have a test engineer or a test technician diagnose defects on the board tester led to a two-pass approach being used, especially in the higher yield cases. A low-skilled operator would perform a 'go/no-go' test on a batch of boards and simply sort them into two piles of good and defective boards. The high-skilled operator with the logic diagram would then come along and diagnose the defects on the failing boards. This two-pass method of operating led to the development of a formula that was used in an economic analysis of the test process, and became widely used. Unfortunately the formula is still widely used today even though this two-pass approach is rarely used because of the automation of the diagnosis. This formula is discussed in the next chapter.

The first automated diagnostics appeared in the form of 'guided probes' from Teradyne and GenRad in late 1972 and early 1973. The Teradyne probe used a database of nodal logic states which were learnt by probing a reference 'known good board'. However, the GenRad probe obtained its nodal data from a digital simulation of the UUT performed on the tester. The main purpose of this simulator was to help with the test program preparation problems and the guided probe was simply a side benefit. The simulator was called CAPS (computer aided programming software) and was a major breakthrough for functional testers. It meant that the test engineer only had to be concerned with determining the input vectors to apply to the UUT since the simulator would calculate all of the logic states for every node at every test step. The simulator could also produce for the first time a measure of the fault coverage of the test program and show which areas of the board had poor fault cover because of poor controllability or observability. Some limited 'design rules checking' also enabled the simulator to highlight design errors that could lead to marginal performance of the UUT. The large amounts of data generated by the simulator meant that the GenRad testers had to be equipped with disk drives rather than the low capacity, and very slow, magnetic tape drives that were the usual bulk storage peripheral in use. This prompted many concerns about using the testers in 'dirty' production environments, but the disk drives proved to be quite reliable and within a few years all of the major players in the market had simulators running on their testers.

Developments over the next few years were more evolutionary than revolutionary. The testers became faster and the guided probes became 'look-ahead guided probes' which no longer had to begin their probing sequence at an edge connector pin. This kept the diagnosis time down by reducing the number of probed points and also minimized probing errors. Diagnostic resolution enhancements were added and major improvements were made to the simulators. Analogue test capabilities had been added to these basically digital testers in the 1973/1974 time frame to produce what became known as hybrid (analogue/digital) testers. There was a big need for such testers because many boards had both circuit types on them. There was an alternative, however, for the fully analogue board in the shape of the 'in-circuit' tester. These testers tended to be much cheaper than the digital functional testers and they were also easier to program. They were essentially impedance measurement devices with a reed relay scanner to enable the measurement hardware to be switched around the various nodes on the board. There were three major manufacturers, Zehntel (later acquired by Teradyne), Faultfinders (now part of Schlumberger) and Marconi Instruments, in the United Kingdom. These testers required access to all electrical nodes on the board to perform measurements on each individual component while it was still connected (in-circuit) to all of its neighbouring components. The effect of the parallel impedance of the network of other devices was eliminated to a large degree from the measurement of the device under test (DUT) by various guarding techniques that effectively 'open-circuited' the parallel paths. The measurement technique was not new. I can remember using something called an 'in-situ' impedance bridge in the United Kingdom in the late fifties. This was essentially an in-circuit tester without the relay scanner and the computer control.

A small company based in Florida, Testline, then introduced a small digital tester which was aimed at the field-service depot test market and the low-volume end of the manufacturing market. The tester tested one digital IC at a time by clipping it and running a pre-defined test program. This was the digital IC equivalent of the 'in-situ' impedance bridge mentioned above. It isolated the DUT from the surrounding ICs by overriding the natural logic states of the surrounding devices. The technique is now well known and is usually called 'backdriving' because current is fed back into the output stage of any device that is normally driving an input on the DUT. Testline had done some research into the potential damaging effects of this forcing current and concluded that it was safe for a short period of time. It was quickly recognized that this approach could be applied to an in-circuit tester to give it a digital test capability. However, the leading ATE manufacturers were sceptical about the potential damage from backdriving. I was working for GenRad at their headquarters in Concord, Massachusetts, at the time and was familiar with the research done there by Jim Skilling and Aldo Mastrocola. They had developed a clever technique for measuring the rise in junction temperature within a backdriven IC. I am sure that other companies also performed their own research. In the meantime the complexity of digital boards had been rising at an alarming rate and the functional testers were falling behind in terms of program preparation. Simulation times were beginning to get very long and it was also difficult to impossible to get simulation models for the more complex devices. Diagnosis times were also increasing due to the large number of points that needed to be probed. As a result of these lengthening times the annual production testing capacity on these systems had diminished considerably.

The electronics market needed a new approach to testing and many people felt that in-circuit would be this new approach, but early attempts to add a digital capability to existing analogue testers had been disappointingly crude and limited. Then GenRad launched the digital version of their 2270 in-circuit tester and a new era began for board test ATE. This tester had memory-backed driver sensors that could output lengthy test patterns within the time limits imposed by the temperature rise caused by the backdriving currents. It also had built-in hardware and software safety measures to ensure that the possibility of damage to the UUT was minimized. The automatic test generation system had the ability to automatically correct for different wiring configurations of the same IC type and many other innovations. This was the breakthrough needed and the basic architecture of in-circuit testers has remained essentially the same up to the present time.

Within two years the unit sales of in-circuit testers had surpassed that of the well-established functional tester. It took a little longer for the dollar volume to get there because the in-circuit testers (ICTs) typically sold for a lower price. The philosophy of the two approaches differed considerably but they are actually a lot more alike, especially in the digital domain, than many people realize. An ICT tests a board using the assumption that:

> If the components are all of the correct type and value, if the active devices are all functional, if they are all present and inserted correctly and if there are no shorts or opens, then the probability is very high that the board will function correctly.

An FNT tests a board using the assumption that:

> If the board behaves as expected when it is driven with a set of test vectors that have been developed to have a high degree of fault coverage, then the probability is very high that the board will function correctly in the end product even if the test data rate is lower than the operational data rate.

Obviously both of these assumptions have some potential flaws; however, in practice things have worked out quite well.

The similarity between the two approaches mentioned above comes about in the following way. A digital in-circuit test is actually a functional test of each individual IC on the board, the main difference being that the ICT tests individual components using a set of pre-defined test programs that are referred to as the 'component library'. This is actually a library of test programs. It is this library approach that results in the much lower test program generation costs for ICT as opposed to FNT. In essence an ICT test program for a board consists of a set of smaller test programs picked out of the library. It is actually a lot more complex than that but this simplification serves to illustrate the difference. Since a functional test program has to be generated for a whole collection of ICs, taking account of how they are interconnected with every board being different, it is easy to see that this is a much more complex task. Because of this complexity a common approach to programming is to break down the task into smaller pieces and test them one by one. In this respect the test plan differences between ICT and FNT can become blurred. There is a big difference in the way that the test patterns are applied and sensed, but the actual test may be very similar. The same could also be said about boundary scan testing concepts.

In the early days of their availability ICTs were generally regarded as being inferior to functional testers (FNTs) in terms of the comprehensiveness or the fault coverage of the testing. For this reason many companies simply used ICTs to pre-screen their FNTs and so improve their efficiency and their capability. The ICTs eliminated the 'easy' manufacturing process faults, leaving the FNT to handle the more 'difficult' problems. It did not take long for these companies to determine that the overall fault coverage of the better ICTs was not too different from that of the FNTs. Depending on the yield and the fault spectrum the ICT could even be better. What was obvious was that the fault coverage of the two approaches was different for different types of defect. This meant that the best fault coverage would be obtained if both approaches were used. This in turn led to the development of testers that combined both test approaches in a single cabinet. These testers became known as combinational testers. This then broadened the choice of the board test part of the overall test strategy enormously. You could use ICT alone or FNT alone, with similar levels of fault coverage, albeit with higher program preparation costs for the FNT approach. You could use ICT and FNT as separate systems in a variety of configurations or you could opt for one of the newer combinational testers. Some companies, however, had little choice because their customers banned the use of ICT. Many military, and some telecommunications, organizations were still sceptical about the potential damage and reduction in operational life that might be caused by the 'backdriving' technique of ICTs and therefore did not permit the use of ICT on their equipment.

In the mid eighties doubts began to be expressed about the future viability of ICTs. They had enjoyed a remarkable success in the ATE market but a number of industry trends were causing concern. The first of these was the packages used for ICs. Through-hole (dual in line) packages had an almost universal pin spacing of 0.1 inch. This was usually expressed as 100 mils (one hundred thousandths of an inch) in the United States. Building the 'bed-of-nails' fixtures for this 'pitch' was relatively straightforward, but the new surface mount devices were forecasted to go down to pitches of 25 mils and even 20 mils with pins on all four edges of the package. Even if smaller probes could be made the build-up of pressure in some areas of boards would be tremendous. Components mounted on both sides of the board created additional probing problems, and the use of 'pin grid array' packages, with contacts underneath the package, would be impossible to probe. Designers were reluctant to give up 'real estate' on the boards to add test points to improve the testability, because the whole point of moving to SMT was to make the products smaller so they could pack in more functionality and be more competitive.

Another problem was the increasing use of application specific ICs (ASICs). The main reason for the rapid test program preparation for ICTs was the relatively straightforward method of using standard test programs stored in the component library. This is only effective, of course, if all of the components on your board are available in the library. This is why many number games are played by ATE manufacturers to make it look as though they have the most extensive library. By definition there will not be a program in the library for a custom component, so the need to create one was beginning to increase the overall programming costs. All of these problems resulted in the formation of the 'joint test action group' (JTAG) which formulated the 'boundary scan' specification. One reason for the success of ICTs was that they were less affected by any poor testability of the board design. With a test pin available on every electrical node of the board they had 100 per cent controllability and 100 per cent observability. This situation would degrade dramatically as a result of the potential problems outlined above. In theory the boundary scan approach would get us back to this unity testability situation while at the same time simplifying the tester requirements.

As things turned out, the fixturing problems were not as bad as had been expected;

clever techniques plus a gradual coming together of design and test meant that good designs could be probed from one side of the board with most probes still on a 100 mil grid. There are exceptions of course, but we did not reach the situation of utter chaos that had been predicted. At the same time the boundary scan implementations were not quite the panacea that had been hoped for. At the present time boundary scan is mainly used for connectivity testing (opens and shorts) rather than the full functional testing of the chips that it had been conceived to do. Boundary scan capabilities have been added to many existing and new in-circuit testers and the two techniques are happily coexisting to provide the best of both worlds.

In the early eighties a new name appeared on the list of tester types. This was the manufacturing defects analyser (MDA), a name that I believe was first used by Fluke. Loaded board shorts testers had been very common in the seventies because of the desirability of repairing any shorts before the board was powered up for testing. Since the FNT was the predominant method in use, and this required power on the UUT, the presence of a short could cause other problems. The MDA could be viewed as being either an enhanced shorts tester or a limited capability ICT. For the military and telecommunications equipment manufacturer, these offered a good economic performance when used as a pre-screener in front of an FNT. They were probably restricted in their ability to use a full-blown digital ICT for the reasons given earlier. MDAs are relatively inexpensive to buy and to program.

At the other end of the scale the mid to late eighties saw the introduction of a number of 'supertesters'. These are very powerful but fairly expensive testers costing between $1 000 000 and $3 000 000. They are usually called multi-strategy testers (MSTs) or simply 'performance testers', since they include many different test methods and set out to test the full rated performance of the UUT. They are architected along the lines of a VLSI component tester and are characterized as having much more flexibility and precision in terms of the placement of drive signals and the sensing of responses. As such they have a much better chance of detecting design induced defects and marginal behaviour of components. From an economic point of view they are usually justified on the basis of reduced diagnosis costs at later test stages or by the possible elimination of the sub-system test stage altogether.

This overview has briefly related the development of the commercial automatic board test system from the 1960s to the present time (1992) to serve as background information for anyone new to ATE and to illustrate the wide range of choices available. If you also bear in mind that each of the tester categories described contains a wide choice of equipment, with widely varying prices and performances, you begin to see how complex it can be to decide on what is best for any given situation. This is why the following chapter on test strategy optimization and modelling is so important. Further tester types such as vision systems, infra-red systems, laser-based solder joint testers, etc., further add to the complexity, as do such issues as burn-in or vibration screening. The following brief comments about each of the major options should help to highlight the main pros and cons involved in any comparisons.

7.3 Testing options

The major options at the board test stage of the overall test strategy are:

1. No testing. Go directly to the sub-system test stage.
2. Test with a dedicated manual test rig.
3. Test with your own custom ATE, possibly using VXI modules.
4. Test with a commercially available ATE system.

The decision here will in most cases be fairly obvious, based upon such factors as volumes, yields, fault spectra, time available and equipment costs. When the direction is not so obvious then a quick and simple economic analysis will usually point to the best alternative. This is not the case when you get down to the more specific issues involved in the choice of tester. This requires a much more detailed analysis, as is explained in the next chapter.

No testing

If we choose to do no testing of individual boards then the boards would pass to the next assembly stage and a decision would have to be made about how to test at this stage. No testing at the individual board stage would most likely be appropriate when:

1. The yield of the boards is very high.
2. The board would be difficult to test unless the next assembly stage is completed.
3. Any condition where it is not economic to test.

Visual inspection

Some degree of visual inspection is usually performed between the soldering process and the first test stage. This may be a cursory look at the board mainly to see if there are any cosmetic problems or any gross assembly or soldering problems. The depth of any inspection will tend to depend on what follows. If the next test stage is an ICT or an MDA there is usually no economic benefit to performing such an inspection. If the next stage is a manual test operation or a functional test operation with no prior shorts test, then it would probably be prudent to make the visual inspection more thorough. To some degree this visual inspection may well be a part of the assembly process itself, in which case the inspection would only be of the soldering. Automatic visual inspection systems exist (vision systems) and might be considered. They tend to be rather slow and because of their grey-scale visual analysis methodology they may run into problems if there is not good colour consistency in the board substrate material and the components. Vision systems based on x-ray are also emerging and these promise to be able to detect problems beneath components on a board.

Bare board testing

The bare board can be considered to be a component since most companies buy in their production requirements from specialized companies. There is usually only a prototyping facility in-house. Specialized bare board testers are available and vision systems can also be used to inspect the bare board surfaces. There are even specialized inspection systems for looking inside via holes on the board. The economics of performing any incoming inspection on bare boards can be approached in a similar manner to component test economics as described in Chapter 6. Essentially it will come down to the defect rates and the cost of any defects that escape detection in terms of the higher costs that will be incurred at later test stages. This can be very high when a complex multi-layer board is fully assembled and the fault is virtually unrepairable because it is in a buried layer.

Shorts testing

Systems designed specifically for shorts testing on loaded boards were commonly employed when the dominant form of loaded board tester was the FNT. When ICTs and MDAs came along there was less of a need for such specialized (limited?) testers. For most applications that need it, the low cost of an MDA may make that a better buy. However, if the dedicated shorts tester can also test for open circuits then it could be preferred to an MDA, which usually has limited opens testing capabilities. With the general improvements in manufacturing quality and the increased use of surface mount technology (SMT), shorts have tended to become less of a problem. In fact with SMT boards opens are likely to be a bigger problem than shorts.

The manufacturing defects analyser

The MDA has enjoyed a lot of popularity in recent years because of its low cost and its easy program preparation. The use of MDAs was originally conceived as a pre-screening for more powerful and more expensive systems and as such was usually an economically viable thing to do and could generate a good return on the investment. Unfortunately, MDAs have occasionally been oversold in terms of their real fault coverage capabilities and some companies have used them as the primary board tester as opposed to a pre-screener. Whether this is a viable strategy or not is a function of the yield, the fault spectrum, the fault coverage of the following test stages and the end use of the product. In some cases it may be the best strategy but in many cases it most clearly is not. One of the main problems is that surprisingly few companies have reliable information about their yields, their fault spectra and other important information that is relevant to test strategy decisions. An important shortcoming of MDAs at present is their limited ability to detect open circuits. Since this can be one of the major defect categories on SMT populated boards this is a bit of a drawback. However, a number of companies are working on a solution to this problem so it could well be solved by the time this book is published.

In-circuit testers

Without doubt the most widely used tester type in operation today is the in-circuit tester. Testers are available in a wide range of capabilities and prices and the price/performance ratio has been improving steadily. The better testers are capable of very high levels of fault coverage coupled with very high throughput rates, which makes them very cost effective if the production volume is fairly high. However, even at fairly low production volumes they often prove to be the most cost effective solution, which is why there are so many of them about. As mentioned earlier, the increased use of ASICs has caused some increase in programming costs because these devices are not available in the standard libraries. However, there are a number of solutions to this problem. All ASIC designs require a test program to verify the ASIC itself and this is usually generated as part of the design process. In some cases this data can be used to create an ICT library model. Another simpler choice is to only perform a rudimentary test of the ASIC on the board and many companies choose to do this. This may sound like an easy way out but it is not entirely against the basic philosophy of in-circuit testing. If the ASIC was made right, if it is mounted correctly, soldered correctly and basically functioning, then there is a high probability that the board will function correctly. Yet another solution is to have the ICT use any scan paths or 'built-in test' within the ASIC itself to test it.

ICTs tend to have a high fault coverage for the majority of detectable manufacturing process defects. I say 'detectable' because there are a number of defect types that cannot be detected. A good example would be the 100 nF capacitor in parallel with the 100 µF decoupling capacitor. The presence or the absence of the 100 nF component cannot be detected by a simple impedance measurement. The fault coverage for component-related defects will typically be lower than for manufacturing defects and the coverage for design induced defects will be rather poor unless it is a fairly blatant problem. Component interaction problems would not normally be tested for, but some testers do have the ability to test a small group of components at one time. Diagnostic resolution and accuracy are good but the ICT is more likely to indict a non-existent defect than a functional tester is. This is particularly true with the lower end testers, which usually have a poorer measurement range and accuracy. At the extremes of the guarding systems capabilities, measurement accuracy can suffer to the point where a measurement appears to be out of specification when the component itself is actually all right. This results in the 'detection' of a defect that does not exist.

Some testers are equipped with additional diagnostic tools that can improve the diagnostic resolution. This can be important from an economic point of view because it leads to a more efficient repair process. Every board repaired incorrectly as a result of poor diagnostic resolution will have to pass through the tester a second time and will have to be repaired a second time. This costs money and reduces the capacity of the tester and the repair station.

Functional testers

Although the ICT has dominated board test ATE sales for the past ten to twelve years there are still quite a lot of functional testers sold. There is still enough scepticism about backdriving to cause many military users to prefer functional test and most 'in-house' built ATE is functional in nature. Program preparation times, and therefore costs, still tend to be the main problem, especially now that 'time to market' has become such a big issue. However, there are many examples of complex functional test programs being generated in very little time as a result of good design and test integration. Automated test pattern generators (ATPGs) can be quite effective if the design is reasonably well structured and if the design is verified by simulation. Much of the needed data is present in the design database and a number of products exist to take this data and produce it in a format needed by the test system.

Combinational testers

Most of the combinational testers available today are essentially in-circuit testers with added functional capabilities. Both analogue and digital capabilities may be optionally available, as may specialized memory testing and boundary scan hardware. Boundary scan requires a very different configuration to a standard ICT. Typically an ICT will have between 1000 and 3000 test pins with a memory depth (pattern depth) behind each pin of between 8 and 16k bits. A typical boundary scan requirement may be for four test pins and tens of megabytes of memory depth. The architecture of ICTs is such that this capability has to be provided by a custom designed module rather than any reconfiguration of the tester's pin memory.

Combinational testers are capable of providing a very high degree of fault coverage, but as usual what you actually get will depend on the fault spectrum that you have or expect to get in the future for your particular production situation. Since these testers cost a lot more money than a simple ICT it is well worth checking that the added test functionality really will produce results for your situation. If the added functional testing capabilities are modular they could always be added later if needed.

Performance test systems

This category of testers represents the current top of the range of commercially available testers in terms of performance, complexity and price. They are known as 'multi-strategy testers', 'performance testers', 'cycle-based testers' and possibly even by other names. They contain the most flexible functional testing capabilities that have ever been available, an in-circuit capability equal to the best ICTs, along with a wide range of other specialized testing capabilities such as algorithmic pattern generation, very fast clock generators, and analogue stimulus and measurement features. They have powerful graphical user interfaces, multiple computers, masses of memory and masses of mass storage. The key feature that differentiates this type of tester from earlier functional board tester designs is the 'cycle-based' nature of the driver sensor sub-system. In most functional testers changes in the input vectors applied to the UUT will occur at fixed time intervals, so a tester operating at 10 mHz will apply the test vectors at a 100 nanosecond (ns) rate unless there are any programmed delays. Similarly the sensor system is able to monitor the logic state of the output pins every 100 ns. Such systems are usually asynchronous and suffer from 'dead-time'. Dead-time is the term used to describe any time in the test sequence when a driver cannot change its logic state or a sensor is unable to read in a logic state. In the cycle-based tester the tester's clock synchronizes to the UUT's clock and the driver edges can be positioned at different times within the test cycle. Such a tester operating at a nominal 10 mHz testing rate will have a test cycle time of 100 ns but the driver can be told to change states in 10 ns steps within the 100 ns cycle, or even within the next cycle. Similar timing flexibility also applies to when you sense the response to an input change. This degree of flexibility enables the tester to emulate the complex timing diagrams of the board and the components on it.

What does this all mean in practical terms? Essentially, if you can emulate the performance of the UUT, you can detect performance-related defects. The type of defects that result in performance anomalies are the design induced defects and the component-related performance defects. These kinds of problem are typically referred to as 'speed-related faults', 'a.c. parametric defects', 'dynamic functional faults', etc. The bottom line from an economic point of view will be that you can justify the higher price of these testers if their extra fault coverage can reduce the cost of diagnosis and repair at the following test stages, including out in the field. Alternatively, the ability to detect the dynamic and the performance-related defects may make it possible to eliminate the next test stage altogether. If this can be done the cost savings could well be substantial enough to make even a $1 000 000 plus price tag look like a bargain. These testers can also be very useful for testing the prototypes of a new design. While on the subject of prices, it seems to me that we tend to have some sort of mental money barrier that works in decades. Depending on the item to be purchased these barriers exist at $10, at $100, at $1000 and so on. When these performance testers were first introduced at around $1 000 000 this was a major barrier, even though a few years earlier the top of the range functional testers were costing between $500 000 and $700 000 each. These typically contained a rather simple 12 or 16 bit computer, a few hundred driver sensors and operated at 2 mHz. A performance tester of today typically contains several 32 bit processors, possibly a dedicated high-speed wider word length processor, thousands of driver sensors and operates at 20 mHz with extremely flexible timing control. With price inflation for five years of 5 per cent per year the $700 000 tester would now be $893 000. This is not very different from the modern performance tester. I suppose that this is further proof of 'Moore's law' about electronics (chips) doubling in complexity while halving in price. We all expect the price/performance ratio of any electronics product to go up and up.

Infrared-testers

This is a very specialized form of board tester that uses infra-red imaging to look at the temperature of the components on a board. This can be used for diagnosis purposes since many defects result in a measurable temperature rise or possibly in a lower than expected temperature if a defective device is not drawing sufficient current. Another type of tester that uses infra-red techniques is designed to test solder joints. Each solder joint is heated up with a burst of energy from a laser. Then the temperature of the joint is measured over time with an infra-red sensor. The profile of the temperature decay will then indicate how good the soldered joint is. A correct joint with the right amount of solder will show a smooth temperature decay with a certain decay time. Too much solder on the joint will show a longer decay, too little solder will show a rapid decay and a dry solder joint will show a different profile again. These testers tend to be used in the testing of high-reliability products when the solder joints may deteriorate with time or mechanical shock.

'In-house' ATE systems

Before commercial ATE systems became available in the late sixties most people made their own. In fact, the first commercial computer-based board tester, GenRad's model 1790, was born out of an 'in-house' system that had been used for a couple of years before being offered on the market. ATE in general, as opposed specifically to component and board testers, actually had its roots in the military prime contractors. Most military projects required that the contractor supplied a complete testing capability along with the product itself. The test equipment was an important, and often very lucrative, portion of the contract. In addition to the military contractors, most large companies had their in-house test groups who designed and built special-to-type test fixture, test rigs or even general purpose test equipment. In the early days of the commercial board test market these in-house groups were the biggest 'competitor' to the ATE suppliers. This designing and building of 'in-house' test systems continued into the early eighties, even for testers that were commercially available. In most cases this did not make good economic sense because even the very large companies still did not have the benefits of scale that the dedicated ATE manufacturers had. In virtually all of the in-house built board testers that I encountered the hardware was excellent and the software was awful. The result was that the test program preparation process was long and expensive and there was usually no automated diagnostics so the throughput was very poor. As a result, large numbers of these in-house systems were needed to get the job done. One company I was involved with had over a hundred of their testers deployed for a production volume that could have been handled with five or six commercial testers. The hardware was also expensive. I heard of a tester that was insured for $1 000 000 during transportation to another factory when the average cost of a similar commercial product would have been $300 000. The software systems that drive modern ATE usually accumulate several hundred man-years of development over their product lifetimes. Most manufacturers will produce several new hardware generations that are all driven by one software system. The software tends to grow and grow. On top of that you also have an enormous investment in the development of component libraries needed for the program preparation system. There is no way that this effort can be matched by any company who only wants to build testers for their own use. For special-to-type equipment, for which there is no commercial equivalent, the situation is very different. You either have to build it yourself, have it custom built or have a commercial tester modified.

In the early days in-house testers were usually built from a stack of commercially available instruments and some form of controller. The equipment was usually built into a standard 19 inch rack and so the terms 'stack and rack' or 'rack and stack' were usually applied to describe these test systems. Today there is a wealth of programmable 'smart' instruments to use as part of a home-made tester, and also a very wide range of modules that comply with the VXI bus standard. Putting together the hardware has never been easier. However, the economic issues have not changed. The amount of software development needed to make a system perform efficiently is almost always underestimated, as is the effort needed to get all of the hardware to play together. If there is a commercially available solution to your needs it will almost certainly be a better economic solution. If there is no commercial equivalent then you really have little choice other than to go ahead and design your own. If this is to replace existing test rigs an economic justification (ROI) is likely to be needed so make sure you fully understand the cost of putting it all together. The alternative might be to use a third party systems house to do it for you. Even if you have the expertise in-house it may be that their time is better spent performing their usual duties rather than developing a custom tester.

Manual board test

In many cases, particularly where volumes are very small, the board test stage of the test strategy is still performed manually in some way. There is a bit of a grey area here in defining this as being the 'board test' stage or the 'sub-system test' stage and this will usually depend on the number of boards in the product. A common method is to use a substitution method of using a fully functional sub-assembly, with one of the boards replaced by an untested unit direct from manufacturing. Another method is to use some form of test fixture that may contain test sequences built into ROMs (read-only memories), but with all measurements and diagnostics being performed manually. Yet another approach for digital boards may be to use word generators built into a personal computer and a logic analyser for measurements. Whatever the method employed there will be some point at which it becomes economically viable to invest in a more automated solution. ATE costs tend to fall as labour costs increase so in general the break-even point at which the cost of the manual approach is the same as the cost of the automatic approach gets lower all the time. The manual approach will still tend to have the same major cost areas to be considered as the automatic approach. For a 'life cycle cost' analysis these would be:

1. The purchase cost of any equipment required.
2. The maintenance cost for the test equipment.
3. The cost of developing the test plan.
4. The cost of test diagnosis and repair.

5. The cost of defects escaping to the following test stages.
6. The cost of defects escaping to the field.

7.4   'Process' testing and 'performance' testing

Around the time when the decade of the seventies ended people in the ATE world (both suppliers and users) began referring to in-circuit testers as 'process testers' and to functional testers as 'performance testers', the logic being that ICTs mainly addressed problems caused by the manufacturing process, while FNTs mainly addressed the functionality or the performance of the board in total. In reality the functional testers of that era were also 'process testers' in that they did little more in the way of fault detection than the better ICTs. FNTs could detect manufacturing process defects such as shorts, opens, mis-inserted components, etc., but they did little in terms of testing the board at its rated speed or in a manner that could detect performance-related defects. They were too slow and they had very limited control over the timing of driver/sensor edges. The FNT typically applied complex functional test vectors to a board from the edge connector and so its performance was heavily constrained by the testability of the board's design. The ICT, on the other hand, could force its way through testability problems and test each device with a more comprehensive set of vectors than would get through to the device pins from the edge connector driven by the functional tester. Therefore, for functional problems related to the functionality of the devices the ICT could do a better job. The FNT scored when the functionality problems involved multiple devices.

True 'performance' testing did not really emerge until the second half of the eighties in the form of the cycle-based performance testers referred to earlier. Even these systems cannot match the full operating speed of the UUT but they do not really have to since control of timing and the lack of any 'dead-time' are more important than sheer speed when it comes to detecting performance-related defects. In other words, it is more important to be able to emulate the way the UUT actually functions than to simply match its speed. These testers are capable of performing some design verification testing that could not be done with a flat timed tester regardless of how fast it can run.

The reason for addressing this here is to warn against being misled by the terminology that some ATE suppliers apply to their products. The ATE market is a very competitive one and has been going through a bad patch for some time. Under these conditions there is a danger that suppliers will be tempted to make their products sound a bit better than they really are. This has certainly been the case with the word 'performance'. There is nothing illegal or unethical about referring to a system as a 'performance' system. Performance is a relative thing, so if a supplier has two products then the better one can quite reasonably be referred to as a 'high-performance' tester. However, their definition of performance may be different to that of other suppliers. The practice is to be frowned upon if you accept that the term 'performance' as applied to a board tester means something specific, such as a need to be cycle-based.

At the end of the eighties yet another term began to be used to describe testers of an in-circuit nature, i.e. ICTs and MDAs. This is the term 'structural tester'. I can only assume that this is referring to the ICT philosophy that says 'if it is built (structured) correctly, using the correct components, with no shorts and opens, then it will probably work'. I do not particularly care for this terminology since 'structural testing' is already widely used to describe the testing of the mechanical integrity of an object, including printed circuit boards. Using the same term to describe a tester that tests electrical connectivity, parametric values and functional truth tables is likely to cause confusion. Having said that, however, it is probably preferable to calling them 'bed-of-nails testers' since the main purpose is to test the board and not the fixture. In the same vein we should probably also stop calling FNTs 'edge connector testers'.

7.5 High speed versus performance

In many cases the confusion about performance testing has been connected with the confusion about testing speeds. It is quite normal in any selling situation to emphasize the importance of any problem that the sales person feels they have a solution for. This obviously gives them a competitive advantage if they can convince the potential customer that this is a vital issue. This is where I would classify the test speed issue. The vendors with high-speed testers claimed it was essential and the vendors with the lower-speed testers claimed it was not needed. 'The higher speed will not improve the fault coverage but it will increase your fixturing costs', etc. The silly thing about this argument was that the difference in speed between the slow systems and the fast systems, in audio terms, was one octave when you really need a decade to see any substantial differences!

The implication was that the higher speed would catch more faults of a 'performance' nature, but without the control of timing edges this is not true. You can detect certain performance defects with a single measurement if you have sufficient control over timing; you do not need to make 20 000 000 tests in a second to do this. From an economic point of view you have to ask yourself what the real benefit of a faster test rate is, especially with an ICT. Does the extra cost of the tester, and the extra cost and time involved in building and debugging complex fixtures that will work consistently at the higher test rates, result in a lower overall cost or not? These questions have to be related to the fault spectrum that you expect to get in practice. Answering this type of question requires a detailed analysis of the test strategy, using the available alternative testers, such as described in the next chapter.

7.6 The fault spectrum

The fault spectrum is one of the most functional concepts in the analysis of testing strategies. It is also the most abused concept. I have seen numerous magazine articles and conference papers where the virtues of one tester or a specific strategy have been promoted based upon the author's definition of 'the industry typical fault spectrum'. This industry typical fault spectrum will vary enormously depending on the product being promoted. The economic argument in favour of MDAs has always gone something like this: 'Since manufacturing defects account for $X\%$ of your fault spectrum, why spend another \$50 000 to find just $Y\%$ more?' The economic argument for the more powerful testers has always been along the following lines: 'The cost of detecting defects at the later stages of test is so high that you must maximize the fault coverage at the board test stage to minimize your overall life cycle costs. This requires coverage for other types of defects.'

I tend to go along with the second argument, not because I was associated for a long time with the marketing of 'high end' testers, but because I have been involved for an equally long time in performing economic analysis on alternative strategies. Unless there is a vast difference in capital cost the 'better' fault coverage tester nearly always wins the economic argument. The more important point here however, is that it does not matter who is closest to the real industry average fault spectrum, you cannot determine *your* test strategy based on such an average because your situation may be very different. The overall fault coverage of any tester will depend to some degree on the fault spectrum of the boards it tests. This spectrum will vary from board type to board type, from production line to production line, and even from day to day on the same line producing the same board type. There are so many variables involved. It actually gets worse as your yields improve. At low yields the spectrum often tends to be more stable because there are a number of underlying consistent common causes for the defects. At very high yields, where the process is well controlled, the small variations that result in defects can be much more random.

Any analysis of your test strategy will require the best possible information about the fault spectrum that you can obtain or predict. Crude analyses that simply apply a global fault coverage percentage to a given test stage are completely inadequate for anything other than a rough estimate of costs and capacities. This is the subject of the next chapter so I will not dwell on it here. The main points I would like to get across here are the importance of knowing your fault spectrum and the potential danger of being misled about the performance of a tester.

7.7   The major cost areas

The performance of the board tester will affect the cost of the operation at the board test stage, at the following test stages and in the field. Therefore these 'life cycle' costs have to be compared when comparing alternative testers. If we simply look at the direct costs involved with buying and operating the tester we can be misled into accepting a low-cost solution for the board test stage that raises costs more at other stages and also damages our quality reputation. As defined in Chapter 1 the life cycle costs for test decisions fall into six main areas:

1. The cost of the equipment. For ATE this is usually just the purchase price of the tester but it could also include any additional equipment that is needed for the

operation of it, especially if it is equipment that can be capitalized. Vacuum pumps, air conditioners and furniture are examples of this extra equipment. For the more manual test stages this would include the cost to develop, build, test and document any special purpose equipment or jigs. The cost of any general purpose instrumentation such as digital voltmeters, oscilloscopes, signal source, power supplies, logic analysers, etc., should also be included, along with at least an inventory carrying cost for any 'products' used for test purposes.

2. The cost of maintaining the equipment (both the hardware and the software elements). For ATE the cost of any maintenance contracts and software support charges. Some companies would include any additional training requirements in here also. For the other test stages similar hardware service costs will be incurred and any software involved may also require some updating.

3. The 'set-up costs' for program preparation and the production of fixtures. For ATE the cost of generating the test programs and the design, construction and debugging of fixtures. For manual operations the cost of developing a test plan, writing it, editing it and debugging it, plus the cost of any interfaces and jigs not included in item 1 above.

4. The operational costs for testing, diagnosis and repair (TDR costs). The cost of actually using the equipment for testing and for diagnosing defects and the cost of repairs or re-work that is usually performed off-line at dedicated repair stations. It can be useful to keep these three cost areas separated because it can sometimes be surprising as to which is the biggest. It is quite common to exclude repair costs from these calculations under the assumption that they will be the same regardless of which tester is selected. This is not true. Repair costs will be dependent on the effectiveness of the diagnostic system and this can vary greatly from tester to tester.

5. The costs (all of the first four categories) for the following test stages beyond the board test. There will be costs associated with the following test stages that fall into the above four areas. The diagnosis and repair costs at these stages will vary depending on the performance of the board test stage. This is where the 'rule of tens' concept starts to work since the cost of each diagnosis and repair action will be a lot higher here than at the board test stage.

6. The field-service costs. The cost of field service will vary with the yield from manufacturing and the fault coverage of the overall test strategy. An additional field cost will be incurred as a result of early-life failures, but this is not usually considered when comparing board testers. If one of the strategies being considered could trap any potential early-life failures before the product leaves the factory then this should of course be taken into account. Even though relatively few defects should escape, the cost of field-service operations is very high. This can often be the biggest cost category of the six.

In case you skipped over Chapter 6 it may be worth reading Sections 6.4 to 6.6. Each of the following sub-sections takes a look at the six major cost areas listed above with some suggestions on what to look for. I have avoided any specific cost information or references to any specific equipment examples because this kind of data becomes out of date very quickly. These sections are a brief checklist of where the costs are and possibly how to avoid some pitfalls.

7.8 Equipment costs

The commercial ATE market is a highly competitive one. For this reason, prices cannot vary over a very wide margin for equipment with similar capabilities. It is unlikely that you would find price differences of more than 30 per cent between competitive offerings; in fact 20 per cent is more likely to be the limit. These price differences can usually be accounted for by differences in the design philosophy or such things as import duties where the equipment originates outside the country. The use of different types of computer peripheral is a good example of where price differences can arise. The nature of mass-storage devices—magnetic tape, floppy disk, cartridge disk—and their sizes can account for some differences, as can printers and terminals, etc. On the more positive side, it is likely that faster storage media or peripherals will also produce cost savings due to more efficient operation.

Price differences between competitive products are therefore not likely to be a major factor. The main reason why anyone ever paid more than they needed to for an ATE system is that they bought more than they needed. It is a great temptation, and one that is difficult to avoid, when buying a large expensive system, to purchase an optional capability, because you may need it at some time in the future; it is a kind of insurance. Another reason why this happens is that if middle management is going to upper management, cap in hand, to ask for a large sum of money, it might as well be 10 to 20 per cent more. This will also avoid the embarrassment of having to go back in six months' time to ask for more money. Before being tempted to fall into this trap, remember a fundamental rule of capital investment: *an investment in a piece of equipment cannot provide an ROI unless the equipment is used.* If you buy an option and never use it, then you have a certain amount of money just lying idle; it is not even earning interest—instead it is depreciating.

If the possibility exists that a feature may be needed and that the feature can be added to the equipment at some later time, it is far better to say so in your proposal to upper management. They will feel you are doing your job better than if you buy something you never use. There will be times when the decision will not be so easy, such as when the feature cannot be retrofitted or when the choice has to be between two different models. In this situation, a careful look at probabilities of needing the extra capabilities, and the risks if you do not buy them, will be necessary.

It should almost go without saying that when comparing prices be sure you are comparing apples with apples. By this I do not necessarily mean comparing the contents of the system, but their performance. For example, it is meaningless to compare how much memory you have in the computer or microprocessor controlling the system when the performance is more dependent on how that memory is utilized by the software.

7.9 **Maintenance costs**

As with the purchase prices for similar equipment, maintenance costs for service contracts or hourly rates will not vary much between suppliers. Service contracts are typically in the range of 8 to 10 per cent of the initial purchase price and hourly rates are also usually competitive. The biggest variables when it comes to maintenance costs will be:

1. How often will it be needed?
2. How quickly will it be fixed?

In other words, reliability and response time to service calls. The best way to evaluate such things prior to purchase is to ask for details of the supplier's quality assurance program (do they do burn-in, do they do a soak test, etc.) and to talk to other users of their equipment. 'How many service engineers do they have?' and 'What spares stock do they hold?' are other questions that should be asked.

Many companies now charge for software updates and other software support capabilities. Some also have optional rapid response applications support that is on-line to your tester via a modem. The vendor's engineers can operate your system remotely to diagnose problems, and software 'patches' or updates can be downloaded via the modem.

7.10 **Set-up costs**

The set-up costs are essentially the test program generation costs and the cost of the test fixture required to interface the UUT to the tester. The actual steps you go through to develop a test program will vary depending on the type of system involved. ICTs usually use an automatic test generation system, which can go a long way to creating the program. Analogue component tests are either selected by trial and error from a library of tests or, in the case of the more sophisticated testers, they may be determined as a result of an automatic analysis of the UUT's circuitry. The diagnostics on an ICT are generally a fall-out of the method of test, so there is little need for any additional work to create any diagnostic database unless the tester has some special diagnostic additions.

Functional test programs are usually developed with the aid of a digital simulator. These can vary quite a lot in terms of their capability, their speed, their ease of use and so on. Much of the evaluation time can be taken up comparing the various test program generation software capabilities. This is just as it should be, however, because the hardware is usually quite easy to compare but the software is not. Since the overall performance of the tester will be more dependent on its software than on its hardware you should be prepared to put quite a lot of work into comparing these capabilities.

Most of the testers from the bigger vendors have the ability to take data automatically from the design database to eliminate the unnecessary duplication of effort and to minimize the mistakes that can occur when this duplication has to take place. Almost all testers require a circuit description of the UUT for the program preparation process and for the diagnostics. This can usually be produced automatically from the circuit description needed for the design. This is usually referred to as the 'physical' data and most available conversion packages provided by the ATE suppliers can support conversion from the formats used by several of the computer aided engineering (CAE) vendors.

In addition to this 'physical' data it may also be possible to take in some of the 'behavioural' data from the design simulator. This can save a lot of time in developing functional test programs and can also be used to develop test program models for ICTs. These are of course functional tests. This can be particularly useful for ASICs. There are also products available for this design to test integration from independent companies who do not themselves manufacture CAE or ATE products. These typically take data in from a variety of commercial simulators or even a physical UUT and convert it to their own native format, which retains a definition of the logic waveforms within the UUT. They can then output the waveform data through a series of converters into the required format for different testers to create programs for component testers, in-circuit library elements, and functional board test programs and probing databases. The key advantage claimed is that the same set of tools can be used at all stages of the process.

The generation of test programs is a critical area for the operation of any ATE. It is also one of the areas where the greatest differences exist between competing products that may at first sight appear to be very similar. The primary objective with any board tester is to get a program with a very high fault coverage in the shortest possible time. This is essential if you are to meet the four main objectives. With short lifetime products you rarely have the luxury of going back to the test program to enhance it once the product is in manufacturing. If a review or a modification of the test program is required as a result of an engineering change order (ECO) you may get the chance to incorporate some changes but frequently you have to live with what you had first time. It is important then to compare competing testers with this in mind. Some testers will get you up to a certain level of fault coverage very quickly but then they may be very slow when it comes to interactively trying to improve upon this. Some, however, behave in the opposite manner. They may take a bit longer to get to some reasonable degree of coverage but then enable you to improve to a much higher level quite quickly.

7.11 Operational costs

The operational costs are the test, diagnosis and repair costs (TDR). The test and diagnosis costs will obviously be directly related to the performance of the tester in terms of its absolute test times and the efficiency of its diagnostic methods. The repair costs will also be partially determined by the tester's performance because the accuracy and the resolution of the diagnosis will determine how long it takes to perform a repair and how often an unnecessary repair action takes place. Figure 4.2 shows that there are four possible outcomes to a test. The two 'passing' cases are of no consequence to diagnosis and repair since neither will take place if the board passes. However, the board may fail the test because it contains a defect that is detected or it may fail the test even though no defect is present. This can happen, for instance, when a measurement is near the end of the tester's range of capabilities. Measurement accuracy tends to fall off rapidly at the extremes of the range and a measurement can therefore appear to be bad when it is in fact all right. Obviously the wider the tester's measurement range, the higher its accuracy, and the slower this accuracy falls at the extremes of the range, the less likely it will be that defects get indicted when they should not be.

This indictment of non-existent defects is only one of the reasons for unnecessary repair actions. Other reasons include incorrect diagnosis caused by limitations of the diagnosis system and incorrect repair action caused by misinterpretation of the diagnostic message. Since every diagnostic action will be followed by a repair action these kinds of problem will increase the workload on the repair stations and also reduce the capacity of the test system.

The 'test time' is usually defined as the time taken to test a defect free board. Depending on the formulae being used to determine costs this time may or may not include the 'handling time' taken to connect the board to the tester for testing and to remove it once the test is complete. My own preference is to combine the testers 'run time' and the 'handling time' and to call this the 'good board test time'. Others prefer to show the run time and the handling time separately. As long as the formula is correct it does not really matter too much how this is done.

The diagnosis time is usually defined as the average time taken to diagnose a defect on the test system. We have to be careful here because of the different nature of the diagnostic methods employed on different tester types. On an ICT the diagnosis is a fall-out of the testing process. You test one component at a time and if it fails the test you have your diagnosis automatically because you know which component was being tested when the failure occurred. The exception to this rule is that some testers have some additional diagnostic tools which are additional to the in-circuit tests. If this is the case then some time should be added to the average to account for the fact that these additional techniques will be used occasionally. However, for a pure in-circuit test of the components the test time and the diagnosis times will be much the same. There may be a small additional time required to print out the diagnostic message. For the diagnosis of a short circuit things are a little different. The shorts test is normally performed before the component tests within the ICT's test program, and if any are detected the testing stops and the board is sent to the repair station for repair. This is because the presence of a short will affect the tester's ability to make other measurements accurately on the nodes that are involved in the short. It is quite possible that a short between two nodes will completely short out one or more components. If the short remained on the board then when the tester tests these shorted components it would measure a very low impedance. Therefore the diagnostic message may well refer to a short and one or more bad components, all of which will be 'repaired' even though there is actually only one defect present. It is also possible that the short could cause damage to the board, when power is applied, if it is allowed to remain. The shorts test on an ICT usually takes much less time than the testing of all the components. It should take a few seconds rather than a few tens of seconds, so this difference has to be accounted for in any economic analysis. How this is done is described in Chapter 8.

7.12 The cost of the following test stages

Each of the test stages that follow the board test stage will have the same set of four cost areas that we have just looked at. There will be some equipment cost, some maintenance cost, some programming cost and some operational costs. These will need to be considered and determined in much the same way as the costs for the board test stage.

7.13 Field-service costs

The last cost area when looking at the life cycle costs related to the choice of tester or test strategy is the cost of field service. As stated earlier, these costs will be a function of the yield out of manufacturing and the overall combined fault coverage of all of the test stages. There is a discussion of these costs in the previous chapter.

7.14 A simple example

Although it is necessary to perform a detailed economic comparison of test strategies and test tactics in the manner described in the next chapter, there are times when a quick and dirty analysis is needed just to get a feel for the situation. This can be done on paper or by developing a simple spreadsheet model. The following example shows how this might be done for two alternative functional board testers. The evaluation team at XYZ Electronics has narrowed down the field to two testers, A and B. Tester A has marginally better performance in terms of program preparation, fault coverage and operational costs, but it is 50 per cent more expensive at \$300 000 as opposed to only \$200 000 for tester B. At first sight it does not appear that the performance differences justify the additional \$100 000, but they decide to perform a simple analysis anyway. The new tester is required for a new product range that is forecasted to sell about 7500 units per year, with each unit containing four boards of low to medium complexity. The annual board volume will therefore be 30 000 boards per year. The yield from manufacturing is expected to be 70 per cent, which equates to an average of 0.357 faults per board ($FPB = ABS[LN(yield/100)]$); i.e. the average number of faults per board (FPB) is equal to the absolute value (ABS) of the natural logarithm (LN) of the yield, when yield is expressed as a proportion rather than a percentage.

For their expected fault spectrum they have determined that the fault coverage will be 95 per cent for tester A and 90 per cent for tester B. Tester A is likely to have fewer problems with misdiagnosis than system B so a correction factor, known as the 'diagnosis/repair loop number', of 1.1 is applied to system A and of 1.15 for system B. This implies that system A would have 10 per cent more diagnosis and repair actions than the true yield of 70 per cent would indicate. System B would have 15 per cent additional, unnecessary diagnosis and repair actions. This figure is also applied to any diagnoses and repairs that result from defects detected at the final test stage.

The test strategy opted for is to follow the board tester with a single final system test stage since there are no 'sub-assemblies' in the product. Table 7.1 tabulates the above information along with calculations of the number of 'detected faults per board' (DFPB), the number of 'escaping faults per board' (EFPB), the 'apparent yield' (Y_a) and the number of diagnosis and repair actions.

As you will see from the table, the assumption is that 90 per cent of the defects escaping from the board test stage will be detected at the system test stage. Table 7.2 lists the main performance parameters as determined by the evaluation team.

The labour cost of the repair technicians is also $20 per hour. Table 7.3 now shows the calculation of the operational costs.

Table 7.3 provides the cost data for the operation of the board testers, the following test stages and the field. The program preparation costs are $144 000 ($12 \times 12\,000$) for system A and $158 400 ($12 \times 13\,200$) for system B.

Table 7.1 Defect detection and escaping defects for the example

	Board test A	Board test B	Final system test With A	Final system test With B
Average faults per board	0.357	0.357	0.018	0.036
Fault coverage	95%	90%	90%	90%
DFPB	0.339	0.321	0.0162	0.0324
EFPB	0.018	0.036	0.0018	0.0036
Apparent yield	71.2%	72.5%	98.4%	96.8%
Diagnosis/repair loop number	1.1	1.15	1.15	1.15
Number of diagnostic actions per board	0.373	0.369	0.0186	0.0373
Number of repair actions per board	0.373	0.369	0.0186	0.0373

Table 7.2 Principal test stage performance parameters

	System A	System B	System test
Average good board test time	45 s	45 s	2 hr[†]
Average diagnosis time	5 min	6 min	2 hr
Average repair time	10 min	12 min	15 min
Diagnosis/repair loop number	1.1	1.15	1.15
Average cost of test program	$12\,000	$13\,200	$5000
Debug time on tester/program	3 days	4 days	N/A
Labour cost of operators	$20/h	$20/h	$30/h

†Time for one system (4 boards).

Table 7.3 Operational cost calculations

| | Board test | | System test | |
	A	B	A	B
Good board testing costs				
30 000 × (45 s/60) × ($20/60)	7500	7500		
2 h × $30 × 30 000/4			225 000	225 000
Diagnosis costs				
30 000 × 0.373 × 5 min/60 × $20/60	18 650			
30 000 × 0.369 × 6 min/60 × $20/60		22 140		
30 000 × 0.0186 × 2 h × $30			33 480	
30 000 × 0.0373 × 2 h × $30				67 140
Repair costs				
30 000 × 0.373 × 10 min × $20/60	37 300			
30 000 × 0.369 × 12 min × $20/60		44 280		
30 000 × 0.0186 × 15 min × $20/60			2790	
30 000 × 0.0373 × 15 min × $20/60				5595
Field costs				
30 000 × 0.0018 × $1000	54 000			
30 000 × 0.0036 × $1000		108 000		

Table 7.4 Summary of cost calculations

	System A	System B
Non-recurring costs		
1 Purchase price	$300 000	$200 000
Annually recurring costs		
2 Maintenance costs	30 000	20 000
3 Test programming costs	144 000	158 400
4 Operational costs		
Testing	7 500	7 500
Diagnosis	18 650	22 140
Repair	37 300	44 280
5 The cost of following test stages	261 270	297 735
6 Field-service costs	54 000	108 000
Total annual recurring cost	552 720	658 055

The annual cost difference of $105 335 represents the cost savings of system A relative to system B. These savings would be generated by the incremental $100 000 invested in system A, so when viewed over several years this would be a cheaper alternative. See the example in Fig. 4.1 and the accompanying text for a discussion about 'incremental investment'. The numbers down the left-hand side of Table 7.4 refer to the six main cost areas outlined in Section 7.7 earlier in this chapter.

Table 7.5 System time required for the expected work load

	System A	System B
Time needed for test and diagnosis		
30 000 tests × 45 s/3600	375 h	375 h
11 190 diagnoses × 5 min/60	932.5 h	
11 070 diagnoses × 6 min/60		1107 h
Totals	1307.5 h	1482 h
Tester time needed for program debug		
12 programs × 3 days × 7 h	252 h	
12 programs × 4 days × 7 h		336 h
Total tester time required	1559.5 h	1818 h

The next thing that the evaluation team look at is the capacity situation for the two alternatives. XYZ Electronics want to avoid buying two testers or operating more than one shift, so the capacity calculations are important. They estimate that they will have 1680 operational hours available each year. This is based on there being 7 usable hours each day, 20 working days each month and 12 months in a year. They calculate the amount of time required on each tester in the manner shown in Table 7.5.

The results in Table 7.5 tend to settle the issue. With only 1680 hours available system B does not have the capacity to get the job done without some overtime, a second shift or a second tester.

This example illustrates a number of things. It shows how a rough analysis can be performed quite quickly with just a paper, pencil and a calculator. It also shows how first impressions can be very misleading. The apparently small advantages that system A had over system B did not appear to be enough to justify an extra 50 per cent on the capital investment. It also shows the importance of performing a 'life cycle cost analysis' rather than simply looking at the direct costs associated with the testers. The bulk of the savings are made in the stages that follow the board tester. System A does a much better job of helping the company to meet its four main objectives. The operational *cost* is lower, the *quality* of the shipped product is higher, the *time to market* is faster and the more powerful system will cope with new *technologies* more easily. It should be stressed that this is a very simple approach and a number of things are not included in the analysis. However, this kind of analysis can quickly weed out the non-starters and so reduce the effort required for the more detailed study of the alternatives.

Notes on the example. The above example featured a comparison between two functional testers which is probably the simplest case from a mathematical point of view. The reason is that functional testers will tend to find only one defect at a time and the diagnosis time will be fairly similar for different defect types. For these reasons we can simply determine the total number of defects detected and multiply this figure by the average diagnosis time. Since each board will receive one 'passing' test, either because it was defect-free to begin with or because it was successfully repaired, we can simply multiply the 'good board test time' by the total production volume. This calculation of the 'testing' cost part of the 'operational costs' is also true for other tester types since the board will only receive one 'passing test'. However, the diagnosis and repair costs are more difficult to calculate because ICT type systems can detect multiple defects in one operation provided that they are detected in the same part of the test program. Specifically, two or more shorts will be detected in one operation and the test would usually be suspended at the end of the shorts test part of the program. The board would then go off to be repaired. If there are no shorts present then the tests would progress into the 'component testing' part of the ICT test program and any defects detected here will be reported at the end of the test. The ICT's 'one component at a time' technique means that it can also detect multiple faults in this section of the program. It should be clear, however, that if there is a short and another defect present then these will not be detected simultaneously because the testing would normally terminate after the shorts test. This means that the mathematics become a bit more complicated. A full discussion of how this situation is handled appears in Chapter 4 along with an example. Failure to take this operational situation into account when performing even a simple analysis like the above example will result in some significant errors.

The situation gets more complex yet for combinational testers or for in-circuit testers with a 'boundary scan' test capability, which are really combinational testers anyway. Now we have a situation where we have the complication mentioned above *and* we can have some defects detected by the 'functional' or 'boundary scan' portion of the test program. This situation becomes a bit difficult to model in a simple way and will almost certainly need the use of a more complex model such as the one described in the next chapter.

7.15 Another view of board test economics

The classical definition of quality is 'conformance to requirements'. If the product conforms to (or meets) all of the requirements of the customer, it is a high-quality product. This leads to the concept that the overall cost of quality is made up of two parts. The 'cost of conformance' is the cost of all the things we do to ensure that the product will indeed conform to the customer's requirements and the 'cost of non-conformance' is the cost incurred when the product does not conform. In general the cost of conformance is incurred before the product is shipped, and the cost of non-conformance is incurred after it ships to the customer. This concept can be adapted to compare the effectiveness of alternative board test systems. Ideally we would like the board tester to find all of the defects present on the boards regardless of the source of the defects. In practice this 100 per cent fault coverage is unlikely to be achieved and some defects will 'escape' to the following test stages and even out to the field. The costs incurred in using the board tester can be regarded as the cost of conformance for the tester, and the costs generated by the escaping defects can be regarded as the cost of non-conformance for the tester. Clearly the tester with the lowest escape rate will have the lowest cost of non-conformance. In general the cost of conformance rises as you try to achieve a higher degree of test comprehensiveness, and the cost of non-conformance falls as the test comprehensiveness rises. The total cost at any point will be the sum of the two and this cost will be at a minimum somewhere near the point where the two cost lines cross each other. This is illustrated in Fig. 7.1.

Several elements of the cost of non-conformance are difficult to quantify. These are the cost of lost repeat business due to customer dissatisfaction and the cost of

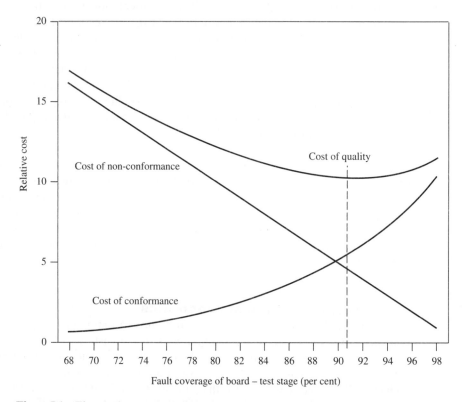

Figure 7.1 The total cost of quality consists of two elements, the cost of conformance and the cost of non-conformance. The cost of conformance will fall as the test comprehensiveness increases, but the cost of achieving this test comprehensiveness will usually rise in a non-linear manner. In this example the cost of conformance rises exponentially by an order of magnitude as the fault coverage of the board test stage increases from 68 per cent to 98 per cent. The cost of non-conformance is made up of additional system test and field test costs. The minimum cost of quality is achieved when the board test fault coverage is a little over 90 per cent

lost sales resulting from a poor quality reputation. Another important area is the added risk of 'product liability' litigation, as referred to in Chapter 2. As a result of this the real minimum may be at a point where the measured cost of non-conformance is well below the cost of conformance.

7.16   The search for cheaper board testing

The test engineering manager is under pressure to reduce the cost of test. At the same time he or she is told that 'testing adds no value'. The only ray of sunshine that can be seen is the increased yields that the quality improvement process has generated. Unfortunately the search for cheaper testing and the high manufacturing yields have occasionally resulted in some less than optimal decisions being made. Cheaper testing does not necessarily come from cheaper testers unless those testers still have good fault coverage, and higher yields do not mean that you can get by with a lower level of fault coverage. In fact the reverse is true. The fault coverage capabilities of a tester are dependent on the fault spectrum that it sees. If all of the faults are simple ones then the coverage can be very high. However, as the fault spectrum moves towards defects that are more difficult to detect the fault coverage will fall. Most quality improvement processes are based upon the Pareto method of eliminating the biggest causes of defects first so there is a tendency for the fault spectrum to become biased towards the most problematic defects as the yield improves. The laws of mathematics cannot be bent. If you attack a 90 per cent yield with a 70 per cent fault cover the escape rate will be the same as attacking a 70 per cent yield with a 90 per cent fault cover. The net result will be no improvement in the shipped quality of the product. All the effort taken to improve the yield will have been wasted.

The cheapest testing will be achieved with a test strategy that is optimized to meet the four main objectives with the lowest possible overall life cycle cost. This and the fact that the board test stage is the cheapest place to detect defects may require that additional costs be incurred here in order to minimize the total. Looking at the board test stage in isolation may give you the wrong answers.

Returning now to the subject of fault coverage, there is a good reason why a higher level of fault coverage is needed in a high-yield situation than in a low-yield situation. The overall fault coverage that is achievable at any test stage is a function of the ultimate capability of the tester, the comprehensiveness of the test program and the fault spectrum. The comprehensiveness of the test program is in turn dependent on the test system, the amount of effort that can be applied and the capabilities of the programmer. In general a test system with a better test generation system, including links to design and a better device library, will enable high fault coverage programs to be produced in less time. The fault spectrum, however, has nothing to do with the tester. This will be a function of the manufacturing technology employed, the effectiveness of the process control, the defect rates of the components, the complexity of the board and so on. The fault spectrum affects the overall fault coverage because different testers will have different levels of fault cover for different types of detect.

In most cases the fault spectrum will change as the yield improves, simply because the quality improvement process will progress at different rates for different types of defect. For example, one reason for the improvement in yields has been the move to surface mount technology. This is a more completely automated process relative to the through-hole process, and a very different one. It would be unreasonable to assume that the fault spectrum of manufacturing induced defects would be the same for these two very different process technologies. For this reason a reasonably accurate analysis of a test strategy, and more particularly for a test tactic, requires that the fault spectrum be broken down to at least eight or ten defect categories. Only then will it be possible to see the different performance of different strategies and tactics when yields and fault spectra change.

7.17 Changing yields and fault spectra

As described in Fig. 1.14, the choice of the optimum test strategy is determined by the external market forces, the internal forces and the current or anticipated performance in terms of yields and fault spectra. Here I have used the plural form deliberately because most companies do not simply have one yield and one fault spectrum. Typically there will be multiple products in production at the same time, all with different levels of maturity, different levels of complexity, incorporating different component technologies and using different manufacturing technologies. This complicates the test strategy selection process considerably. Ideally you may require to be running several different test strategies for different groupings of problems, and this is what many companies will do. Unfortunately not all companies have sufficient production volume to justify the use of multiple strategies, so choosing the best possible compromise becomes the objective. Naturally the highest volume products or those with the poorest yields should probably dictate the direction. For the other products, with different fault spectra, you may well have to live with a poor fault coverage at board test and a higher degree of fault detection at the following test stages. As long as the yields are reasonably high this may not be much of a problem since the bulk of the cost at the following test stages may well be the 'good unit' test costs as opposed to the diagnosis costs. This possibility is illustrated in the results of the earlier example shown in Table 7.3. The annual testing costs at the final system test are $225 000 whereas the diagnostic costs are only $33 480 for the case where board tester A is used and $67 140 for board tester B. In practice the relationship between 'good unit' test costs and diagnosis costs will be a function of the relationship between test times and diagnosis times as well as the yield seen at these stages. Depending on the product, the 'good unit' test time can sometimes be very much longer than the diagnosis time. For example, a complex product that requires numerous adjustments, calibration and a long testing procedure may take several hours or even days to 'test' at the final system test stage. However, if a straightforward logical approach to diagnosis is possible then this may only take an hour or so, possibly less. On the other hand, you can get products that have very short test times and very long diagnosis times. Under these conditions the diagnosis costs will dominate unless the yield is high.

Yields and fault spectra can vary because of the production mix, but they can also vary for one board type or group of board types. This can happen as a result of changes in materials and components from outside suppliers or because of sudden shifts or gradual drifts in some part of your own manufacturing process. Another important reason for changes is the learning curve effect. When a new product goes into production there may be problems causing the yield to be lower than planned for. As these problems become resolved the yield rises and as it does the fault spectrum will also change.

When you buy expensive capital equipment the cost is usually justified on the basis of the equipment having a useful life of several years. Most financial departments seem to favour five years as being the time period that any ROI should be based upon. Five years is a long time in the electronics industry and many changes could occur in that time. For instance, it only took about three years for the sales rate of digital in-circuit testers to catch up with functional board testers. It is necessary then to take a look at the current trends in the industry and to see how these might influence your choice of strategy and your choice of testers. There is some further discussion of this subject in the next chapter.

8. Test strategy analysis

The test engineering equivalent to 'do it right first time, every time' is 'do the right testing right, first time, every time'.

8.1 Overview

This chapter is effectively the core or perhaps the heart of this book because *The Economics of Automatic Testing* involves the analysis of test strategy alternatives. Most of the chapters that precede this one deal with the issues that have to be considered when making test strategy decisions. Some of these cover broader business issues such as 'quality' and 'time to market', while others deal with more specific issues such as 'board test' and 'component test'. Most of the chapters that follow this one deal with the things that have to be done once the strategy decision has been made—most specifically, Chapters 9, 11 and 12 on the evaluation of equipment, financial appraisal and presentation to management.

It is probably wrong today to simply refer to a 'test strategy' because the issue has become much broader than that. The test strategy is really a part of the overall design, test and support strategy that should be developed to be optimal for the specific situation that you find yourself in. This is all described in Chapter 1, with the conclusion that the sub-optimization caused in part by 'departmental' thinking is no longer acceptable. Having said all this, however, this chapter concentrates on the test strategy part of the design, test and support strategy. The chapter is roughly divided into three parts:

1. A discussion about test strategies and an overview of the more commonly implemented strategies.
2. A description of the approach to be taken in developing an economic model for comparing strategies.
3. A description, with examples, of a current test strategy analysis model.

Figure 8.1 shows the concept of how the test strategy will influence your ability to meet the four major objectives, and how the choice of strategy is itself influenced by such factors as yield, fault spectrum, production mix and industry segment. Other influencing factors will be the four objectives themselves, the testability of designs and the degree of design and test integration that exists or is planned.

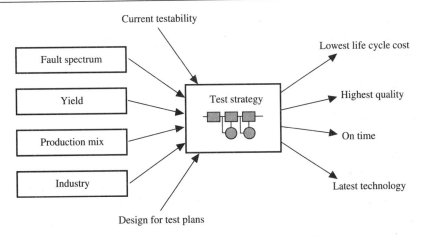

Figure 8.1   Test strategy determinants

8.2   Test strategy determinants

Yield

The proportion, or the percentage, of defect-free units emerging from manufacturing has an important influence on test strategy selection because it affects the emphasis placed on diagnosis times, repair times, throughput and capacity. Low-yield situations tend to need more complex test strategies with several stages of test addressing different parts of the fault spectrum. It was not uncommon in the seventies and the early eighties to see a strategy that included incoming inspection of components, bare-board testing, loaded board shorts test, a full functional test of boards, several sub-assembly tests and a final 'system' test. This strategy complexity was needed to address all of the different fault types in an efficient manner. Also, when functional testing was the predominant test method employed by board testers, a separate shorts test was a prudent tactic because of the potential damage that could be caused by powering up the board with shorts present. At the other end of the yield scale, a very high and consistent yield could result in the very simple strategy of performing the final product test only.

Low yields tend to favour a strategy based on a 'process tester' such as a manufacturing defects analyser (MDA), an in-circuit analyser (ICA) or an in-circuit tester (ICT). This is because low yields are usually the result of manufacturing defects as opposed to the other two major sources of defects, design induced defects and supplier induced defects. There can be exceptions to this of course. I am aware of several low-yield situations caused by supplier induced defects. These are boards that contain a large number of programmable logic devices which tend to have much higher defect rates than high-volume commercial (merchant) devices. In such cases a strategy that includes incoming inspection might be more appropriate. Another reason for low yields favouring 'process testers' is that they tend to have much higher throughput rates at low yields than 'functional testers', mainly due to the long diagnosis times involved with functional test. At high yields this difference tends to diminish because there are fewer defects to diagnose. The 'good board test time' may also be shorter on a functional tester than on a process tester. Therefore high yields tend to result in a wider choice of test strategies.

Fault spectrum

Of all production variables, the fault spectrum probably has the biggest influence on test strategy selection. If you take an extreme example this should be quite obvious. A fault spectrum dominated by solder shorts would be poorly served by a high-speed 'performance' tester, and a spectrum dominated by tricky timing faults would be poorly served by an MDA. A fault spectrum that contains a broader mix of defect types may well require both testers or possibly a 'combinational' system. Figure 8.2 shows a fault coverage profile for two different testers plotted for a specific fault spectrum. This clearly shows that it would be beneficial to employ both testers if maximum fault coverage is required. Ideally it is necessary to apply some form of weighting to such profiles so that the more important (frequently occurring) defects have more impact on the comparison. This can be done in a simple manner by multiplying the percentage number for the occurrence of the defect by the fault coverage for that defect type. This is essentially what a test strategy analysis model will do. Another approach might be to weight the defect types by a factor representing the difficulty of detection at later stages of test.

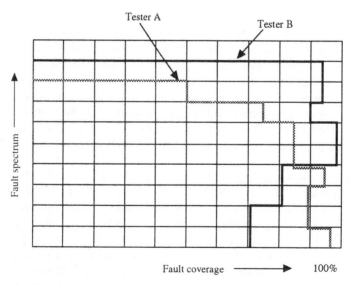

Figure 8.2 Fault coverage profiles for two different testers

Production mix

The 'production mix' refers to the number of product (board) types and the production volumes involved. A 'low-mix/high-volume' manufacturing operation tends to have a different set of problems to a 'high-mix/low-volume' manufacturer. The term is also used to describe the technology mix involved. Some manufacturers produce mainly digital boards or possibly mainly analogue boards, but others may use a whole gamut of technologies in their UUTs—simple digital boards, extremely complex digital boards, analogue boards, combined analogue/digital boards, boards with r.f. circuitry, optical devices, displays and so on. All of these need to be tested and so this will influence the choice of test strategy and tactics.

In 'high-mix/low-volume' situations much of the cost emphasis will be on the 'set-up cost' areas of program preparation and fixturing. In contrast 'low-mix/high-volume' manufacturers will not be too concerned about these costs, but will be concerned about throughput and capacity since they will want to buy as few testers as they need to get the job done. High capacity will need fast run times (the good board test time) and rapid diagnostics. Accurate diagnostics is also a requirement since any additional diagnosis needed to overcome an incorrectly performed repair will reduce tester capacity. High volumes usually make it easier to justify more expensive, and hopefully better performing, testers. Very low volumes, regardless of the mix, may make it difficult to justify any form of ATE.

Unfortunately, many companies find themselves in the 'medium-mix/medium-volume' situation. This situation tends to have more options available and requires more analysis of these options in order to find the ideal strategy. Situations like this will find the development or the purchase of a strategy analysis model most beneficial.

Market segment

The market segment that you operate within will influence the test strategy because of the different customer demands and competitive situations. Military and aerospace customers will demand higher levels of assurance that the product is defect-free and reliable over a wider range of conditions than the customers in some other market segments. The test strategy adopted in these cases will probably involve some environmental stress screening (ESS) as well as routine testing and final product assurance.

Testability

The levels of testability that you experience, or that you expect to experience in the future, will also influence the strategy choice because different test approaches are affected to different degrees by poor testability. At the board level ICTs tend to get around many of the traditional testability problems by grabbing each IC in turn and testing it in isolation. A full functional 'edge connector' tester will tend to have more problems if you still suffer from poor testability. Unfortunately, most measures of testability are only relative. They do not give an absolute measure. Testability tends to fall into the 'I know it when I see it' category.

Accessibility

Accessibility is closely related to testability in that it refers to the 'mechanical' testability as opposed to the electrical testability. MDAs and ICTs need access to all nodes of the board and ideally from one side of the board. If the board designs do not allow for this then these testers will be less effective unless they have some functional test capability and/or some boundary scan capabilities. Functional testers will be less affected by accessibility but they may require a very large programming effort to reach acceptable levels of fault coverage.

Design for test (DFT)

The inclusion of DFT (testability) circuitry into ASIC and board designs is both a part of the overall test strategy and a determinant of the tactics within the strategy. Obviously if the designers have gone to the trouble of including boundary scan into their designs then you have to choose a tester with good boundary scan capabilities for program generation and the application of tests.

The four forces

Since the overall requirement is to find a test strategy that is optimal for meeting the main objectives of *cost*, *quality*, *time* and *technology*, then it is obvious that these have to be uppermost in your mind when comparing strategies. Some companies are setting very ambitious quality targets for themselves. So which of the strategies under consideration is most likely to achieve these goals? Will this have a negative impact on time to market? It may be that the programming times for the better strategy are unacceptably long. Is there an alternative strategy that can meet both goals? What will this do to the costs? Will it be able to cope with new technologies in a smooth and speedy manner? The answers to these questions are crucial to making the right choice and a detailed analysis of the strategies will be needed to get the appropriate data.

8.3   Variation in the key determinants

In quality terms 'variation' is the enemy, as it also is with test strategy selection. Yield and fault spectrum are two of the key determinants of a test strategy and these are prone to vary. A company manufacturing high volumes of a single product in a well-controlled operation may have a fairly consistent yield and a fairly consistent fault spectrum, but this will be an exceptional case. The yield and the fault spectrum will be different for different products, for different production lines, for different manufacturing technologies, for different shifts, for different days of the week, for different batches, etc. There will be variation by product in all three of the main defect categories of design, supplier and manufacturing induced defects. Hopefully you will be able to keep this variation to a minimum and so select the best strategy. If the variation is very high then the process is not in good control and you will need to select a strategy that is flexible enough to cope with wildly varying yields and fault spectra. You will also have to plan for more capacity than would be needed if the process was in better control if you are to cope with the testing bottleneck that will occur when the yield is low and the spectrum is biased towards the more problematic type of defects.

8.4 Industry trends

When looking at alternative test strategies it is also necessary to look at the various trends that are taking place within the industry and to try to determine what their impact will be on the determinants of the ideal strategy. Over the past ten years or so some of the more important trends have been:

1. The quality improvement drive.
2. The increased use of ASICs.
3. The move towards surface mount technology.
4. The increased use of design automation.

These include design technologies, component technologies and manufacturing technologies, with quality covering all three. More recent trends include the increased use of boundary scan (test technology), the multi-chip module (MCM) (packaging technology), the increased use of VXI bus-based equipment (test technology) and so on. All of these have an impact on your test strategy planning process.

The four trends in the list all had a significant impact on both the yield and the fault spectrum. The quality drive that took place in the eighties, and still today, naturally resulted in increased yields. This in turn led to changes in the fault spectrum because the two tend to go hand in hand. Quality improvement usually involves fixing the biggest problem first and then addressing the new biggest problem. This inevitably results in changes to the fault spectrum. The move to surface mount technology led to further increases in yields due to its more automated process. Initially yields dropped because the process was new and contained more process steps than the 'through-hole' method, but once companies moved up the learning curve yields increased dramatically. Fault spectra changed also, not only as a result of the increased yields but also because it was a very different process. Short circuits tended to dominate the fault spectra of through-hole boards but open circuits tend to be more of a problem with SMT. This is bad news for many MDAs because they do not do a good job of testing for opens. Indeed, board testers, either functional or in-circuit, do not perform a test for opens. Open circuits are detected as a result of some other test failing. An open circuit to a digital input causes the pin to behave as if it was 'stuck at one' because the test patterns do not reach the pin. An open circuit in the analogue part of a board being tested on an in-circuit tester will cause a measured value to be excessively high impedance, and so on. Some work is being done to find a workable 'opens' test. If successful this could improve the performance of MDAs for SMT-based boards.

Many of the early implementations of ASICs were to replace large areas of conventional SSI/MSI logic with a single chip. This reduced the parts count, possibly the number of boards in the product, the size and power consumption of the product, and probably increased reliability. Such designs naturally resulted in higher yields and changes to the fault spectrum. The same is true for the trend towards design and test integration and the increased use of design automation techniques for board design as opposed to component design, where it has typically been used. This has led to a decrease in the problems that were caused by the duplication of effort in design and test, and in the number of design induced defects due to the more thorough design process.

Monitoring the trends and trying to predict the effect that they will have is an important element in test strategy planning. Not many people would have been brave enough in 1980 to predict the yield levels seen today in the better run factories, for boards with many times the complexity of their 1980 counterparts.

8.5 Increased yield and fault spectrum changes

The outcome of all of the trends and changes that have been taking place has been the improvement in yield levels into the first test stage and the changes to the fault spectrum. This has been good, but for some companies it has caused some problems because the fault spectrum has changed from being dominated by manufacturing defects towards being dominated by much more difficult defects to detect and diagnose. From a testing point of view the ideal fault spectrum, assuming that there has to be some defects, would consist mostly of manufacturing defects and a negligible amount of supplier induced and design induced defects. This situation is shown in the middle of Fig. 8.3. This is good because the manufacturing defects can be detected easily with 'process' type testers that are relatively quick and easy to program. If the fault spectrum is biased more towards the supplier induced defects, as in the left of Fig. 8.3, or towards the design induced defects, as shown on the right, then a much more difficult testing problem occurs. There is a need for a very powerful functional tester in the real 'performance' tester end of the ATE market if we are to have a reasonable degree of fault coverage at the board test stage. Since this is the cheapest place to find defects we need to maximize the fault coverage here, but an MDA or an ICT will not do a good enough job against this kind of fault spectrum. Many companies found themselves in this sort of position as a result of their quality improvement drives. In the early days of the quality revolution an enormous amount of effort was put into improving the manufacturing process and this dramatically reduced the number of manufacturing defects. Major improvements to the quality of components was also achieved, but this was offset to some degree by the increased number of components on a board and by the increased use of ASICs and programmable devices that tend to have higher defect rates than the high-volume components. In many cases there was less of an improvement in the design area since the majority of board designs were still produced using traditional methods. We have all seen the statistic that 50 per cent of ASICs do not work with the board prototypes. This is partly

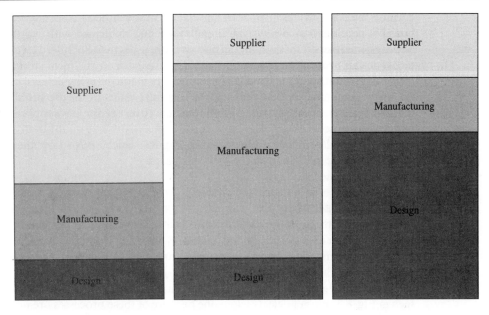

Figure 8.3 Three different fault spectra

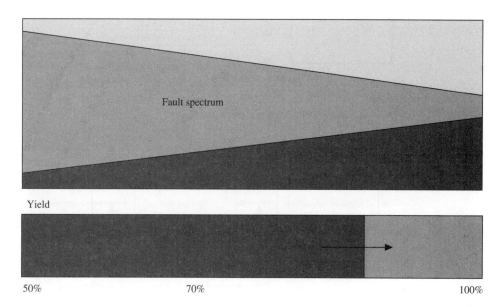

Figure 8.4 Graphical depiction of how the (normalized) fault spectrum might change as the yield increases

because the ASIC is designed with simulation and synthesis whereas the rest of the board is developed as a physical 'breadboard' and debugged with word generators and logic analysers. This is changing now as more use is made of board level simulation in the design process. In theory this could get us back to the ideal situation of high yields and a fault spectrum that is predominantly manufacturing defects. The problem for many companies is that they can be blissfully unaware of the problem because the board testers in use have a very poor fault coverage for the supplier and design induced defects.

This potential problem is illustrated in Fig. 8.4 which shows how the normalized fault spectrum can change as the yield increases.

8.6   The ideal test strategy?

Is there such a thing as the ideal or universal test strategy? In principal there is, but we can implement it with different tactics. If we accept that we have to design the product right first time and then manufacture it right first time using components and materials that are right first time, then we have the simple defect occurrence model defined earlier in this book. This then leads to the simplified fault spectra as seen in Fig. 8.3. We get defects when one or more of these processes are not performed right first time. In some ways the ideal test strategy is one that addresses these three sources of defects as fully as possible. Historically the electronics industry has done just that with the strategy shown in Fig. 8.5. We used incoming inspection to eliminate

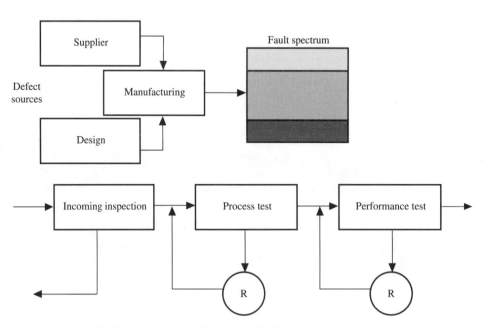

Figure 8.5   The 'ideal' test strategy, addressing all of the defect sources

the supplier induced defects, a board test stage and a sub-assembly test stage to eliminate the manufacturing process defects and a final system test stage to verify the performance of the entire product. With increasing quality levels the trend has been to eliminate the incoming inspection stage, or rather to push the testing responsibility back towards the vendor. This is as it should be. If we are responsible for verifying the performance of our products then why should the component suppliers not be responsible for theirs? If you are in the business of producing sub-assemblies rather than the entire product then the same will be true for your customers. It should also not be necessary for the customer to perform any incoming tests on your sub-assembly.

Over the years this ideal strategy has been implemented in a variety of ways, or with a variety of tactics, mainly as a result of the available testing technology at the time. Occasionally, however, the objectives of the strategy can become rather misplaced. Early objections to the in-circuit test relative to the functional board test were based on the fact that the ICT did not test the entire board operating as it would in the end product. Why should it? If the purpose of the board test stage is to check that the board has been built correctly then it does not matter if it is not tested functioning in its entirety. That test can be performed at a sub-system test stage where any additional problems caused by interactions between boards can also be tested for. There was, and there still is today, a tendency to mix up the objective of making sure that the product has been built correctly with the objective of testing its performance to the design specification. This simplistic view of what each step in the test strategy should be doing does of course have to be tempered by the fact that the board test stage is the cheapest place to find defects; therefore the more types of defect we can find there the better. This is probably the strongest argument for the 'performance' class of board tester: 'Any "performance" related defect, such as design induced and supplier induced defects, that can be caught here will cost less than finding them at later stages.' The only way to see whether this statement is valid is to perform a detailed cost analysis. It may in some cases be cheaper to find these defects at a sub-system test stage. It will depend on the number of such defects and the degree of automation present at the sub-system test stage. The use of VXI based test rigs and artificial intelligence (AI) based diagnostics at this test stage could change the economics of this argument away from the 'performance' tester. On the other hand, if the 'performance' tester can lead to the elimination of the sub-system test stage then the economics can swing back in its favour.

Confusing the testing of the 'process' with the testing of the 'performance' is often very prevalent in analogue functional test systems, where the user will often specify a set of measurement accuracies that are targeted at the performance test. There can sometimes be good reasons to do this, as when adjustments need to be made or if there are 'select on test' components involved. It is often argued that the need for 'select on test' components is the result of poor design, but it is not always as simple as that and it cannot always be avoided. However, it is always well worth taking a careful look at your test strategy to see just what tests are needed at each of the stages and where is the best place to make them. You can easily wind up paying a lot of money for extreme measurement accuracies that are not really needed in the context of the real objective of the test.

8.7 Commonly used test strategies

In this section I will take a look at some of the more commonly used test strategies and comment on their performance, their advantages and their disadvantages. I will also comment on specific issues related to the modelling of the strategy as a lead in to the next section which discusses modelling in general.

Functional board tester based strategies

Throughout the seventies this was the most widely used test strategy for both analogue and digital boards. Analogue in-circuit testers also became very popular in the seventies but there were probably many more 'rack and stack' systems about, many of which were built 'in-house'. The test philosophy was to exercise the entire board with a set of carefully developed tests to prove that the board 'functioned'. Digital testing tended to be the big problem because the complexity was beginning to grow to the point where program preparation was becoming a real problem. This problem was eased by the introduction of digital simulators which helped the programmer and also provided the diagnostic data for the guided probe. Functional testing is still widely used today, especially where there is still concern over the dangers of 'backdriving'. The primary advantage of functional testing is that it gives confidence that the UUT has been exercised in a similar manner to the way it will eventually be operated in the product. The fault coverage of the test program will be known to some degree for digital circuits because of the fault simulation process. I say to 'some degree' because it will only be known for the simulated fault set with one fault at a time simulated. However, in practice this is a reasonable indication. Another advantage of functional testing is that the test fixture that interfaces the board to the tester is usually much simpler than a 'bed of nails' fixture since it only requires wiring to the edge connector sockets. For the same reason it may be possible to use one fixture for a family of board types. In practice this simplicity of fixturing cannot always be achieved. A disadvantage of functional testing is the greater reliance that it places on having a testable design. If the testability is poor it is frequently necessary to resort to a bed of nails fixture to gain access to additional test points. For boards populated with dual-in-line devices IC clips can be used to gain additional test points, but this cannot be done for SMT boards.

The biggest disadvantage of functional testing has always been the complexity and the time taken to develop test programs. The complexity of the designs and the test program generation process has meant that very highly skilled programmers have been needed. Testability problems have, as indicated earlier, just made the problem worse. There were many man-years of time put into the development of automatic, or automated, test pattern generators (ATPGs) but the increase in complexity of the components and the boards kept getting too far ahead for any really practical solutions. However, there is now some hope for this. The big problem with the ATPG systems was that they only understood simple circuitry and it was too complex to build in the recognition needed to handle more complex problems. The development of the boundary scan specification was in many ways an admission of defeat. In effect it recognized that the only way to make ATPGs work was to define a very structured design style with a relatively standardized test procedure. This and the links to design are making it possible to take automatically generated test vectors from the design database and produce them in the native language of a board tester. Companies who have a well-integrated design and test process are today producing high-coverage test programs in days or weeks that would formerly have taken many months to prepare.

A problem that digital functional testers share with their in-circuit counterparts is the need for library models. They are different models, of course. The functional tester requires a simulation model whereas the ICT needs a test program plus some other data. The problem is usually the availability of the information needed to create the models accurately. The use of ASICs also exacerbates this problem because, by definition, there will be no ASICs in the libraries supplied by the equipment vendor. Again the integrated design and test process can supply the solution here in many cases.

The diagram in Fig. 8.6 shows a typical functional test strategy matched against the simplified fault spectrum and the 'ideal' strategy—ideal that is from a fault coverage point of view. This diagram, and the others that follow, attempts to show approximately the areas of the simplified spectrum that the test strategy addresses in the absence of any incoming inspection. As Fig. 8.6 shows, the functional tester

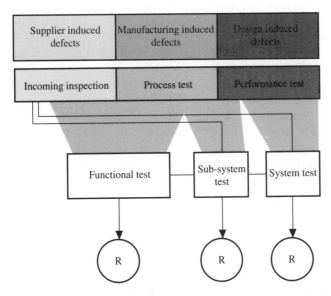

Figure 8.6 What functional board tester based test strategies have attempted to achieve

mainly addresses the manufacturing process defects that would cause the board to malfunction and some portion of the potential supplier defects, namely the functional (truth table) defects. In the seventies when functional failure rates were quite high this capability was important if there was no incoming inspection. Today the defect rates for the simpler devices is very low so this capability has less value. The sub-system test offers the possibility to detect any process defects that escape from the FNT and others that relate to the assembly, such as interactions and problems in other parts of the unit that could not be tested on the board tester. The SST stage may also detect some performance-related defects and some component defects of a parametric nature. However, these would have to be quite bad to be detectable at this stage, especially if the design is tolerant of wide variations in parametric values. There were functional board testers in the seventies that had parametric measurement capabilities, but these had a relatively short life. The capability added cost to the tester, slowed it down and could only address device pins that came out directly to the edge connector, which was of limited value. The final system test stage is usually set up to prove that the product meets its design specification prior to shipment. It is generally assumed that most performance defects would be detected here as a malfunction of some sort. In practice the fault coverage at the final test stage may be quite low. This obviously depends on the nature of the product and the nature of the tests that are performed. However, consider the following two points:

1. The test procedure for this stage is written primarily to prove that the product works—that it meets its design specification. We know from experience with ASIC designs that a set of test vectors that the designer creates to prove a design will not necessarily have a very good fault coverage. Until a few years ago the test program supplied with an ASIC design to the ASIC vendor was usually the design verification vectors. When passed through a fault simulation these vectors frequently had very low coverages in the 50 to 70 per cent area. This often led to disputes between the ASIC vendors and their customers about the performance of the ASIC so that the ASIC vendors eventually insisted on fault simulation and a 95 per cent minimum fault coverage. This same situation is frequently true for final system test procedures. A set of tests to prove that it works will not always indicate a reason why it may not. You cannot test all of the combinations that the end user might apply to the product. There simply is not time.
2. Another way to look at the issue of fault coverage at the system test stage is to consider the testability diagram showing the controllability and the observability into a circuit block on a chip. Then consider this block buried four or five layers deep in several other testability diagrams and you get a feel for the problem. This is illustrated in Fig. 8.7. In a fully bus structured product the number of layers might not matter but there are many other situations where it does.

The moral of all of this is that ideally we should aim to detect most of the defects before the final test stage is reached.

Primary inputs   Primary outputs

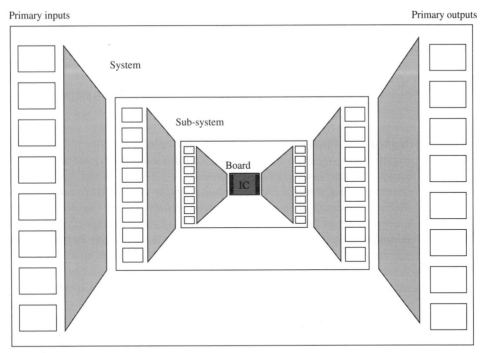

Figure 8.7   System level testability problems

MODELLING ISSUES FOR FUNCTIONAL BASED TEST STRATEGIES

For any accurate economic modelling of a process step it is necessary to determine the average amount of time that each UUT spends on the tester, including the handling time, the test time (for a UUT that passes the test), the diagnosis time for any defects and the number of visits made to the tester. This time, which is referred to as TTT (total time on the tester) is critical to cost, throughput and capacity calculations. The functional board test is easier to model correctly than most other tester types because it is usually a one fault at a time diagnostic approach. This is because most functional testers will stop testing at the first failing test and enter their diagnostic routine. Once the diagnosis is completed the board is removed and sent to the repair station. Depending on the UUT the test time for a board passing the test will often be quite short relative to the handling time. In contrast the diagnosis time, with a guided probe, is usually quite long relative to the handling time so these times need to be handled separately in the calculations. It is particularly important to try to determine the diagnosis time as accurately as possible if the yield is below about 80 per cent because the diagnosis time will probably dominate. As yields become higher the 'passing test time' will dominate the overall time (TTT). The basic formula for a functional test stage is therefore

Total time on the tester = test time + diagnosis time (average faults/board)

$$\text{TTT} = T_t + T_d(\text{FPB})$$

where

$$T_t = \text{test time for a good board (including handling time)}$$

$$T_d = \text{average diagnosis time for a defect (including handling time)}$$

$$\text{FPB} = \text{average number of faults per board (FPB} = \text{LN } Y)$$

Dynamic (high-speed) functional tester based strategies

Figure 8.8 shows what the higher speed functional testers are attempting to do. It is hoped that the faster rate of application of the digital test vectors will enable the tester to detect some of the performance-related faults. In practice it is unlikely that the fault coverage of these testers is very different from the slower testers which are known as *static* functional testers because of their low speed. Some devices have a minimum clocking frequency but this does not necessarily mean that the testers pin drivers need to run at these speeds because many testers have clock generators that run at a higher speed than the regular pin electronics. The key to detecting marginal timing or 'dynamic' defects is to run smarter rather than to run faster, and this is what the true 'performance testers' attempt to do.

MODELLING ISSUES FOR DYNAMIC FUNCTIONAL TESTER BASED STRATEGIES
There are no special requirements for the model of these strategies relative to the static testers. The differences will be in the input parameters to the model that define the performance of the tester, its cost and the cost of program preparation and fixtures.

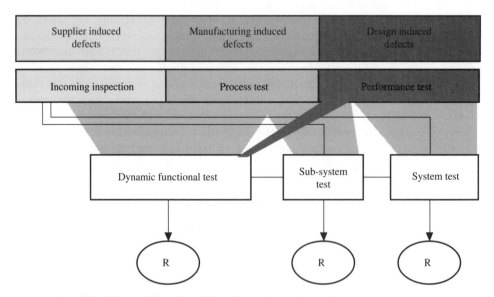

Figure 8.8 What 'dynamic' functional board tester based strategies have attempted to achieve

In-circuit board tester based strategies

This is almost certainly the most widely used strategy at present. By the late seventies analog in-circuit testers were already dominating the market relative to functional analogue testers, but the digital capabilities were rather simplistic and did not really dent the digital functional tester market. Then the technical breakthrough mentioned in the previous chapter occurred and everything changed. In a few years ICTs were outselling FNTs for complex digital board testing. The major advantages were the easier programming, the faster diagnostics and the more precise diagnostic messages. The ability to detect multiple defects in one test, with the restrictions outlined in Chapter 4 (Fig. 4.5), was also welcome because the low yield typical at that time meant that many boards had multiple defects. There were relatively few custom devices around at that time so modelling these for the component library was not the issue that it is today. The main disadvantages of the in-circuit tester are generally regarded to be the elaborate fixtures that are needed and the need to 'backdrive' digital circuits to isolate the component under test. Backdriving was a big issue for certain industry segments such as the defence industry and the telecommunications industry. Both were concerned about the potential damage to devices, the possible 'weakening' of devices causing reliability problems and the fact that the component manufacturers would not guarantee that their products would withstand backdriving currents. I was involved in what was probably the most extensive research program into the effects of backdriving to be conducted. This was performed in the United Kingdom for the Ministry of Defence (MOD) by the Electrical Research Association. The work was funded equally by the MOD, a number of defence contractors and almost all of the leading ATE manufacturers. The tests showed a number of phenomena not previously recognized and confirmed others. However, the results showed that backdriving is fairly safe so long as the ATE manufacturer and the test programmer are aware of the potential dangers and take some precautions in the design of the tester and the test programs. There was an extensive accelerated life test (burn-in) performed as part of the program, which was the first time that this had been done. The results showed no problems with the batches of components that had been backdriven. Indeed, for some component types the backdriven devices performed better. A side benefit of backdriving is that the current is high enough to destroy 'weak' devices and so improve the reliability of the product by weeding out these poor specimens. As a result of this work a DEF-SPEC (defence specification) was issued that outlined the precautions needed and essentially made it possible to use ICT on military products.

When I was involved in selling in-circuit testers and the backdriving issue came up, I would often pose the following questions in a slightly tongue in cheek manner:

1. 'When you remove a solder short from a board after diagnosis on a functional tester, is it also your policy to replace all of the ICs connected to the shorted nodes?' The usual answer of course was 'no', but this represents a backdriving situation that may have lasted for several minutes during the guided probe diagnosis. You cannot connect the outputs of two TTL devices together. When

you do, as with a short, one of them will win the battle and force current into the other. The current will be a lot less than a backdriving current from a tester but it will go on for about 50 000 to 100 000 times as long.

2. 'If you accept that burn-in procedures that operate devices at 150°C (junction temperature) for 168 hours improve reliability, why is it that 100°C for 5 milliseconds is bad?' This is, of course, a simplistic view because there are other mechanisms involved than simply the temperature. However, it did tend to put the issue into perspective.

I have no hard data but I would estimate that well over 90 per cent of all digital boards manufactured in the past 12 years have been tested on ICTs. In that time there has been no explosion of increased field failures which you would expect if the testing method was dangerous. On the contrary, the service reliability of electronic equipment has increased substantially in that time, especially when you consider the increase in complexity that has taken place.

Figure 8.9 shows the typical ICT based strategy matched to the simplified fault spectrum and the 'ideal' strategy. If you compare this to Fig. 8.6 for the FNT based strategy you will see that they are essentially the same. The ICT will detect a high percentage of manufacturing process defects and some percentage of supplier induced defects such as out-of-tolerance passive components and functional problems with linear and digital devices. The sub-system test stage will hopefully detect any component interactions and board-to-board interactions as well as defects associated with other parts of the sub-assembly. This stage may also detect some of the supplier

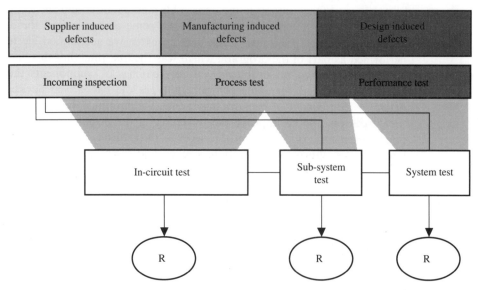

Figure 8.9 What in-circuit board tester based test strategies have attempted to achieve

induced defects and some design induced defects that manifest themselves as performance problems. The final system test stage should then determine whether the product meets its design criteria. If it does not then there must be some remaining defects, from all three sources, that will need to be diagnosed. Some of these will be interactions between sub-assemblies that will not have been tested for previously. The comments about the potential fault coverage of this final test stage in the previous section about functional based strategies still apply.

The simple diagrams in Figs 8.6 and 8.8 cannot show the subtle differences between the fault coverage of the FNT and the ICT that do exist. The FNT can find defects that the ICT cannot, such as component interactions, and the ICT can find defects that the FNT cannot, such as an out-of-tolerance or completely missing pull-up resistor. What the diagrams do show, however, is that the two strategies are not too dissimilar in terms of their overall testing performance, even though the test philosophy of the approaches is different.

MODELLING ISSUES FOR IN-CIRCUIT BASED TEST STRATEGIES

Modelling an ICT test stage accurately is a little more complex than for functional testing. The fundamental need is still to determine the average time each board spends on the tester (TTT), but this is complicated by the fact that an ICT can find multiple defects in one operation. Specifically it will find all shorts in one operation and all other defects in one operation. However, if there is a short and some other defect present then two diagnosis operations and two repair operations will be required, so the board will make three visits to the tester. This is all described fully, along with some examples, in Sec. 4.8 of Chapter 4 on statistics. Since the number of diagnosis and repair actions will be a function of the proportion of boards failing the tests we have to use the yield numbers in the formula rather than the average number of faults. Specifically we need to use $(1 - Y)$, which is the proportion of failing boards. Therefore the formula for ICT becomes

$$\text{TTT} = T_t + T_{ds}(1 - Y_s) + T_{do}(1 - Y_o)$$

where

T_t = test time for a defect-free board (including handling time)

T_{ds} = diagnosis time for a short (including handling time)

Y_s = yield for shorts

T_{do} = diagnosis time for all other defects (including handling time)

Y_o = yield for other defects

As can be seen from the formula, it is necessary to have separate terms for the diagnosis of shorts and the diagnosis of other defects. This not only handles the number of visits to the tester correctly but it also accounts for the fact that the diagnosis of a short will take less time than the diagnosis of some other defect. The shorts test is usually the first test of an ICT program and it is usually quite rapid, depending on the number of electrical nodes on the board. The remainder of the testing usually takes a lot longer because of the nature of one component at a time of in-circuit testing. There may also be a requirement for time delays to allow for capacitors to charge-up, etc. If a defect is detected within this part of the test program the testing will still continue to the end of the program, unlike a functional tester which will stop at the first failing test. As a result of this the diagnosis time for a defect in the 'other' category will take about the same time as testing a defect-free board. Some additional time may be required when a diagnosis occurs, to allow for the printing of the defect ticket and to attach it to the board. Also, if the tester has some additional diagnostic tools to improve accuracy then some time will need to be added to the average diagnosis time to allow for the occasions when these diagnostic resolution tools are needed.

High-speed in-circuit based test strategies

Similar comments apply to this type of strategy that applied to the dynamic functional tester based strategy. It is not very clear that there are any substantial improvements in fault coverage that result from an increase in the test rate. The primary advantage would be that more tests could be applied in a given time and this may be beneficial in terms of keeping the backdriving times short. To get more tests, as opposed to the same number in a shorter time, would also require more memory in the pin electronics, but most of the current testers have similar amounts of memory regardless of their test application speed. Another area where higher speed is an advantage is the application of boundary scan test vectors. These tend to be very long sequences over a few pins whereas ICTs are mainly structured to have relatively small memories (typically 4 to 16k bits) over a large number of pins. For this reason some dedicated hardware, with a lot of memory, is used for the boundary scan tests, so this has no relationship to the test rate of the standard in-circuit driver/sensors. I am quite sure that it is possible to identify a fault condition that can be detected at 15 but not at 12 MHz, but I would suggest that this type of defect is a relatively small portion of the overall fault spectrum. However, this is exactly what this chapter is all about—the analysis of alternative strategies and tactics to find the optimum solution for your particular situation. Only with a fairly detailed analysis will it be possible to determine whether any marginal performance improvement is worth the extra costs that might be involved. High-speed programs and fixtures usually take longer to prepare and to debug. Bearing in mind the comments in other parts of this book about the need to consider escapes rather than fault coverage a small improvement may well be worth a lot.

Figure 8.10 shows what the high-speed ICT based test strategy is hoping to achieve. Relative to the lower speed testers it is trying to pick up some additional fault coverage. However, the speed of these testers and their functional counterparts is nowhere near the actual operating speed of the boards they are testing. Numerous

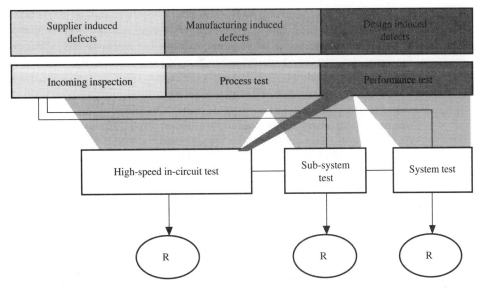

Figure 8.10   What 'high-speed' in-circuit board tester based test strategies have attempted to achieve

'benchmark' tests comparing commercial testers frequently result in lower speed testers performing better than the faster ones. This probably supports the contention that the theoretical fault coverage may well be set by the performance of the hardware, but what you can actually achieve in a given amount of time is more dependent on the test generation software provided with the tester. There is an undeniable market demand for faster test rates so most of the major vendors offer high-speed versions whether they believe them to be necessary or not. Indeed, there are very few low-speed testers on the market at the time of writing.

MODELLING ISSUES FOR HIGH-SPEED ICT BASED STRATEGIES
There are no real differences in the modelling process or the formulae required relative to a lower speed tester. The differences will be defined in the input parameters to the model rather than the model itself.

Combinational tester based test strategies
As I indicated in the previous chapter, digital/analog in-circuit testers were originally assumed to have an inferior fault coverage relative to functional testers. For this reason they were often used to pre-screen the functional tester that was performing the 'real' tests. Users quickly realized that the fault coverage could be similar or even better than the FNT, but more importantly the coverage was for a slightly different set of faults. Even though the fault coverage of each tester may have been, say,

90 per cent, the combined fault coverage could be 95 per cent or even better depending on the fault spectrum. The big problem, of course, was the need to develop two test programs and two fixtures. However, ICTs were much cheaper than the high-performance functional testers in the early eighties and the program development was virtually automatic, so this added cost could often be justified for a few per cent of extra coverage. Remember that this could mean a massive reduction in the escape rate. Having two tester types also added a great deal of flexibility to the test strategy options. Some board types could be tested on the ICT some on the FNT and some on both, depending on their fault spectra and the needs of the end product.

ATE manufacturers were quick to spot an opportunity and started to develop 'combinational' testers that combined both ICT and FNT capabilities in one system. Some of the early testers were not particularly well integrated in that the ICT and FNT drivers came to separate pins on the interface, but this deficiency was soon overcome by better scanner design. Now only one test program and one fixture were needed, although both increased in complexity quite a lot. ICT fixturing was not entirely compatible with the needs of the functional tests, especially if any high frequencies were involved. Therefore 'dual-level' fixtures were used which enabled the bulk of the test probes to be withdrawn after the in-circuit tests were completed. This left fewer probes in contact with the board to reduce circuit loading for the functional test. A major advantage of combining the two techniques in one system was the reduced amount of board handling time involved. With separate systems two handling operations are needed, even for defect-free boards, whereas only one is needed on a combinational tester. This increases the throughput of the strategy considerably because handling time can be a major part of the total time spent on the tester (TTT).

With combined strategies of this type, whether on separate testers or a combinational system, it is usual to perform the in-circuit tests first. This is because of its better capability of identifying manufacturing process defects quickly and precisely. Also, if there is a short circuit on the board you will want to have it removed before powering-up the board. An alternative to this normal approach has been used whereby the functional test is performed first and the in-circuit test is only used if the functional test fails. The ICT capability is therefore only used for diagnosis and there may be no diagnostic capability on the functional tester. Some companies have used an in-house built functional test rig for the functional go/no-go test and a commercial ICT in the diagnosis and repair loop. This strategy is illustrated in Fig. 8.11. This strategy is effective where the yield is high and confidence in the fault coverage of the functional test is high. One problem with this approach is that boards that pass the functional test do not go through the ICT, so you lose the benefit of maximizing the fault coverage. Another problem is that some of the boards that fail the functional test may contain defects that the ICT cannot detect. These would go back to the FNT for diagnosis, but if this stage has no diagnostic capabilities you are faced with diagnosing the board at the next test stage.

The situation with combinational testers is a little different today from that in the early eighties. Today most combinational testers are basically ICTs with some

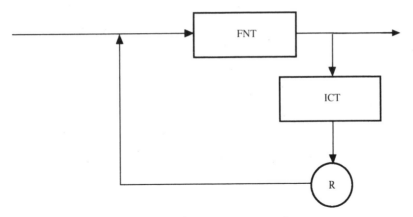

Figure 8.11 A functional test, with in-circuit diagnostics test strategy

additional functional capabilities. These are designed to overcome accessibility problems. Essentially, most of the testing is performed as an in-circuit test and the functional capability is used to test areas of the board where accessibility for probing is inadequate. In general most of the testers with a full ICT and FNT capability fall into the cycle based 'performance tester' category. These have digital/analog in-circuit test capabilities, digital/analog functional test capabilities and a variety of test generation techniques. Since any tester that combines two or more testing techniques can be called 'combinational' you have to be careful to fully understand what capabilities the tester really has in terms of test application, test generation and diagnostics. For example, some combinational testers have a functional test capability but no functional diagnostic tools.

Figure 8.12 shows the combinational strategy relative to the simplified fault spectrum. Basically it is trying to do the same things as the other strategies but to do it better in terms of the fault coverage or to overcome problems such as poor access for probing.

MODELLING ISSUES FOR COMBINATIONAL TESTER BASED STRATEGIES

This is a more complex modelling situation than the ICT model because defects can be detected in more ways. It is necessary to know the test times and the diagnosis times for defects detected in the shorts test part of the test program, in the remainder of the ICT part of the program, in the functional part of the program or in a boundary scan test if used. It is still not a complex problem because it will simply be necessary to add terms to the ICT formula as appropriate. The following example shows how the formula might develop for a tester with ICT, FNT and boundary scan capabilities:

$$\text{TTT} = T_t + T_{ds}(1 - Y_s) + T_{do}(1 - Y_o) + T_{df}(\text{FPB}_f) + T_{db}(\text{FPB}_b)$$

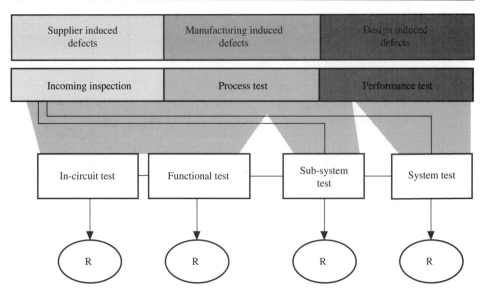

Figure 8.12 What combined functional and in-circuit board tester based test strategies have attempted to achieve

where

TTT = total time on the tester

T_t = test time for a defect-free board (including handling time)

T_{ds} = diagnosis time for a short detected by ICT (including handling time)

Y_s = yield for shorts at ICT

T_{do} = diagnosis time for other faults detected by ICT (including handling)

Y_o = yield for other defects at ICT

T_{df} = diagnosis time for fault detected in the FNT part of the test program (including handling time)

FPB_f = average fault per board detected functionally

T_{db} = diagnosis time for faults detected in the boundary scan test (including handling time)

FPB_b = average faults per board detected by boundary scan

Some refinement and breakdown of the input parameters may be needed to correctly model certain situations. For example, if most of the board is being tested by in-circuit methods and a section is being tested by boundary scan due to access problems, you may have a situation where some shorts would be detected by ICT and some by boundary scan. (You may have noticed that, in deference to American readers, I have not used an abbreviation for boundary scan.) This situation can be handled in two ways in a model. You can have additional defect categories for shorts and split them between ICT and boundary scan or you can have a single defect category for shorts and alter the fault coverage of one of the testers to suit the situation. Therefore, instead of assigning a 99 per cent or a 100 per cent fault coverage to the ICT for shorts you might only assign a 70 per cent coverage if only 70 per cent of the board is tested by ICT. The remaining shorts would then be picked up by the boundary scan test with its full fault coverage assigned. This example assumes that the in-circuit test is performed first. If the boundary scan test is performed first you would assign a 30 per cent fault coverage to it and the full coverage to the ICT. It can often be easier to adjust the modelling by changing the input parameters than to try to build every possible situation into the model in the first place.

Performance tester based test strategies

The modern cycle based performance tester was described in the previous chapter. Essentially it is a combinational tester with a very powerful and flexible functional testing capability. The in-circuit testing capabilities are similar to those found in 'high end' in-circuit testers or combinational testers. It is usually the functional capability and other testing approaches that set the 'performance tester' apart from the high end combinational tester. The price is something else that sets them apart.

Figure 8.13 shows what the performance tester sets out to achieve. It is basically covering the same portion of the fault spectrum that the combinational tester addresses with a higher fault coverage and the ability to detect some performance-related defects. One argument in favour of performance testers is that they can eliminate the need for a sub-assembly or a 'hot mock-up' test stage. This may well be valid for some products but if the sub-assembly is fairly complex, with several boards and a backplane, etc., then it is unlikely that you could realistically eliminate this stage. It will, of course, depend to a large degree on the turn on yield of the sub-assembly. Should it be possible to eliminate the sub-assembly test stage then the economic benefits could make it quite easy to justify the high price of such systems. One important benefit of these testers, with their multiple test tactics, is that different parts of a board can be tested in ways most appropriate to the circuitry involved. Some of these testers will even segment the board automatically and recommend the type of tests to apply.

MODELLING ISSUES FOR PERFORMANCE TESTER BASED STRATEGIES

There are no real differences between modelling such testers and modelling a combinational tester. The operational and performance differences will be defined in the input parameters to the model.

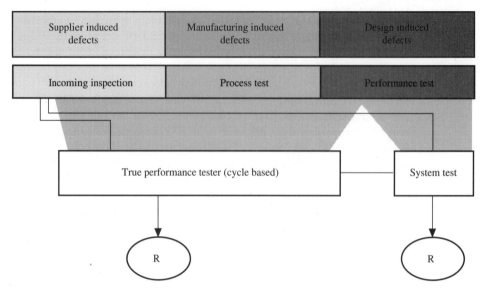

Figure 8.13 What true 'performance' board tester based test strategies have attempted to achieve

8.8   The economic analysis of test strategies

A question that needs to be addressed is whether or not an economic analysis is the right way to compare the effectiveness of alternative test strategies. I think that this is an easy question to answer. If I did not believe that this was the right way to do it I would not have sat down for many hours pounding the keyboard of a PC to write this book. I am biased, so let us look at the issue in a logical manner. If you set up two identical production units, manufacturing the same products, with identical equipment and the same personnel operating both, but with two different test strategies, what would be the differences? Assuming also that the design for test implementation is the same for both lines then the differences would be found in:

1. The capital cost of establishing the two operations.
2. The set-up costs involved (programming and fixturing).
3. The time taken to get set up.
4. The running costs (test, diagnosis and repair).
5. The quality costs involved (in-house and in the field).

If you accept that these will be the key differences then an economic comparison seems to be the logical route to take. Another view of this is that the common denominator of the four market forces of cost, quality, time and technology is cost. The lowest costs will usually be achieved when the quality is highest, time to market will influence profitability and new technologies are needed to remain competitive, keep down costs, improve quality and get to market on time. This is a convoluted system of servo loops.

Selecting the right test strategy is important because it also impacts other areas. Once in place the strategy may impose constraints on new designs, constraints on manufacturing and affect your achievable quality. This in turn will affect customer relations, repeat sales, new accounts, field service costs and profitability. A primary objective of the strategy will be to achieve a high level of fault coverage at the lowest possible cost. This will be achieved with an optimum strategy, testable designs and an effective quality improvement process.

8.9 Modelling the test strategy

Figure 8.14 shows the basic structure of an economic model for test strategy analysis. The model is a set of calculations that represent, or model, the actual production test process. The calculations are performed on input data that defines the unit under test, the production variables, the fault spectrum and the performance of the various test stages. The results of the calculations provide information on the costs involved, the capacity requirements and the quality impact of the strategy.

The production variables include data on the production volume, the number of shifts operated, the labour costs of the various operators involved, etc. The product variables include data on the complexity of the UUT and the fault spectrum. The yield will be computed from the fault spectrum data. The test stage capabilities include data on the capital cost of equipment required, the programming workload, the test times, the diagnosis times, the repair times and the fault coverage of the test stages for the defined fault spectrum. Some of the supplier induced parts of the fault spectrum can be computed from a knowledge of the number of each component type present

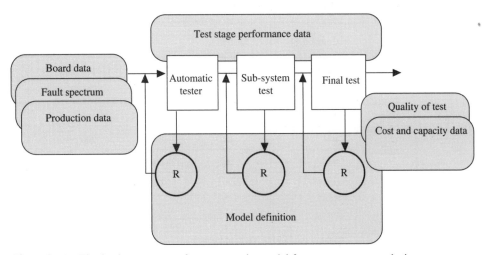

Figure 8.14   The basic structure of an economic model for test strategy analysis

and the experienced defect rate for that category of component. This is shown in Fig. 8.15 as a partial listing of the component types. One of the biggest problems with this or any other type of analysis concerns determination of the input data, especially if this is a future projection. If you have no first hand knowledge of the defect rates you may have to rely on the supplier's quality reports or specifications.

The manufacturing process induced part of the fault spectrum and is then determined as shown in Fig. 8.16, followed by the design induced defects as shown in Fig. 8.17. The three categories can then be summed and the yield computed in the usual manner:

Supplier induced defects	0.053
Manufacturing induced defects	0.120
Design induced defects	0.040
Total	0.213 faults per board

$$\text{Yield} = e^{-0.213} \times 100\%$$

$$= 80.82\%$$

The next step is to define the fault coverage performance of the various test stages versus each of the fault types defined in the fault spectrum. This is usually done in the form of a 'fault coverage matrix', as shown in Fig. 8.18. A section of the model is then set up to take the actual defect data from the fault spectrum and apply it to the test strategy to determine how many of each defect type are detected and how many escape, from each of the test stages. This I call the 'defect detection matrix' and an example is shown in Fig. 8.19. The escaping defects from one test stage become the input defects to the next stage. Some of these are detected and some escape to become the input defects to the next stage. This continues until the 'next test stage' becomes the customer. From this it is possible to determine the quality aspects of

Component	Quantity	Defect rate		FPB
Resistors	150	0.01%	100 ppm	0.015
Capacitors	80	0.02%	200 ppm	0.016
MSI	50	0.01%	100 ppm	0.005
LSI	40	0.02%	200 ppm	0.008
VLSI	10	0.05%	500 ppm	0.005
ASIC	4	0.10%	1000 ppm	0.004

Total of 0.053 FPB (yield of 94.84%)

Figure 8.15 Determining potential supplier induced defects

	Defect	FPB
	Missing components	0.01
	Reversed components	0.03
	Wrong components	0.01
S	Track short	0.01
	Track open	0.01
M	Solder short	0.05
D		

Total of 0.12 FPB (yield of 88.7%)

Figure 8.16 Determining potential manufacturing process induced defects

	Defect	FPB
	Timing/speed	0.01
	Interactions	0.02
	Memory problems	0.01
S		
M		
D		

Total of 0.04 FPB (yield of 96.1%)

Figure 8.17 Determining potential design induced defects

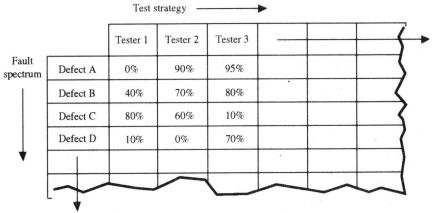

Figure 8.18 The fault coverage matrix for each test stage

225

	Test strategy						
	Tester 1					Tester 3	
			Tester 2				
Fault spectrum	FPB in	Detected	Escapes FPB in	Detected	Escapes FPB in	Detected	Escapes
Defect A	0.1	0.08	0.02	0.01	0.01	0.005	0.205
Defect B							
Defect C							
Defect D							

One per strategy

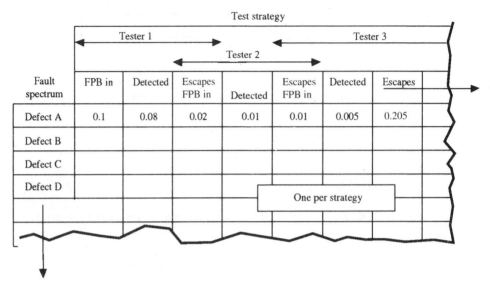

Figure 8.19 The defect detection matrix

	From Manufacturing	Tester 1	Tester 2	Tester 3
FPB in	---	0.86	0.36	0.11
Yield in	---	42%	70%	90%
Detected FPB	---	0.50	0.25	0.07
Apparent yield	---	61%	78%	93%
Fault coverage	---	58%	69%	64%
Escaping FPB	0.86	0.36	0.11	0.04
Yield out	42%	70%	90%	96%

One per strategy

Figure 8.20 The testing result of the analysis shown as a yield progression

the strategy and summarize them in some manner. An example of this is shown in Fig. 8.20. This I call a 'yield progression' because it shows how the yield progresses throughout the test process.

The number of defects detected at a test stage will determine the number of diagnostic operations and repair operations at that stage. From this and the input data on test, diagnosis, and repair times, as well as the labour costs of the operators, it is possible to calculate the costs for the various test stages. The TTT parameter discussed earlier also determines the throughput at each test stage. The throughput figure, the total available time, and the lost time due to program preparation and other factors will determine the capacity. This and a knowledge of the production volume will enable the model to determine how many test stages of each type, and how many repair stations, will be required. A calculation of the available capacity at each stage can also be performed.

Getting the formulae right first time, every time

In the first edition of this book, which appeared in 1982, 1 highlighted the fact that an inaccurate formula was being used in a number of test strategy models. Many of these were being published by ATE vendors in the form of application notes or manual worksheets. In some of these the crime was amplified by using the same formula for both functional and in-circuit test stages. Unfortunately the formula is still appearing in various forms, all of which can be reduced to the most common form, which is shown below:

$$\text{TTT} = T_t(2 - Y) + T_d(1 - Y)$$

There are two terms in the formula. One defines the time spent testing defect-free UUTs and the other defines the time spent diagnosing defects. They are both wrong for both the ICT and the FNT cases. The multiplier for the test time term $(2 - Y)$ implies that some boards are tested twice with a passing result. This does not happen in practice. Once the board passes the test at a given stage it will migrate on to the next test stage. Each time the board visits the test stage with a failing result it will get a diagnosis. Once all of the defects have been repaired it will pass the test and then migrate on. There is no reason to test the board twice. For the diagnosis term the $(1 - Y)$ multiplier indicates that the proportion of failing boards will receive one diagnosis action. This is not correct for functional testing, which will find only one fault at a time, and it is not correct for in-circuit testing because it does not model the fact that boards with shorts will not get a full test on their first visit to the tester. A simple example will illustrate the errors.

Example

Assume a yield of 70 per cent which equates to 0.36 faults per board, of which 0.2 are shorts and 0.16 are other defects. For simplicity we will assume that the testers have a 100 per cent fault coverage. The following times have been determined:

For the functional tester

$$T_t = 35 \text{ s}$$

$$T_d = 360 \text{ s}$$

For the in-circuit tester:

$$T_t = 60 \text{ s}$$

$$T_{ts} = 32 \text{ s}$$

$$T_{to} = 65 \text{ s}$$

Taking the functional case first, the old formula produces the following result:

$$\text{TTT} = 35(1.3) + 360(0.3)$$

$$= 153.5 \text{ s}$$

The formula defined earlier gives

$$\text{TTT} = T_t + T_d(\text{FPB})$$

$$= 35 + 360(0.36)$$

$$= 164 \text{ s}$$

These errors for the functional case will diminish as the yield increases because, as Y tends to unity, $(2 - Y)$ will tend to unity and $(1 - Y)$ will tend to zero.

For the in-circuit case the old formula gives

$$\text{TTT} = 60(1.3) + 65(0.3)$$

$$= 97.5 \text{ s}$$

The formula defined earlier gives:

$$\text{TTT} = T_t + T_{ds}(1 - Y_s) + T_{do}(1 - Y_o)$$

where

$$Y_s = (e^{-0.2}) = 0.82$$

$$Y_o = (e^{-0.16}) = 0.85$$

Therefore

$$\text{TTT} = 60 + 32(0.18) + 65(0.15)$$

$$= 75.51 \text{ s}$$

The incorrect formula was partly correct in the very early days of ATE when the testers had no diagnostics. At that time the testing operation usually involved two people. A low-skilled operator would perform a go/no-go test sorting the good boards from the bad boards. Then a skilled technician would diagnose the bad boards with a logic probe, a logic diagram and skill. The defective boards therefore had two tests. They had a test when the low-skilled operator did the sorting and they had a second test after being repaired—the so-called re-test. This makes the $(2 - Y)$ multiplier correct for this type of operation. The $(1 - Y)$ multiplier in the diagnosis time term was still wrong because functional testers still only tested for one fault at a time. Therefore if you had an average of one fault per board (pretty good at that time) you would diagnose an average of one fault per board. The yield for the 1 FPB case is 0.37 so $(1 - Y)$ would be 0.63. The formula would therefore understate the workload considerably. The 63 per cent of the boards that were defective would contain an average of 1.59 defects on each board, all of which would require a diagnostic action.

Using flow diagrams to check formula accuracy

Flow diagrams of the type shown in Figs 4.10, 4.11b and 4.12b can be very useful for testing the formulae used in a model. In can be a tedious process but it is worth getting right. An example is shown in Fig. 8.21 where a batch of 1000 boards with a yield of 70.5 per cent is tested. The upper part of the diagram shows how the defect distribution is calculated as per the method defined in Chapter 4. On the first visit to the tester 705 boards pass the test because they are defect-free. A further 25 also pass even though they contain defects due to the 85 per cent fault coverage for faults in the 'other' category. Therefore 270 boards fail the test, 139 failing the shorts test (100 per cent fault coverage for shorts) and 131 failing for other reasons. These are repaired and visit the tester for a second time. Now the 25 boards that contained both a short and some other defect fail the test and the remainder of the repaired boards pass. The 25 failing boards go to the repair station and then on to the tester for their third visit before passing and migrating on to the next stage. The number of escaping defects in the 'other' category was determined as follows:

0.20 other defects per board and an 85 per cent fault cover
Therefore 0.17 detected faults per board
Y_o is therefore $(e^{-0.17}) = 0.84$
$1 - Y$ is therefore 0.156

Thus 156 boards out of the batch will have 'other' defects detected but 25 of these will be on boards with a short and therefore will not be seen on the first visit to the tester. Therefore only 131 will be detected on the first visit. The flow diagram shows what will happen in practice and the numbers agree with the $(1 - Y)$ figures for the two defect types.

This example assumes that all of the diagnoses are correct, that the repair actions will be performed correctly and that no non-existent defects will be 'detected' by the tester. In practice all of these problems can occur. The four possible outcomes of a test defined in Fig. 4.2 need to be accounted for in some manner. This is most easily

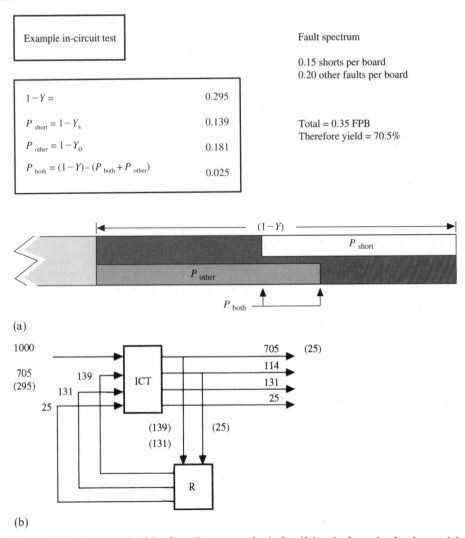

(a)

(b)

Figure 8.21 An example of the flow diagram method of verifying the formulae for the model

done by applying a correction factor to the number of diagnoses and the number of repair actions. The number of passing tests is not affected. This correction factor is known as the 'loop-number multiplier' or simply the 'loop number', since it is applied to the diagnosis/repair loop of the test process. The loop number simply adds a percentage to account for the number of unnecessary diagnosis and repair actions. For most cases this is likely to be in the 10 to 20 per cent range so the loop number will be between 1.1 and 1.2. This gives the basic functional formula:

$$TTT = T_t + T_d(FPB)LN$$

Where LN is the loop number (also sometimes abbreviated to LP). The basic in-circuit formula becomes

$$\text{TTT} = T_t + T_{ds}(1 - Y_s)\text{LN} + T_{do}(1 - Y_o)\text{LN}$$

If you want to be really precise you should apply a different loop number to the 'shorts' term and to the 'other' term because it can be argued that there is less likelihood that shorts will be misdiagnosed, wrongly indicted or incorrectly repaired. Figure 8.22 shows what happens to the flow diagram of Fig. 8.21 when the loop-number effect is added in. Since there will potentially be some unnecessary diagnoses and repairs for each batch of boards visiting the tester, the number of visits increases. If we assume that the probability of an unnecessary diagnosis/repair action is the same for each group of boards visiting the tester then the simple loop-number approach is not really correct. For a probability of 0.2 that there will be an unnecessary diagnosis/repair the loop number is set to 1.2 (i.e. 20 per cent additional actions). The more correct calculation of LN would be

$$\text{LN} = 1 + P_{DR} + (P_{DR})^2 + (P_{DR})^3 + (P_{DR})^4$$

where P_{DR} is the probability of an unnecessary diagnosis/repair action (0.2 in this example). However, the error relative to simply using $1 + P_{DR}$ is quite small, as is shown in Table 8.1.

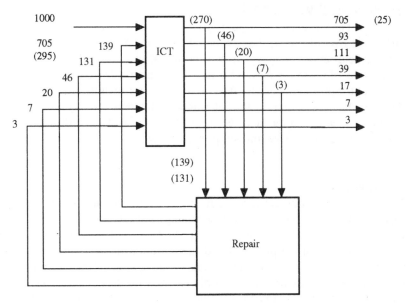

Figure 8.22 Adding the loop number correction factor to the example in Fig. 8.21

Table 8.1 Loop-number errors if the simplified $(1 + P_{DR})$ formula is used

P_{DR}	Error, %
0.2	4
0.15	2.2
0.1	1.0

8.10 An example of a test strategy analysis model

The remainder of this chapter is an edited version of parts of the user manual for a model that I developed towards the end of 1992 called EVALUATE. The model was developed as a result of suggestions from several people who had recognized that many companies did not analyse their test strategy options in sufficient detail because of the amount of work involved in developing a model. They felt that if such a tool was available commercially then more people would undertake the necessary analysis. Since the following material is taken from the manual, which out of necessity is fairly tutorial in nature, there is a little bit of repetition of material covered in earlier parts of this book. However, since this is an important subject I hope that this will not prove to be too much of a problem. At least it makes this section readable without constant cross-references to earlier chapters.

8.11 Overview of EVALUATE

EVALUATE is a decision support tool mainly for the manufacturing test strategy part of the overall test strategy, which would also include the design to test strategy and the field-service strategy. It can, however, give an indication of the effect that different decisions in the design for test (DFT) area will have on the cost and quality aspects of the product. Since EVALUATE produces an indication of the likely number of defects that will 'escape' to the field for each different strategy analysed, it effectively provides a measure of the field-service workload also. This information will be useful in establishing the service and support strategy.

The testing decision process

Test strategy decisions are just a part of an overall hierarchy of strategic decisions beginning with the overall corporate strategy for the company. They are, however, a very important part because the decisions made will have a significant impact on the life cycle cost of the products and therefore on the profitability of the company. The cost of test has become a much bigger percentage of the overall manufacturing cost of most complex products, mainly because the test generation tools needed to create effective tests have not been able to keep pace with the growth in complexity of the products. New manufacturing technologies such as surface mount and multi-chip modules have increased the difficulty and therefore the cost of test. The industry is being forced to do what it should have been doing since day one—design for testability. Testability decisions made in the engineering department will impact the time to market for the product and this in turn will determine whether the revenues and profit targets can be achieved. These decisions will also determine how much fault coverage can be achieved in the time available and this in turn will affect the quality of the product and therefore its life cycle cost.

The decision process follows the route of objectives—strategy—tactics. The objectives will obviously be unique to each company, but one important element of them is common throughout the electronics industry. This is the need to consistently develop new products that cost less, that have higher quality, that are brought to the market on time and that contain the latest technology. Meeting these objectives with the maximum effectiveness is a difficult and complex task. This is made even more difficult by the fact that these objectives can be in conflict with each other if the overall strategy is not well thought out. Selecting the most effective test strategy is part of this overall process and by itself is a complex problem because of the large number of variables involved. This complexity has frequently led to companies resorting to using a simple approach to the analysis of alternative strategies. Unfortunately any simplification of a complex problem leads to inaccuracies and a lack of resolution. The simple models usually make too many assumptions which limit their use to a quick overview. Most of the models that have been made available over the past ten to fifteen years have been produced by the vendors of automatic test systems. As such they have been aimed mainly at helping the potential customer to develop some cost data for inclusion in a financial justification for a tester. These models have been produced in the form of brochures containing a worksheet form which you complete and then perform the calculations manually, and more recently in the form of spreadsheet models. Most of these have suffered from one or more of the following problems:

1. They make too many simplifying assumptions.
2. They may only model one type of strategy.
3. They may use incorrect formulae.
4. They lack the detail needed to compare tactics.

This last point is an important one. Since we need to optimize the strategy and the tactics in order to maximize the effectiveness, the model needs to have enough detail to be able to see differences in the performance of different testers (the tactics). This is actually a strange omission for the ATE vendors to make in their models because to some degree they are trying to use these tools to convince customers that they have a more cost effective solution. The point about the incorrect formulae is also important. Since much of the input data will often be an estimate or a forecast you need the model to be fairly precise so that it does not introduce further errors. The core of the model is the calculation of the average total time that each tested unit spends at each test stage, including any re-tests after repair and the effects of incorrect diagnosis and incorrect repair actions, etc. From this time value the model calculates the cost information, the throughput at each stage, the capacity of the test stage, the number of testers needed, the available surplus capacity and so on. If this time calculation is in error then all of the cost- and time-related results will be in error. There is a formula that has been around in the test industry for many years that does not accurately model what really happens in a test stage. This formula has frequently been used in these simplified models for both in-circuit and functional test stages. Even if it was correct for one it could not be correct for both, since the two test techniques operate in different ways.

EVALUATE has been developed to address these limitations. It provides electronics manufacturers with an accurate model of all of the commonly used test strategies for board test, sub-assembly test and final system test. It provides the level of detail needed to analyse both the test strategy and the test tactics without the need to spend many man-months developing your own detailed models. EVALUATE is independent ot any equipment vendor.

EVALUATE has been developed using ObjectVision. This is an application development program for Microsoft Windows that has been created by Borland International Inc. ObjectVision uses a 'forms' metaphor so EVALUATE is used by completing several forms with the necessary input data. The results are then presented on other forms which may be viewed on the screen or printed. You navigate through the various forms by clicking on-screen 'buttons' with a mouse. This can also be done from the keyboard by using the usual Windows keyboard commands to get the 'focus' to the relevant button and then pressing the 'enter' key. These methods take you through the forms in a logical sequence; however, you can also move in a random manner from form to form by selecting **Form|Select** from the pull-down menu bar at the top of the screen.

8.12 Test stages modelled

- *MDA* Manufacturing defect analyser. In EVALUATE an MDA is modelled as an in-circuit tester but with reduced capabilities as defined by the input parameters.
- *ICT* In-circuit tester.
- *FNT* Functional tester.
- *COMB* Combinational tester. This tester type is used to model combinational systems which are basically ICTs with some added functional capabilities, ICTs with added boundary scan capabilities and on up to full 'performance' testers.
- *SST* Sub-system test.
- *SYS* Final system test.

8.13 Forms description

The following is a brief description of each of the forms used for a normal analysis. More details of the individual input parameters appear in Sec. 8.21 on input parameter reference.

Header (Fig. 8.23)

The first form you come to when you start up EVALUATE is called 'Header'. This is the title form which enables you to enter information and notes about the analysis being performed.

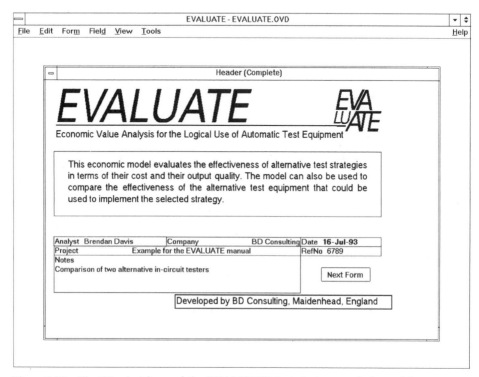

Figure 8.23 The 'Header' form of the EVALUATE test strategy analysis model

Board Data (Fig. 8.24)

This is where you specify the complexity of the board, or boards, that the analysis is being performed for. EVALUATE can be used to analyse strategies for a single board type, a group of boards or your entire production. To analyse a strategy for more than one board type you simply enter data that is an average for the boards in question. You enter the quantity of each component type present on the board and the defect rate for those devices in parts per million (ppm). EVALUATE then calculates the average number of faults per board (FPB) that this combination of quantities and defect rates will produce, and sums the results into several sub-groups:

1. Passive components.
2. Linear active components.

3. Mixed signal devices.
4. Digital devices.
5. Memory devices.
6. Custom devices.
7. Other.

Clicking on the 'Next Form' button then takes you to the 'Fault Spectrum' form.

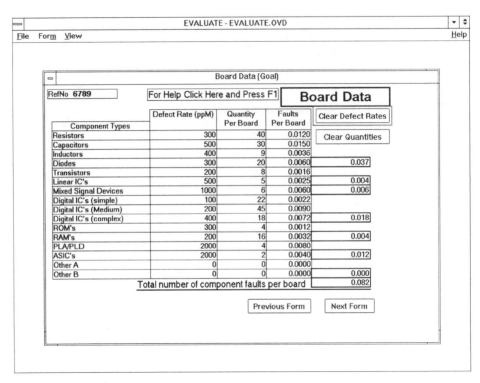

Figure 8.24   The 'Board Data' form of the EVALUATE test strategy analysis model

Fault Spectrum (Fig. 8.25)

The sub-totals from the 'Board Data' form are brought forward to the 'Fault Spectrum' form. Here you now add in all of the other defect classes that go to make up the overall fault spectrum. These fault classes are grouped into three main categories.

1. Design induced defects.
2. Supplier induced defects.
3. Manufacturing process induced defects.

The component defect sub-totals from the 'Board Data' form are included in the 'supplier induced' category. Each of the categories has one 'other' defect class so that you can define a defect class for each of the categories if you suffer from defects that are not included in the lists. Selecting the 'Next Form' button takes you on to the 'Production Data' form.

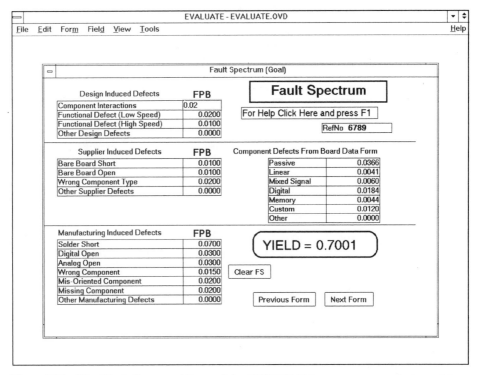

Figure 8.25 The 'Fault Spectrum' form of the EVALUATE test strategy analysis model

Production Data (Fig. 8.26)

Here you supply information on the production volume, the time available per shift, the number of shifts and the labour costs for the various operators, test engineers and programmers. Selecting the 'Next Form' button takes you on to the 'Test Stage Data 1' form.

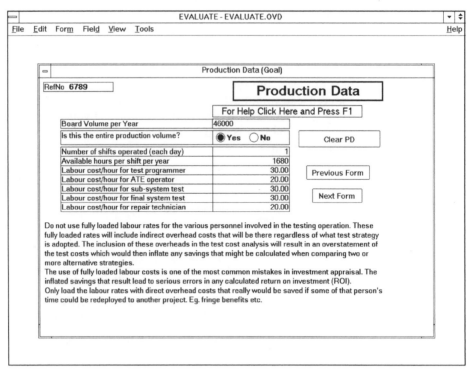

Figure 8.26 The 'Production Data' form of the EVALUATE test strategy analysis model

Test Stage Data 1 and 2 (Figs 8.27 and 8.28)

These two forms are used to provide data about the performance of the various test stages. On the first of these forms you provide data for up to two in-circuit board testers and two functional board testers. The reason you can specify two of each is to make it easier to make comparisons. For example you can compare two different strategies such as 'in-circuit test (ICT) followed by sub-system test (SST) followed by final system test (SYS)' versus 'manufacturing defects analyser (MDA) followed by SST followed by SYS'. As far as EVALUATE is concerned the only differences between an MDA and an ICT lie in the prices and the performance levels of the testers as defined on these and other forms.

Alternatively you may wish to compare two different tactics for the same strategy. For example, you can compare a high-performance tester with a similar system that has lower performance and a lower price from the same vendor, or you can compare the products of competing vendors. Having access to two in-circuit type testers (ICT-A and ICT-B), two functional type testers (FNT-A and FNT-B) and two combinational type testers (COMB-A and COMB-B) speeds up this type of comparison process.

The data supplied on these forms relates to the test, diagnosis and repair times; the number of test programs; the cost of fixtures; the number of engineering changes

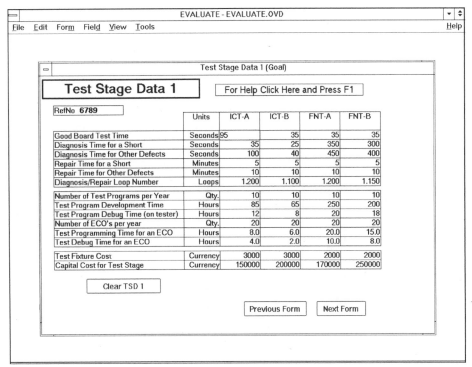

Figure 8.27 The 'Test Stage Data 1' form of the EVALUATE test strategy analysis model

and the programming time needed to accommodate them and so on. Selecting the 'Next Form' button takes you to the 'Test Stage Data 2' form where the same information can be added for the combinational testers, if required, and for the sub-system test stage and the final system test stage.

The 'Test Stage Data 2' form is a little different to the 'Test Stage Data 1' form because there are a few more pieces of information required for accurately modelling a combinational tester. Most current combinational testers are essentially in-circuit testers with some added functional capabilities. There are, of course, exceptions to this but if the system combines both capabilities then the in-circuit tests would be performed first so that any shorts can be repaired before power is applied to the UUT. An in-circuit tester with an added boundary scan capability would also be modelled using the COMB tester. The additions to this form provide the ability to specify the diagnosis time for defects detected in the 'functional' test portion of the test program and the diagnosis time for defects detected during the boundary scan tests. Since some combinational testers have no automatic diagnosis for functionally detected defects it is also possible to specify an 'off-line diagnosis time', so the cost of this can be included in the results. This time, however, has no impact on the time spent on the tester and therefore does not affect the throughput and capacity calculations.

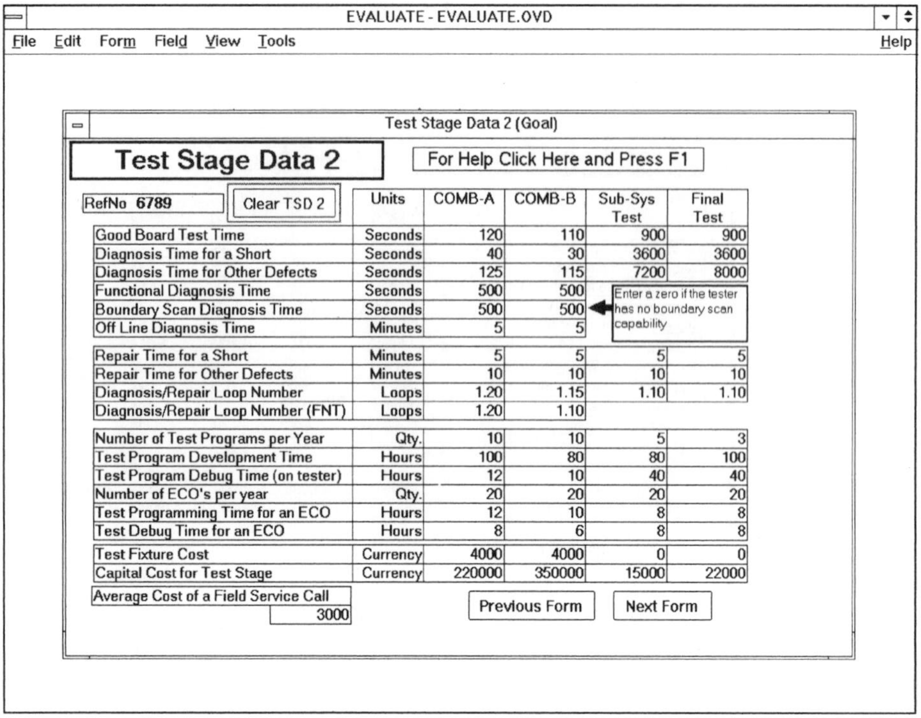

EVALUATE - EVALUATE.OVD

File Edit Form Field View Tools Help

Test Stage Data 2 (Goal)

Test Stage Data 2

For Help Click Here and Press F1

RefNo 6789 | Clear TSD 2

	Units	COMB-A	COMB-B	Sub-Sys Test	Final Test
Good Board Test Time	Seconds	120	110	900	900
Diagnosis Time for a Short	Seconds	40	30	3600	3600
Diagnosis Time for Other Defects	Seconds	125	115	7200	8000
Functional Diagnosis Time	Seconds	500	500		
Boundary Scan Diagnosis Time	Seconds	500	500		
Off Line Diagnosis Time	Minutes	5	5		
Repair Time for a Short	Minutes	5	5	5	5
Repair Time for Other Defects	Minutes	10	10	10	10
Diagnosis/Repair Loop Number	Loops	1.20	1.15	1.10	1.10
Diagnosis/Repair Loop Number (FNT)	Loops	1.20	1.10		
Number of Test Programs per Year	Qty.	10	10	5	3
Test Program Development Time	Hours	100	80	80	100
Test Program Debug Time (on tester)	Hours	12	10	40	40
Number of ECO's per year	Qty.	20	20	20	20
Test Programming Time for an ECO	Hours	12	10	8	8
Test Debug Time for an ECO	Hours	8	6	8	8
Test Fixture Cost	Currency	4000	4000	0	0
Capital Cost for Test Stage	Currency	220000	350000	15000	22000

Enter a zero if the tester has no boundary scan capability

Average Cost of a Field Service Call: 3000

Previous Form | Next Form

Figure 8.28 The 'Test Stage Data 2' form of the EVALUATE test strategy analysis model

The other input parameter collected on this form is the 'Average Cost of a Field Service Call'. This information is used to apply a cost to the defects that escape from the overall test strategy. It provides a means of including this important cost element into the life cycle cost of the strategy. This can also be regarded as the 'cost of non-conformance' for the test strategy because it relates a cost to the 'escapes' whereas the other costs can be regarded as the 'cost of conformance' for the strategy or the cost of defect detection. Selecting the 'Next Form' button on this form then takes you to the first of three 'Fault Coverage Matrix' forms.

Fault Coverage Matrix Forms (Figs 8.29, 8.30 and 8.31)

This is where you specify the expected fault coverage for each of the test stages for each of the defect classes in the fault spectrum. This level of detail is required in order to make sure that the overall fault coverage is reasonably correct for widely varying

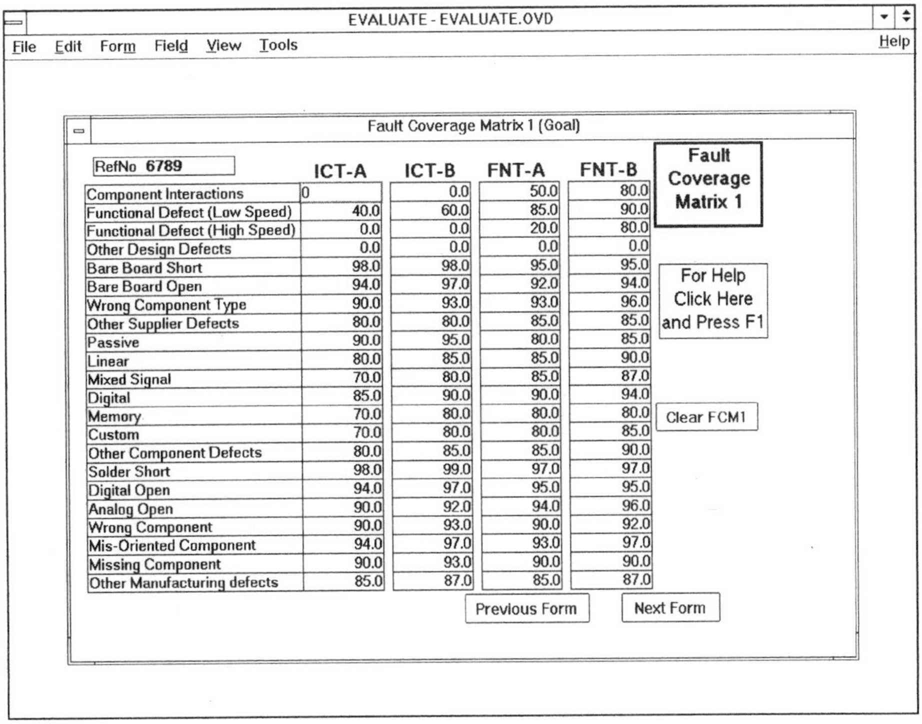

	ICT-A	ICT-B	FNT-A	FNT-B
Component Interactions	0	0.0	50.0	80.0
Functional Defect (Low Speed)	40.0	60.0	85.0	90.0
Functional Defect (High Speed)	0.0	0.0	20.0	80.0
Other Design Defects	0.0	0.0	0.0	0.0
Bare Board Short	98.0	98.0	95.0	95.0
Bare Board Open	94.0	97.0	92.0	94.0
Wrong Component Type	90.0	93.0	93.0	96.0
Other Supplier Defects	80.0	80.0	85.0	85.0
Passive	90.0	95.0	80.0	85.0
Linear	80.0	85.0	85.0	90.0
Mixed Signal	70.0	80.0	85.0	87.0
Digital	85.0	90.0	90.0	94.0
Memory	70.0	80.0	80.0	80.0
Custom	70.0	80.0	80.0	85.0
Other Component Defects	80.0	85.0	85.0	90.0
Solder Short	98.0	99.0	97.0	97.0
Digital Open	94.0	97.0	95.0	95.0
Analog Open	90.0	92.0	94.0	96.0
Wrong Component	90.0	93.0	90.0	92.0
Mis-Oriented Component	94.0	97.0	93.0	97.0
Missing Component	90.0	93.0	90.0	90.0
Other Manufacturing defects	85.0	87.0	85.0	87.0

Figure 8.29 The 'Fault Coverage Matrix 1' form of the EVALUATE test strategy analysis model

fault spectra. This is one area where a simpler economic model usually suffers from some limitations. Some only enable you to specify a global level of fault cover or, at best, the fault cover for a highly simplified fault spectrum. Since the achievable fault cover for a given tester varies widely with defect types, you need to have a reasonable level of detail in order to see differences between different testers or to see how the same tester will cope with different fault spectra. The fault spectrum may well be quite different for different products within the same manufacturing operation or it may change as a result of a process change or an improvement in the yield brought about by quality improvement activities.

The first form collects the data for the two in-circuit testers and the two functional testers. The second form collects the data for the two combinational testers and the third form collects the data for the sub-system test and the final system test stages. This completes the collection of the input data, so selecting the 'Next Form' button takes you to the 'Test Strategy Selection' form.

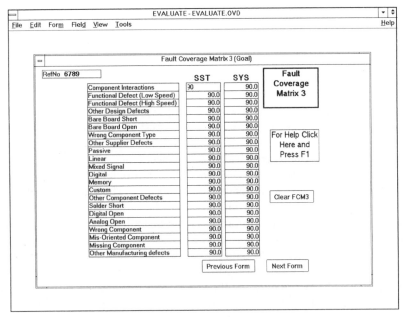

Figure 8.30 The 'Fault Coverage Matrix 2' form of the EVALUATE test strategy analysis model

EVALUATE - EVALUATE.OVD

File Edit Form Field View Tools   Help

Fault Coverage Matrix 3 (Goal)

RefNo 6789	SST	SYS	Fault Coverage Matrix 3
Component Interactions	90	90.0	
Functional Defect (Low Speed)	90.0	90.0	
Functional Defect (High Speed)	90.0	90.0	
Other Design Defects	90.0	90.0	
Bare Board Short	90.0	90.0	
Bare Board Open	90.0	90.0	
Wrong Component Type	90.0	90.0	For Help Click Here and Press F1
Other Supplier Defects	90.0	90.0	
Passive	90.0	90.0	
Linear	90.0	90.0	
Mixed Signal	90.0	90.0	
Digital	90.0	90.0	
Memory	90.0	90.0	
Custom	90.0	90.0	
Other Component Defects	90.0	90.0	Clear FCM3
Solder Short	90.0	90.0	
Digital Open	90.0	90.0	
Analog Open	90.0	90.0	
Wrong Component	90.0	90.0	
Mis-Oriented Component	90.0	90.0	
Missing Component	90.0	90.0	
Other Manufacturing defects	90.0	90.0	

Previous Form   Next Form

Figure 8.31 The 'Fault Coverage Matrix 3' form of the EVALUATE test strategy analysis model

Test Strategy Selection (Fig. 8.32)

This form is the only one of the standard set of forms that is not usually printed. No data is entered on this form. It simply consists of a graphical representation of each of the test strategies with a column of buttons alongside them to select the strategy you wish to analyse. If data has been input for all test stage types EVALUATE actually performs the analysis of all of the strategies simultaneously. This selection form is simply a convenience and time-saving feature since it allows you to view and print only those strategies you are interested in rather than having to wade through all of the others as well.

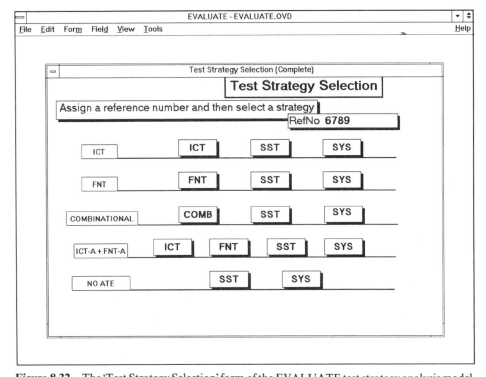

Figure 8.32   The 'Test Strategy Selection' form of the EVALUATE test strategy analysis model

Defect Detection Matrix (Figs 8.33 and 8.34)

Selecting a strategy takes you to the 'Defect Detection Matrix' form for the strategy. There will be two of these forms for the strategies with the direct comparison feature

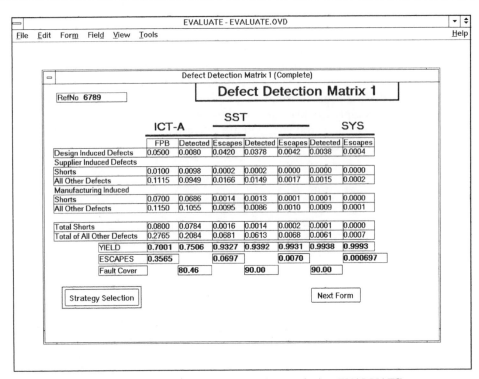

Figure 8.33 The 'Defect Detection Matrix 1' form of the EVALUATE test strategy analysis model

and one each for the other strategies. These forms show a summary of the defects that are input to the first test stage and then show how the defect detection progresses throughout the test strategy. For each test stage you can see the input defects, the detected defects and the escaping defects. The escaping defects become the input defects to the next test stage and so on. The defects escaping from the final system test stage become the escapes to the field and as such represent a measure of the quality of the strategy and an indication of the field service workload that the strategy will create.

For each test stage the input yield (Y_{in}), the apparent yield (Y_a) and the output yield (Y_{out}) are calculated. The apparent yield is the yield seen by the test stage. This will be higher than the actual yield entering the test stage because the tester does not detect all of the defects. Y_{out} for the first stage becomes the Y_{in} for the second stage and so on. The escapes for each stage are summed together and the overall fault coverage for each stage is computed. Selecting the 'Next Form' button then takes you to the 'XXX Results' form.

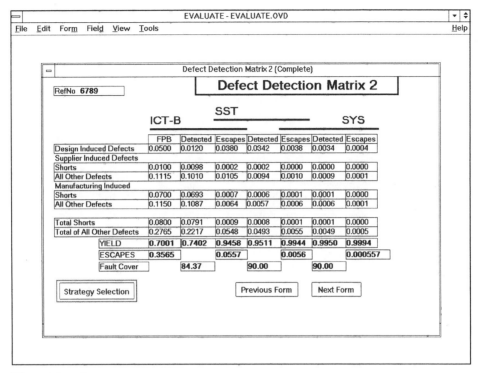

Figure 8.34 The 'Defect Detection Matrix 2' form of the EVALUATE test strategy analysis model

XXX Results (Fig. 8.35)

This form shows the overall results for the current strategy (XXX). It provides a breakdown, sub-totals and totals of costs for test, diagnosis, repair, program preparation, fixtures, ECOs and field service. It also adds up the capital cost of the investment required to implement the strategy. Also listed is the throughput in boards per hour through each test stage and the number of each of the test stages required to handle the volume. The available spare capacity of each test stage is also calculated.

If you entered the input data for all of the tester types on the 'Test Stage Data' and the 'Fault Coverage Matrix' forms, you can now return to the 'Test Strategy Selection' form and select another strategy. This will take you through the 'Defect Detection Matrix' and the 'XXX Results' forms for this new strategy. Alternatively, if you want to re-run the current strategy with some changes to the input data you can return to the 'Header' form and then through various input data forms, change the relevant data, select the strategy and immediately see the effect on the results.

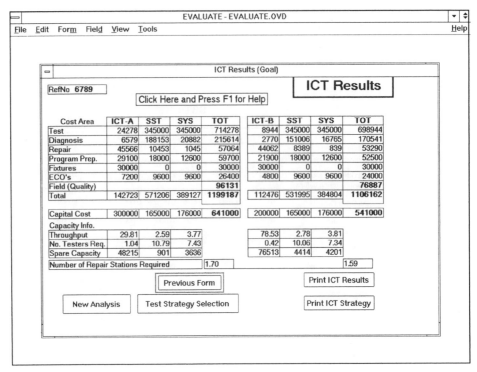

Figure 8.35 The 'XXX Results' form of the EVALUATE test strategy analysis model

8.14 Using EVALUATE

When you first start EVALUATE you will find that it contains some sample data. This is included so that you can browse through the various forms and familiarize yourself with the model. When you are ready to try your first analysis with your own data you can clear the sample data from the various input forms, input your own data and view the results.

If you wish to retain each analysis that you perform you will need to use the **File|Save** As command and give each analysis a unique name.

8.15 Test strategy determinants

With few exceptions all electronics manufacturing operations should have a similar overriding objective. This is to develop products that cost less, with higher quality, brought to the market on time and incorporating the latest technologies. If you cannot do this consistently you cannot be consistently competitive. Other company-specific objectives such as 'to become the leader in ...', or 'to improve our customer satisfaction levels ...', or 'to increase profitability by ...', etc., will all be dependent on meeting this superordinate objective of cost, quality, time and technology.

The selection of the ideal test strategy should support the corporate strategy established to meet these objectives, but the ideal strategy will be different for different situations. Each situation can be broken down into two areas, each of which have two elements which will be the primary determinants of what will be the ideal strategy. The first area, and possibly the most important in most situations, is the current or expected performance of the manufacturing process. This performance can be summarized by two major elements, the yield and the fault spectrum. The second area is the nature of the production operation and this can be split into two major elements, the nature of the markets served and the production mix. The following list summarizes just a few of the possible variations of these four elements.

Yield

1. All products have high yield.
2. All products have low yield.
3. There is a mixture of high yield and low yield products.
4. The yield is stable.
5. The yield is improving.
6. The yield is declining
7. The yield is improving for the low-yield products and declining for the high-yield products.
8. etc.

Fault spectrum

1. The fault spectrum is consistent for all products.
2. The fault spectrum is different for different products.
3. The fault spectrum is stable.
4. The fault spectrum is changing.
5. Manufacturing defects are becoming less of a problem.
6. Manufacturing defects are becoming more of a problem.
7. etc.

The markets served

1. The military and aerospace market.
2. The automotive market.
3. The telecommunications market.
4. The computer, peripherals, consumer and industrial markets.
5. We serve only one market.
6. We serve many markets.
7. We serve the military market but we are moving into other markets.
8. etc.

The production mix

1. High volume, few products.
2. High volume, many products.
3. Low volume, few products.
4. Low volume, many products.
5. Production mix is stable.
6. Production mix is changing.
7. Board volume is increasing due to diversification.
8. Board volume is declining due to the use of ASICs.
9. etc.

8.16 Types of analysis

There will be a number of different kinds of situation where an EVALUATE analysis will prove useful, but these can be broadly classed into three groups. The first is where you want to analyse your current test strategy and compare it with some alternative approaches. This would be particularly important if there has been some significant changes since your current strategy was implemented. Possibly one of the most common changes in recent years will have been a substantial improvement in the yield from manufacturing, even in the face of increasing complexity. Another common change will have been the move to surface mount technology. These changes, which will inevitably be accompanied by changes to the fault spectrum, can often reduce the efficiency of what was once an ideal strategy. In this kind of situation you will be working with data that should be known to some reasonable degree of certainty.

The second situation is where you want to analyse a future situation—a new product range, a major new technology, a product type that is quite new to the company, a big increase in production volume, etc. All of these, and other, situations will involve a move into the unknown. You will not know any of the key data with any degree of certainty. You will be in the area of 'forecasting, estimating and guesswork' (described in Chapter 1 of this book). These situations tend to require a lot more analysis work. You need to perform a variety of 'what if' scenarios and determine which are the most critical parameters. Then you take the time to determine the most likely range of values for these critical parameters. When you have done that, you do the analysis all over again. It is in this kind of situation that economic models have the most value. They enable you to perform a large amount of work in a relatively short time.

The third situation is the analysis of tactics rather than strategies. By this I mean the selection of the best testers with which to implement your chosen strategy. There are two fundamental requirements here. The first is to select the testers that will result in the best implementation of the strategy in terms of meeting the overall objectives, and the second is to determine the cost savings that will be plugged into an ROI (return on investment) calculation to justify the purchase of the equipment.

8.17   Sensitivity analysis

Although EVALUATE does not contain any specific features to enable an automatic sensitivity analysis to be performed, it is in fact quite easy to do. Most sensitivity analyses will involve changing one parameter at a time, with the others held at their nominal values, and observing the change in the results to see if the answers are 'sensitive' to the changed parameter or not. It is normal to plot the results for several parameters on a graph to see which of them display the steepest (most sensitive) slope. If you plot three points, the low end of the range, the nominal (most likely) value and the high end of the range, it will be obvious whether the results have a linear or a non-linear relationship with the varied parameter. If the relationship is non-linear it may be necessary to plot a few additional points in order to plot the curve more accurately. This is very quick and easy to do. You simply change the value of the parameter in question and then go to the 'Results' form for the strategy of interest by using the **Form|Select** command from the pull-down menu. If you are analysing more than one strategy you then go to the 'Results' form for the next strategy of interest.

Once you have established which of the parameters are the most important it is normal to take a bit more care over establishing, or predicting, the most likely nominal value and then repeating the analysis using the most likely values of all of the high-sensitivity parameters.

8.18   Entering data

You can enter data for a new analysis by editing the input parameters of the master version of EVALUATE or by editing an earlier analysis that you performed. So long as you remember to file the new version with a different name the older versions will be retained. If the input data for this new analysis is very different to the version you are editing then it will be quicker to clear the old data from the forms before entering the new data. This can be done for all the forms simultaneously by using the **Edit|Clear All** command from the menu bar. Alternatively the **Form|Clear** command will just clear the current form. In addition to these menu commands each of the input parameter forms has an on-screen button that will clear the current form data cells. The 'Board Data' form has two clear buttons so that you can clear only the defect rate data, or only the quantities of components data. The reason for this is that once you have established your component defect rates they are unlikely to change much from board type to board type. Therefore this data will be common to a number of analyses and you can save time by not having to enter the data each time. For this reason it is probably not a good idea to use the **Edit|Clear All** command unless you really do want to start from scratch. The **Form|Clear** command will also clear the defect rate data if this is issued when the 'Board Data' form is the current form.

When a form, or a section of a form, is cleared, the data entry cursor will move to the first parameter and then progress from left to right and down the form as you enter the data. You can, however, position it on any cell by using the mouse. If you only need to change a few values for your new analysis then it will be quicker to edit the individual cells rather than clearing the forms. EVALUATE will not perform calculations if any of the relevant cells on a form remain blank. This is why data entry is faster if you clear the forms first. If you change the value in any input parameter cell then EVALUATE will re-calculate the effect of that change for all strategies before you can move to another cell to edit. However, if you are only changing the contents of a few cells this will still be much faster than clearing the entire form.

8.19   Interpreting the results

Most of the results are fairly self-explanatory. They are made available on two types of form, the 'Defect Detection Matrix' and the 'Results' forms. The defect detection matrix shows a summary of the main defect categories, how many of these are detected at each stage and how many escape to the next stage. The input yield, the apparent yield and the output yield for each test stage is computed and displayed along with the escape rates in 'faults per board' and the fault coverage achieved by each test stage. The results forms show the primary economic data broken down into the three areas of operational costs, capital costs and capacity information. All costs are calculated separately for each stage of the test strategy and totalled for each stage and for each cost category.

Test cost

This is the cost of testing defect-free units. The formulae used in EVALUATE make the assumption that each UUT, at each test stage, receives only one test that passes. This will occur because there was no defect present or because there was a defect that did not get detected. The passing test may occur on the UUT's first visit to the test stage or it may occur on a subsequent visit that follows a repair operation. Any test that fails is classed as a 'diagnosis' rather than a 'test' since some diagnostic action will usually take place. For some tester types the diagnosis may take less time than the test of a defect-free UUT, as with the diagnosis of a short on an in-circuit tester. The general assumption is that once a UUT passes a test it will migrate to the next stage of the process. If it fails then a diagnosis will take place and will take a different amount of time; the cost of this is included in the 'diagnosis' cost category. The cost in this 'test cost' category is effectively the minimum cost for the test stage since this cost would be incurred even if there were no defects detected.

Diagnosis cost

This is the total cost of diagnosing all detected defects at each of the test stages. It includes the cost of multiple defects that might require more than one diagnosis, as with a functional test that can only find one fault at a time. The formulae used also correctly account for the fact that an in-circuit test can find multiple faults in one operation unless one of the defects is a short. Under these conditions the test would be halted after the shorts test part of the program, the short would be repaired and the board would then be returned to the tester when any other defect types present would be detected. The 'diagnosis/repair loop number' input parameter on the 'Test Stage Data' forms is also used to account for the fact that some defects will not be repaired correctly the first time and so result in an extra visit to the tester and an increase in the average diagnosis time per UUT. This correction factor also accounts for the fact that a tester may indict a non-existent defect which will also result in unnecessary diagnosis time.

Repair cost

This is the cost of the repair actions that are required to remove detected defects from the UUT. The 'diagnosis/repair loop number' input parameter on the 'Test Stage Data' form is used to account for the fact that not all repair actions will be performed correctly the first time. This will result in the UUT failing the test on its return to the tester and so an additional visit to the repair station. This correction factor also accounts for the fact that a tester may indict a non-existent defect and this will also result in additional repair time.

Program preparation cost

This is the cost of preparing and debugging the test programs or test procedures for the non-automated test stages. The cost is derived from the number of programs, the stated times for development and debugging, and the labour cost of the programmers.

Fixture cost

This is the cost of the fixtures needed for the various board types. The cost of each fixture is input to the 'Test Stage Data' forms and should include the cost of the fixture kit and the cost of assembly and debug. EVALUATE uses the following simple logic to determine the number of fixtures required. If the calculated number of testers required to handle the production volume is greater than the number of new programs, EVALUATE assumes that each tester will require a fixture. If the number of test programs is greater than the number of testers needed then EVALUATE assumes that only one fixture is required for each board type. In calculating the number of testers required to handle the production volume EVALUATE allows for the time needed to debug the test programs and the debug of program changes caused by Engineering Change Orders (ECOs).

ECO cost

This is the cost of program changes and the debug of test programs caused by engineering changes to the board design.

Field cost (quality)

This is the cost of field service that results from the escape rate of the strategy and the average cost of a field repair as input to the 'Test Stage Data 2' form. EVALUATE assumes that all escapes will result in a product failure at some point in the warranty period. If you feel that this is a pessimistic view you could alter this assumption by varying the 'average cost of a feld-service call' parameter. For example, if you feel that only 80 per cent of the escapes will result in a failure in the field, reduce the cost per call by 20 per cent. Bear in mind that this data is only considering defects that could have been detected in the factory prior to shipping the product to the customer. Early life failures caused by component degradation or misuse of the product are not included here. This figure of field cost is intended to be a measure of the cost of quality (or the cost of non-conformance) and is a part of the life cycle cost of the strategy. See Sections 6.5 and 6.6 for further discussion of field service costs.

Capital cost

This is the cost of capital equipment purchased to implement the strategy. If the 'YES' button is selected as the answer to the 'Is this the entire production volume?' question on the 'Production Data' form then the calculated 'number of testers required' is rounded up to the nearest whole number before calculating the cost. If the 'NO' button is selected then the 'number of testers required' is not rounded up since it is assumed that some of the tester's time will be used for other UUT types. In other words, the capital cost is allocated in proportion to the tester time required.

Throughput

This is the throughput rate of the various test stages in boards per hour.

Number of testers required

This is the total number of each of the test stages required to handle the production volume as well as the debugging time required for test programs and changes to test programs caused by ECOs.

Spare capacity

This is the available production capacity at each test stage. It is derived from the proportion of testers available and the production throughput rate. It is assumed that there will be no additional programming load.

Number of repair stations required

This is the total number of repair stations needed to cope with the repair work from all test stages. One of the simplifying assumptions made by EVALUATE is that the repair work is performed at a central facility regardless of where in the test strategy the defect was detected. This is a reasonable assumption for most situations because of the specialized equipment that is often needed to perform the re-work of the boards. Specialized de-soldering and re-soldering equipment is required for all board manufacturing technologies but it is especially pertinent to SMT boards.

8.20   The hidden forms

In addition to the main results forms that are displayed each time you run EVALUATE for one or more strategies, there are some other forms that remain hidden unless you specifically call them up. This is achieved by using the **Form|Select** menu and selecting the name of the desired form from the list that is displayed. The hidden forms are not printed when you click on the various print option buttons on the 'Results' forms; instead you have to access the **File|Print Form** command. These forms are normally hidden because they contain a level of detail that is not usually needed for most analysis work. The first type of hidden form is a more detailed version of the 'defect detection matrix' which shows the complete fault spectrum for all test stages. There is one of these for each of the test strategies. The extra information contained on these forms would be useful if you need to know in detail where certain defect types are being detected. These forms are named with an abbreviation of the relevant test strategy. The other hidden form is the 'Intermediate Results' form. This also contains information for all eight of the strategy sets that may occasionally be useful.

8.21   Input parameter reference

Any computational model such as EVALUATE can suffer from GIGO problems. GIGO is an acronym for 'garbage in garbage out'. Quite obviously the accuracy of the results will depend both on the accuracy of the model and the accuracy of the input data. Most of the time models such as EVALUATE will be used as part of a forecasting system to determine what the problems might be in the future. As such a lot of the input data will not be known with any great degree of certainty, even if you want to analyse your current strategy much of the data needed will not be known with any degree of certainty. As a result some of the data may have to be estimated.

8.22   Forecasting, estimating and guesswork

The comments in this section of the EVALUATE manual about forecasting are taken from Sec. 1.12 with the same name in Chapter 1 of this book so they are not repeated here. One of the most important of the techniques that are referred to in this section is the use of modelling and other decision support tools.

This is what EVALUATE is all about. It will not give you absolutely precise numerical answers because you are unlikely to be able to supply it with totally accurate input data. What it will do, however, is save you a lot of time in making the large number of analyses that will be needed to determine the critical variables and observe the impact that changes in yield, or fault spectrum, or fault coverage, or whatever, will have on the life cycle cost of your products.

Another requirement for reliable results is to understand just what the various input parameters are and how the model uses them. The next section describes the input parameters in more detail.

8.23 The 'Board Data' form

The input parameters here are fairly straightforward. The number of each component type on the board, or an average across a group of boards, is input along with the expected defect rate in parts per million (ppm). EVALUATE then uses this data to calculate the average number of faults per board (FPB) for each of the component types. These are then summed into similar groupings and transferred to the 'Fault Spectrum' form. Since the defect rates are not likely to change much from board type to board type this form has two 'CLEAR' buttons, one to clear data from the 'quantity per board' column and one to clear the 'defect rate (ppm)' column. This enables you to retain the defect rate data when you switch to the analysis for a different board type.

The term 'parts per million' is actually an abbreviation for 'defects per million opportunities for a defect'. If you have 200 devices of one type on a board then you have 200 opportunities for a defect. A defect rate of 500 ppm would mean that each group of 200 components selected for a board would contain 0.1 defects $(500 \times 200/1\,000\,000)$.

8.24 The 'Fault Spectrum' form

The component faults determined on the 'Board Data' form are transferred to the 'Fault Spectrum' form, where you then add in the other defects that make up the fault spectrum. These are divided into three groups: design induced defects, supplier induced defects and manufacturing induced defects. The design induced group includes the following fault classes.

Design induced defects

COMPONENT INTERACTIONS

These are interactions between two or more components that will cause incorrect functioning of the board. Such defects are not usually detectable by in-circuit testers because of their 'divide and conquer' methodology of testing one device at a time. However, there are some testers that have the ability to test a small portion of the board as if it were one component. This is usually referred to as 'cluster testing' or 'function testing'. These testers would normally be referred to as being combinational testers since they are combining in-circuit and functional testing in one unit. This can be a bit of a grey area but it is worth remembering that a digital in-circuit tester is in fact a functional tester if it applies functional tests to the digital components. It simply applies a number of smaller test programs as opposed to one large test program. The fault coverage is likely to be quite low for this type of defect because only a limited set of component clusters are tested. A fully functional (edge connector) test is likely to have a better fault coverage for this type of defect, especially if it runs at a reasonably high speed.

FUNCTIONAL DEFECT (LOW SPEED)
These are functional problems, caused by a design error, that might be detectable with a functional test performed at a relatively low speed, relative to the normal operating speed of the UUT.

FUNCTIONAL DEFECT (HIGH SPEED)
These are functional defects, caused by a design error, that would only be detectable if the test was performed at or near to the operating speed of the UUT. Alternatively these defects might be detectable by a tester with good control over the timing of the input application and the sensing of the response.

OTHER DESIGN DEFECTS
These are any design induced defects you anticipate that do not fit into any of the above categories.

EVALUATE treats all of these design induced defect classes in the same way from the point of view of the calculation. Therefore you can assign your own definition to these four classes if you wish.

There is an important difference between design induced defects and the other two major groupings. Supplier induced defects and manufacturing induced defects are random in nature, but a design error will be present on all boards. Fortunately they rarely cause all boards to fail because they usually result in marginal behaviour. It is because of this marginal behaviour that these are the most difficult and most expensive of all defects to detect. Since component performance may change (drift) with time and temperature these defects, which frequently result from an incomplete understanding of a components performance limits, have a higher than usual probability of escaping to the field.

Supplier induced defects

BARE BOARD SHORTS
Short circuit problems associated with the blank printed circuit board as opposed to shorts caused by the assembly or soldering process.

BARE BOARD OPENS

An open circuit connection on the blank printed circuit board as received from the vendor.

WRONG COMPONENT TYPE

Wrong type or wrong family of device as supplied by the vendor. Mis-marked or wrongly coded components would fall into this category. This used to be a common source of problems but it is now less of a problem as component manufacturers have tightened up their internal procedures in the pursuit of quality. A particularly problematic fault of this type would be the wrong family type of IC. The device may well function in exactly the same manner as the correct family type but with lower or higher speed or drive capability.

OTHER SUPPLIER DEFECTS

Any supplier induced detects not covered by the above classes.

Manufacturing induced defects

These are all of the defects that result directly from the manufacturing process, including the loading of insertion and placement systems, the kitting of manually inserted components, the insertion and placement process, soldering, cleaning, handling, etc.

SOLDER SHORTS

Short circuits caused by the soldering process as opposed to any shorts on the blank PCBs as received from the vendor.

DIGITAL OPEN CIRCUIT

Open circuits associated with the digital circuitry on a board. Since opens associated with the blank PCB are listed in the 'supplier induced defects' section, these opens will be caused mainly by insertion problems (bent legs), placement problems on SMT boards (skewed devices), the soldering process (poor joints and 'tombstoning') and thermal shock (bond wire problems). While solder shorts tend to be one of the largest fault classes on traditional through-hole technology boards, open circuits tend to be more of a problem on surface mount technology boards. It is worth noting that very few board testers actually make tests specifically for open circuits. Opens are usually detected as a by-product of some other form of test. On an in-circuit tester the test of an individual component will show up an open circuit but the diagnosis may not be very precise. As a result, boards with open circuits may pass around the diagnosis/repair loop more times than they should because the wrong repair action was taken. A functional tester will detect an 'open' in a similar manner. The test results will be incorrect and the guided probe will isolate the defect to the relevant node and even to the correct pin. In some cases the probe will be able to measure the results on both sides of the 'open' and so make an accurate diagnosis, whereas an in-circuit tester will usually only have one test probe on each node. Bear in mind that similar board behaviour will occur if the open is on a track, at a via hole (a connection between track layers), at a component pin, inside a device, in the test fixture or inside the tester itself. Some testers have additional diagnostic tools to differentiate between these various possibilities. At a 'functional verification' (hot mock-up or hot bed) sub-assembly test the same problems arise. There is no specific test made for opens so they will show up as a functional fault. The one exception to the rule of 'no specific tests for opens' that is becoming more commonly used is boundary scan tests. These can be set up to make sure that the same logic states are present at all terminal branches of an electrical node. There is a lot of research going on to develop specific tests for opens, since this is a major limitation at present with the manufacturing defect analyser type of tester.[1]

ANALOGUE OPENS

Open circuits associated with the analogue circuitry on the board. Again, since very few testers have any direct tests for 'opens', these will be detected as part of some other measurement, either as a direct measurement of a component value, as in an in-circuit test, or as a functional malfunction for a powered-up test.

WRONG COMPONENT

The wrong component type or value placed on the board. This can occur in both manual and automatic insertion or 'pick and place' assembly. The wrong component may be included in the assembly kits or the wrong reel or stick of components may be loaded on to the insertion or placement machine.

MIS-ORIENTED COMPONENTS

Any component incorrectly inserted or placed on the board, e.g. reversed diodes, capacitors, transistors, etc., reversed ICs, misaligned or skewed surface mount devices, etc.

MISSING COMPONENTS

Any component or link that was not inserted/placed on the board prior to soldering, such as a surface mount device on the underside of the board that fell off during the soldering process because of improper application or curing of the adhesive, or because of poor silk screening of the solder paste, etc.

OTHER MANUFACTURING DEFECTS

Any manufacturing process induced defects that do not fall into any of the above.

8.25 The 'Production Data' form

BOARD VOLUME PER YEAR

The total annual production volume of the board or boards that are the subject of this particular analysis. If there is a substantial number of field service returns to the

---

[1] At the time of writing several manufacturers have introduced 'opens' test techniques. These can currently be grouped into three areas. Some measure the parasitic diode or transistor present at the input to a device, some measure the capacitance between the lead frame and a probe plate on top of an I-C, and some measure if current flows when pins are driven.

factory, these could be included here also. Alternatively these could be the subject of a separate analysis.

IS THIS THE ENTIRE PRODUCTION VOLUME?

Here you simply click on a 'YES' or 'NO' 'radio button'. EVALUATE uses this information to determine how to allocate the capital cost of the test stages. If this analysis is for the total annual production volume then all of the capital cost of the equipment needed, rounded up to the next whole number, is included. If this is not the total volume then only the proportion of capital equipment required is counted. For example, if 1.5 testers are required for the board test stage at a cost of $200 000 each then the 1.5 would be rounded up to 2.0 and the total cost of $400 000 would be added to the capital cost if this analysis was for the total annual volume. The available capacity on the tester would be based on the 0.5 of a tester that is available, and the calculated throughput. If the analysis is not for the total annual volume then only the capital cost of 1.5 testers would be allocated since it is assumed that there will be other board types to test. There is some potential for error here since not all of the capital cost will be allocated unless the total volume requires an exact whole number of test stages (which is unlikely). This is not really critical since the capital cost calculation is only intended to be a comparative figure when comparing alternative strategies or tactics. If a return on investment (ROI) is calculated based on the cost differences between two strategies, then the actual investment capital cost for the various alternatives would be used for the calculation.

NUMBER OF SHIFTS OPERATED (EACH DAY)

Usually either one, two or three shifts.

AVAILABLE HOURS PER SHIFT PER YEAR

This figure should be the actual available time for each shift per year. It would typically be calculated in a similar manner to the following examples.

Example A:

$$(7 \text{ usable hours/day}) \times (20 \text{ usable days/month}) \times 12 \text{ months} = 1680 \text{ h}$$

Example B:

$$[365 \text{ days} - 104 \text{ days (weekends)} - 20 \text{ days (vacations)}] \times 6.5 \text{ h} = 1566.5 \text{ h per shift}$$

Allowances can be made here for lost time and downtime, so if the shift is actually 8 hours this figure can be reduced accordingly. In the examples it is reduced to 7 hours and 6.5 hours respectively. Another reason to reduce this time might be for frequent program changes on the testers. With many operations moving to JIT, with its inherently small batch sizes, this becomes an increasing source of lost time. For example, changing the program and fixture ten times during a shift at ten minutes per changeover will lose 1 hour and 40 minutes of production time. The loss of production test and diagnosis time that results from test program preparation and debug is accounted for automatically as a result of other input parameters.

LABOUR COST/HOUR FOR Xxxxxx

There are five labour rate categories for test engineers (programmers), ATE operator, sub-system test engineer, final system test engineer and repair technician. There is also a warning on the form against using fully loaded labour rates when performing an economic analysis. This is a fairly common mistake that even appears quite frequently in articles, technical papers and application notes on the subject. Most companies determine their loaded labour rates for manufacturing operations by 'loading' or 'burdening' (USA) the direct labour costs with all of the overheads that relate to the manufacturing operation and some of the company-wide overheads. This is done, quite correctly, to determine the true cost of operations and to establish a minimum selling price for a product. However, in most cases it is quite wrong to use this loaded cost when determining the cost savings that would be made by introducing a new piece of equipment or a new test strategy. Indeed, for any analysis that will result in the calculation of an ROI it would probably be incorrect and will of course overstate the savings and therefore the return.

The reason is quite simple. The changes being considered will usually only result in direct labour savings and will not save any of the overheads. Replacing some old board testers with one that has five times the throughput will obviously save direct labour costs, but it will not save any rent or leasing costs on the building. It will have no noticeable effect on the cost of heating, lighting and other utilities, or on the cost of the supervisory staff and the management. It will probably have no impact on any of the other elements that go to make up the overheads. For an accurate assessment of the savings you should only load the direct labour costs with those cost elements that would indeed be saved if the people involved could be re-deployed to other activities. There are occasionally some exceptions to this general rule but they are quite rare.

8.26 The 'Test Stage Data' forms

There are two such forms which collect data about some aspects of the performance of the various test stages and the test programming workload.

GOOD BOARD TEST TIME

This is the complete test time tor a defect-free unit, including the handling time required to load the UUT on to the tester and to remove it after the test is completed. Any other time that the operator or the test engineer needs to fill out forms, etc., should also be included. For sub-system and final system test stages where more than one board may be tested at a time, the time per board should be determined by dividing the total time by the number of boards present in the UUT at these points in the test process.

DIAGNOSIS TIME FOR A SHORT

This is the average total time to perform the tests and diagnose the presence of a short circuit including the handling time, etc. Testers that make a specific test for shorts, such as MDAs, ICTs and testers with a boundary scan capability, will usually perform the shorts test first. Since this is usually a fairly quick test, and since the tester will usually be set up to stop testing if a short is detected, this time will be less than for other defect types. For example, an in-circuit tester with a thirty second test time and a thirty second handling time will have a 'good board test time of sixty seconds'. If the shorts test takes only two seconds of the thirty second test time then the time to diagnose a short, including handling, would be thirty-two seconds. The reason for this 'stop on shorts' approach is that the presence of the short will impair the tester's ability to diagnose other defects in the vicinity of the short. Taking a simple example, the short may be directly across a component whose value will be measured. This will make such a measurement impossible and may result in two diagnostic messages for the same fault: one to identify the short and one to say that the component has the 'wrong value'. The presence of a short could also cause additional defects to occur when the board is powered-up for the active tests. It is important to differentiate between shorts and other types of defect because this time difference will affect the average test time per board. This in turn will affect the costs and the throughput calculations.

DIAGNOSIS TIME FOR OTHER DEFECTS

This is the average diagnosis time for all defects other than shorts. On an MDA or an ICT this will usually mean completing the entire test program. Functional testers (FNT) do not usually differentiate between these defect types in that all defects will be detected as a result of incorrect functional behaviour. Unless the functional tester has any special shorts test capability it will take the same time, within reason, to diagnose all fault types. In actual fact most functional testers will stop testing at the first detected fault. That defect would then be repaired and the board would be re-tested. If a second fault exists this will now be detected at some test later in the program relative to the first defect detection. If the run-time of the program is lengthy there will be some defects that will take more time than others to detect. However, unlike an MDA or ICT, this is relatively unpredictable so an average time has to be used. In most cases the run-time for an FNT will be considerably shorter than the handling time. If it is not, then it might be reasonable to take an average value for the diagnosis time in the following manner.

As an example, on a functional tester the handling time for each board averages thirty seconds. The test run-time (excluding the handling time) for a defect-free board is twenty seconds. Assuming that defects occur randomly some will be detected early in the test program and some will be detected towards the end. The simple approach might be to average the run-time for diagnosis to ten seconds. In this case the average diagnosis time, including handling, will be forty seconds.

FUNCTIONAL DIAGNOSIS TIME (ONLY ON THE 'TEST STAGE DATA 2' FORM)
This parameter only appears on the 'Test Stage Data 2' form. It enables the diagnosis time to be specified for defects detected in the functional test portion of a combinational test program, as opposed to defects detected in the in-circuit test part of the program. Some combinational testers have a functional test capability but no automated diagnostics. If this is true for the tester in the analysis then set this time to zero. The diagnosis would then be performed 'off-line' of the test system.

BOUNDARY SCAN DIAGNOSIS TIME (ONLY ON THE 'TEST STAGE DATA 2' FORM)
If the tester has a boundary scan test capability then any diagnosis time for defects detected by the boundary scan test should be included here. The boundary scan test time for a defect-free board should be included in the 'Good Board Test Time' parameter. Again, if there is no automated diagnostics for the boundary scan test then set this parameter to zero. This diagnosis time applies specifically to any connectivity tests performed using boundary scan. The fault coverage for these tests should be entered in the FNT column of the 'Fault Coverage Matrix 2' form, for the relevant COMB tester (either COMB-A or COMB-B, or both if they both have this type of boundary scan test), for all of the 'opens' and 'shorts' defect types. The reason for this is that most current use of boundary scan testing is to determine that all connectivity is correct on the board. If the boundary scan tests are also being used to test individual components or functional blocks then this can be treated as a functional test and the relevant fault coverage data should be entered into the FNT column of the COMB-X tester for the relevant defect types. Any defects not tested for should have their fault coverage data set to zero.

OFF-LINE DIAGNOSIS TIME (ONLY ON THE 'TEST STAGE DATA 2' FORM)
If the functional test or boundary scan test capabilities of a combinational tester do not have automated diagnostics then any defects detected during these tests will need to be diagnosed away from the tester (off-line).

REPAIR TIME FOR A SHORT
This is the average time required to find the physical location of the short and to repair it. This input parameter is one that is frequently underestimated because it can sometimes take a long time to find the short. Removing it usually presents less difficulty. Bear in mind that the tester will usually only identify the electrical location of the short, not its physical location.

REPAIR TIME FOR OTHER DEFECTS
This is the average time to repair defects other than short circuits.

DIAGNOSIS/REPAIR LOOP NUMBER

This is a correction factor to account for the fact that not all repair actions are necessary. For every diagnosis there will be a repair action and for every incorrect repair action there will be an additional diagnosis and yet another repair action. When a board is tested there are four possible outcomes of the test:

1. The test may pass because there is no defect present.
2. The test may pass because a defect that is present is not detected.
3. The test may fail because a detect is detected.
4. The test may fail even though there is no real defect present.

The diagnosis/repair loop number is concerned with the fourth of these possible outcomes, with the fact that the diagnostic message may be imprecise or mis-interpreted and with the fact that the repair action may create another defect. For these reasons more diagnosis and repair actions will take place than the theoretical number that would be calculated from the yield data. In general MDAs and ICTs are more likely to indict a defect that does not exist due to some of the limitations of how they isolate the device under test from the other devices connected to it. The analog guarding techniques have their limitations and if you are operating near the limits then measurements may be in error enough for the device to appear to be out of specification when it is not. The absolute range and accuracy of the measurement techniques will also affect the performance of the tester in this area. On the other hand, the diagnostic resolution and accuracy will tend to be better for an ICT than for an FNT. Therefore it is more likely that the FNT diagnosis will lead to an incorrect repair action. It is important to try to estimate this loop number and not to ignore it because it will affect the cost and the capacity results.

NUMBERS OF TEST PROGRAMS PER YEAR

This is the number of totally new test programs or test procedures that will have to be developed for each of the test stages. For board test stages a test program will be required for each board type, but for the sub-system test and the final system test stages (SST and SYS) the number of test procedures that have to be written will be a function of the number of boards in the sub-assemblies and in the final product. For example, if you have to develop test procedures for three new products that contain 21 board types in 7 sub-assemblies then you would enter 21 for board test stages, 7 for sub-assembly test stages and 3 for the final system test stage. This assumes that the sub-system test stage tests each sub-system in its entirety. If this stage effectively tests one board type at a time, with the other boards in the sub-assembly being 'known good units' then you may have to develop 21 test procedures for the SST stage also.

TEST PROGRAM DEVELOPMENT TIME

This is the average time taken to develop the test programs excluding the time taken to debug the test program on the actual test system or test stage. This time is considered in the next input parameter so that EVALUATE can see how much tester time is taken up with program preparation.

TEST PROGRAM DEBUG TIME (ON TESTER)
This is the amount of time required to debug the test program and the test fixture on the tester (or the test stage in the case of the manual test stages). This time is lost time from a production testing point of view.

NUMBER OF ECOs PER YEAR
This is the average number of engineering change orders expected in a year. In general each engineering change will require that the test programmer examines the test programs and test procedures to see if their effectiveness will be adversely affected by the design change. If it is then the programs, and possibly the fixtures, may have to be modified.

TEST PROGRAMMING TIME FOR AN ECO
The average time spent examining and modifying test programs and test procedures as a result of an ECO.

TEST DEBUG TIME FOR AN ECO
The average time spent debugging the results of a program or fixture change on the tester. This will be lost production time as far as the tester is concerned.

TEST FIXTURE COST
The cost of the fixture kit, the design and the construction of the fixtures. Fixture test and debug time should be included in the 'test program debug time' (see above) so that any lost production time on the tester is taken into account.

CAPITAL COST FOR TEST STAGE
This is the capital cost for one tester or test stage for each of the types under consideration. For the commercially purchased board testers this will be the purchase price of the systems. For the SST and SYS test stages this will be the cost of any commercial test equipment required, such as oscilloscopes, DVMs, logic analysers, VXI equipment, etc. If the development cost for a VXI-based solution is to be capitalized then this should also be included here. If not, then the cost could be included as programming time. Be careful of including any one time costs in with the fixturing costs because this will be multiplied by the number of fixtures needed. The calculation of the capital cost for each test strategy presented on the 'Results' form is only intended to be a comparison guide between the test strategies. Once the short-listed strategies and tactics have been decided on the true capital costs will need to be determined more accurately for the development of an ROI. For example, there may be other capital costs involved that cannot be applied to the EVALUATE analysis. One example of this would be the cost of air conditioning equipment that may be needed for the test facility.

AVERAGE COST OF A FIELD-SERVICE CALL

The average cost of a field-service call can be calculated or estimated in a variety of ways. For some purposes you may well need the actual direct cost of the average service call in terms of the engineer's time cost, the cost of materials and the cost of travel and lodging expenses if appropriate. However, when looking at the broad effects of a test strategy it is necessary to include much more in the cost calculation. This cost is used to determine an important part of the life cycle cost of a particular decision on test strategy. If one strategy allows twice as many defects to escape to the field then you can expect that over time much of your field-service costs will double. This is a major part of the cost of quality and one approach to this that is widely used is to make the assumption that if your quality was perfect then there would be no requirement for any form of field-service organization. The common way to determine the average cost of a field-service call is therefore to determine the total cost of the field-service operation and then to divide this figure by the average number of calls made in a year. In this manner the overhead elements of the field-service operation are also taken into account. A large part of this overhead is the inventory of spare parts that are needed to support the operation, including those parts that are in the 'pipeline' going back to the service depots or the factories for repair. The problem with using this total cost approach is that it will not necessarily reflect the short-term savings to be made by improving the field quality because of the inertia of such an operation. You cannot halve its size overnight even if the quality doubles overnight. Thus, in many cases it will be appropriate to take some intermediate figure between the actual direct cost of a call and the fully loaded cost. There is an important difference here between the inclusion of the service overheads and the non-inclusion of manufacturing overheads as discussed in the comments about the 'labour cost/hour of Xxxxxx' parameters. For a manufacturing operation you cannot possibly eliminate the manufacturing overheads. You can try to minimize them, but you cannot manufacture without space, electricity, machinery, supervisory staff, management, etc. JIT (just in time) methods have reduced inventory levels drastically, but they are not always eliminated. Even in an ideal world there will be some level of manufacturing overheads. However, in this ideal world there could be zero field-service overheads. Some quality gurus argue quite convincingly that regardless of your quality levels there is always room for improvement and the improvement will save costs. If we could reach perfection, field service could be eliminated in just the same way that many test and inspection operations have been.

This then raises the question about the possible elimination of the manufacturing test. Will we ever be able to produce complex products without the need to test them? The advances that have been made in quality over the past twelve years have been tremendous, but so have the increases in complexity. If we can continue with the same rate of defect elimination that we have experienced it might be possible to eliminate everything except the final test stage, but there are problems with this approach. If complexity and customer expectations about quality continue to rise, then it may be difficult to develop a final test procedure that is good enough to verify that the system is defect-free. The final test procedure as typically implemented today can have a rather poor fault coverage (see the comments about the fault coverage of SST and SYS in the 'fault coverage matrix' parameter in the next section), so the only real hope for this kind of utopia lies in further development of built-in self-test capabilities. However, this simply increases the complexity and also has to be tested. It will be interesting to see how things develop over the next ten to twelve years.

8.27   The 'Fault Coverage Matrix' forms

The 'Fault Coverage Matrix' forms are used to define the remaining operational effectiveness of the various testers and test stages. Completing these forms is likely to be the most difficult part of the analysis requirements, but it is an important area because it establishes, more than any other element, the life cycle cost of the product or products under review. The cost of testing can be split into two main areas. The cost of all the testing we do in the factory before we ship the products can be thought of as the cost of conformance of the test strategy. This is what it costs to make sure that the product really does conform to its design specification. The cost of supporting the product after it has been shipped to the customer can be thought of as the cost of non-conformance, the cost of correcting problems that we failed to eliminate earlier in the process. This concept can be extended to look at the effectiveness of individual test stages. For example, if your test strategy is ICT + SST + SYS > FIELD, then when you are comparing individual test tactics, such as which ICT to use, you can consider the operational cost of the ICT as the cost of its conformance and the cost of all other test stages as the cost of non-conformance for the ICT. Ideally we need to minimize the cost of conformance and the cost of non-conformance, but the real issue is to minimize the sum of the two, the life cycle cost of the test strategy. This can often only be achieved by spending a bit more on the cost of conformance since the cost of non-conformance will usually be the larger of the two elements. A simple example will clarify this and also introduce an important concept.

Example

You have the choice of two testers. One has a fault coverage potential, relative to your current fault spectrum, of 90 per cent. The other, more expensive, tester has a potential for 95 per cent fault coverage. The usual view of this choice is that one system is 5 per cent better than the other but it is possibly 30 per cent more expensive to buy. It does not sound like a good deal. However, the escape rate from the better tester is 5 per cent as opposed to 10 per cent from the cheaper system. That is a 100 per cent improvement. This is the important concept referred to above. We need to think in terms of 'escape rate' rather than 'fault coverage' when comparing testers and test strategies. The life cycle cost, and certainly the cost of the non-conformance part of it, will be determined heavily by the escape rate. In this simple example we can expect to see a two-to-one difference in diagnosis and repair costs at the following test stages, and a two-to-one difference in field-service costs for defects other than early life failures.

The important thing to remember here is that the choice of which ICT to buy is unlikely to affect the way that the SST and SYS test stages are set up. Therefore we can assume that the fault coverage of these stages will be the same regardless of which ICT is selected. As a result the two-to-one difference in escapes will pass through to the field. If the following stages detect 90 per cent of all remaining defects then the field escapes would be either 1 or 0.5 per cent depending on which ICT is used. If the following stages detect 99 per cent of all remaining defects then the escape rate would be either 0.1 or 0.05 per cent. This is still two to one.

The effective overall fault coverage for any test stage will be a function of the performance of the tester and the fault spectrum. For a commercially available test system the achievable fault coverage will depend on the performance of the testing hardware and the software for both the testing functions and the test program generation. The hardware will determine what the maximum possible coverage could be but the software will determine how likely it is that you will reach the theoretical limit set by the hardware. How far you get will depend on how far any automatic program generation can take you, how easy it is to improve on this, how much time you can devote to the job and the skill of the programmer. The tester's software will also determine what level of skill the programmer needs to have. Another major issue that will affect how far you can get in a given time will be the availability of library models. In-circuit testers require libraries of tests for passive devices, linear devices and digital devices. Functional testers that rely on simulation techniques for their program preparation and diagnostic data also need device libraries, but of a different type. The completeness and the robustness of the libraries is of major importance. Unfortunately, this is also one of the key areas where the enthusiasm to sell the products can lead to some creative ways of counting the number of devices in the libraries. All of these are areas where commercial testers vary considerably. It is tempting to assume that all testers of a given type are very much the same. Nothing could be further from the truth, but the differences, although important, tend to be subtle. The performance differences are often the result of many small differences rather than one major difference. In order to develop an understanding of the potential differences in the fault coverage that can be achieved, it is necessary to understand how the competing testers work in some detail. However, this is likely to be a requirement for the evaluation of any major investment project.

The fault spectrum will influence the achievable fault coverage because different tester types will have different capabilities when it comes to different defect classes. If your fault spectrum consists entirely of solder shorts and fairly basic manufacturing defects then the actual fault coverage will be similar for an MDA, an ICT, an FNT or a combinational tester. However, as the fault spectrum becomes more diverse the MDA will be left behind. Then the ICT will be left behind, followed by the FNT and so on. As you introduce more exotic defects such as design induced timing faults you have to look at more sophisticated testing strategies. Therefore a good understanding of your actual or predicted fault spectrum is essential for any thorough analysis. It is also necessary that the analysis model has the ability to differentiate between some reasonable variety of defect types in order to see the performance differences between alternative strategies and tactics.

Unfortunately not many companies are fully conversant with their fault spectrum. A lack of fault coverage for certain defect types can lead to a belief that these defects do not exist. When they go on to cause a failure in the field there is either an assumption that this was an early-life failure that could not have been detected in the factory or the information about the defect never gets back to the manufacturing department. Ignorance is bliss. What we do not know about does not hurt us. But it does. Every field failure within the warranty period costs a lot of money, and every failure in the field at any time costs us even more in customer satisfaction, quality reputation, customer loyalty and the risk of 'product liability' litigation. Product liability litigation is when you get sued because your product caused some personal injury or possibly even death. The awards for such cases in the United States typically run into the millions or even tens of millions of dollars, and these cases are on the increase in Europe. The whole issue was highlighted some years ago in the United States when a lady sued the manufacturer of a microwave oven because it killed her pet dog when she put it into the oven to dry it off! She won the case because the judge ruled that there should have been a warning in the operating instructions about the dangers of microwave radiation. This was not even a problem with the product but merely the documentation that went with it.

A few years ago we were involved in an analysis for a company who felt that they had very few component problems—just a couple of hundred defects per year. However, a visit to a sister company who were still performing incoming inspection showed a range of defect rates of between 200 and 500 ppm. Both factories were obtaining their components from the same vendors via a common purchasing department. These defect rates translated into 2300 defective components per year at their current consumption rates. Further investigation showed that the testers used by their sister company for their incoming inspection were fairly old and relatively unsophisticated. Therefore the fault coverage that they were achieving was probably not very high. The real number of defective components was possibly in excess of 4000 per year.

What happens to the undetected defects? The simple answer is that most of them escape to the field. Fortunately they do not all cause the products to fail since the designs may tolerate some marginal performance. However, many of them will cause failures, and many of these could have been detected in the factory. While on the subject of design tolerance to out of specification devices, consider the following. The more tolerance the design has, the less likely it is that the defective device will be detected in the factory if there is no incoming inspection. If the device's parameters drift, then there is a reasonable chance that it will cause a field failure at some time. The failure mechanisms of many devices do result in drifts. There is a wealth of data on the effects of stress screening programs to prove this.

Design induced defects

The degree to which this class of defects occurs will largely be a function of the complexity of the design, the newness of the technologies involved and the design methodologies utilized. Companies that have well-established design procedures using the leading edge of design automation technology may well have very few defects of this type. Their ASIC designs will be thoroughly proven with 95 per cent plus fault coverage test programs verified with fault simulation and worst case timing analysis. They will be using similar techniques as much as possible for the design of the rest of the board to ensure compatibility between the board, the ASICs and the firmware. Design and test integration will be well established and a concurrent engineering process will be operating with effective teamwork between design engineers and test engineers which results in better test programs that are available faster than ever before. Unfortunately there are not very many companies that fit this description. The 'over the wall' approach of handing over a completed design to test engineering and then getting on with the next untestable design is still far too commonplace. All too often ASIC prototypes and board prototypes do not work together when they are first integrated and designers are put under such pressure to get the job done that they overlook or misinterpret a key element of a component specification. The inevitable errors that creep into a new design manifest themselves as incorrect behaviour caused by interactions between components, tolerance accumulation and marginal or intermittent behaviour. As stated earlier, a design mistake will be present on all boards, unlike the other defect groupings that will occur in a random manner. Fortunately these design defects, because of their nature, will not cause all boards to fail but the same comments about the way that component parameters can drift with time (and temperature) will apply. This type of defect along with a.c. parametric failures of devices, which manifest themselves in a similar manner, are the most expensive to find and the most likely to go undetected in the factory.

Supplier induced defects

The defect rates of high-volume commercial components, often called merchant devices, have improved in leaps and bounds over the last twelve years. The component manufacturers were the first of the Western electronics companies to suffer from the threat from Japan. It was this that spurred on the beginnings of the quality revolution in the early eighties. The improvements have been so great in some areas and this improvement has received so much publicity that there is a danger of assuming that it is the same for all component types. Unfortunately this is not the case. Certain classes of components that are made in smaller volumes can have much higher defect rates than the high-volume and less-complex devices. Custom devices, customizable devices and mixed signal devices are all classes that can have relatively poor defect rates. Even some relatively high volume LSI and VLSI devices can have defect rates measured in per cent rather than ppm for the fairly strict incoming inspection procedures of military equipment manufacturers.

Manufacturing induced defects

Tremendous improvements have also been achieved in the defect rates of the various manufacturing process steps such that we have seen major improvements in process yields even as complexity has risen. There has been a large amount of education on quality, what it means, what it does for you and how best to achieve it. This has taken place at all levels in the hierarchy. Process control charts, statistical process control (SPC) techniques and new analytical methods to find the cause of quality problems are all in common use, but again there are almost as many exceptions as there are practitioners of the new quality ethic. Far too many companies pay lip-service to quality. They talk about it constantly because they know that it is right, but they do not commit themselves to it as strongly as they should, especially when fighting for survival in a recession. As Edward DeBono says, 'There is never a good time to implement a good idea.' Therefore we still get manufacturing defects. The robotic 'pick and place' machines that have such low defect rates are still loaded with the wrong components due to human error and the vastly improved soldering systems still fall foul of the continuously reducing dimensions between component pins and tracks.

8.28   An example analysis using EVALUATE

Figures 8.24 to 8.35 not only show the layout of the forms they also contain an actual analysis that was produced as an example for the user manual. The example shows how EVALUATE can be used to compare the effectiveness of two alternative testers. It is assumed that a preliminary analysis has been performed to compare alternative test strategies and that it has been decided that an in-circuit test, followed by a sub-system test and then a final system test will be the most appropriate strategy. Figure 8.24 shows the board data defining the complexity of the board and the defect rates of the components. These defect rates are fairly arbitrary. They are used for the example only and may not be representative of current quality levels.

The sub-totals for the various component defects that are determined on the 'Board Data' form are brought forward to the 'Fault Spectrum' form shown in Fig. 8.25. The other defect types are added here in three groupings: 'Design Induced Defects', 'Supplier Induced Defects' and 'Manufacturing Induced Defects'. Once all of the data has been entered, the overall yield is calculated and displayed. This acts as a quick check that the defect data has been entered correctly.

Clicking on the 'Next Form' box brings up the 'Production Data' form shown in Fig. 8.26, where you enter information about production volumes, available time and labour costs for the various personnel involved. A message on this form warns against the dangers of using fully loaded labour rates when making calculations that will be used to determine cost savings.

Cells that are highlighted, with their contents left justified, such as the 'Component Interactions' cell on the 'Fault Spectrum' form and the 'Board Volume Per Year' cell on the 'Production Data' form, are the active cells that can be edited. When entering data into a form whose cells have been 'cleared' the cursor will move around the form as you complete each cell. If the form has not been 'cleared' and you just want to edit a few cells then you can use the mouse to click on the cell you wish to edit.

The 'Test Stage Data' forms shown in Figs 8.27 and 8.28 are used to define some of the performance parameters of the various test stages, to define the programming workload and to specify the capital cost of the equipment needed. The first of these two forms collects data about two 'in-circuit' and two 'functional' board test systems. These are generic tester types and can be defined to be anything from the simplest MDA to the fastest and most sophisticated testers that use either the in-circuit or the full functional testing techniques. Variations in the ease of program generation and debugging between the testers, as well as their test and diagnosis speeds, are defined here. The diagnostic accuracy and resolution also varies between testers, and this can be defined by varying the repair times and the 'Diagnosis/Repair Loop Number'. The second of these two forms, shown in Fig. 8.28, collects data about the combinational testers and the sub-system, and final system test stages. Data for the combinational testers is not actually needed for this analysis, which is simply a comparison of two ICTs, but is included for completeness. The average cost of a field failure is also entered here so that a more complete life cycle cost can be calculated.

The 'Fault Coverage Matrix' forms define the defect detection performance of the various test stages. The first form shown in Fig. 8.29 enables definition of the performance of the in-circuit and the functional testers. Since this particular example compares the performance of two ICTs the data for the FNTs could have been left blank. In practice it will be normal to open the file for the strategy analysis that would have been performed first and then to edit the data for the specific testers of interest.

Notice that the ICT-B is defined as having a better fault coverage for a number of the defect types. This reflects this system's better performance in terms of its measurement capability, its program generation facilities and the more comprehensive device library. The 'Fault Coverage Matrix 2' form is not required for this particular analysis but is shown for completeness in Fig. 8.30. The 'Fault Coverage Matrix 3' form defines the performance of the sub-system test stage and the final system test stage. Since the main purpose of this exercise is to compare the two alternative ICTs, the fault coverage of these stages has been set to an arbitrary value of 90 per cent for all defect types. For a more precise analysis of the cost of the strategy, some thought should be given to the fault cover that could be expected for each of the individual fault classes because this will have a bearing on the performance of the strategy.

The 'Test Strategy Selection' form shown in Fig. 8.32 is used to select the strategy for which you want to see the results, and operates by clicking on one of the buttons on the left-hand side of the form. This will then take you to several forms that show the results of the analysis for the selected strategy. If the input data has been provided for all of the tester types then you can return to this form to select another strategy and view or print the results for this new strategy also. All strategies are calculated each time that one or more of the input parameters is changed. Strategy results can also be accessed directly using the **Form|Select** pull-down menu.

The first of the results forms seen will be one of the 'Defect Detection Matrix' forms shown in Fig. 8.33. These show a summary of how each of the major defect groups are detected as the boards pass through the test process. The column headed FPB (faults per board) lists the defects coming into the first test stage. This is followed by a listing of the detected defects and then the escaping defects from that stage. The escaping defects from the first test stage become the input defects to the next test stage, where some of these are detected and some escape. This process continues throughout the strategy and is indicated by the overlapping lines beneath the abbreviations of the test stage names above the table.

The yield at each stage of the process is calculated and shown at the bottom of each column. For the 'FPB' column and for all of the 'Escapes' columns, this is the actual yield computed from the actual number of defects present. For the various 'Detected' columns this is the apparent yield (Y_a) as seen by the test stage. This will be higher than the true yield input to the test stage because the test stage does not detect all of the defects and therefore sees a yield that is apparently higher than the true yield.

The number of escaping defects from each stage is also summed and displayed, as is the actual overall fault coverage for the stages based upon the number of defects detected versus the number that entered the stage. The strategies that give the option of directly comparing alternative testers, such as the ICT-SST-SYS strategy shown in this example, have a second 'Defect Detection Matrix' form for the second tester, and this is shown in Fig. 8.34.

The final form displayed for each strategy is the 'XXX Results' form. In this example it is the 'ICT Results' form shown in Fig. 8.35, since each strategy name is abbreviated to the name of the board tester used. This form provides the economic information about the strategy in the form of cost, throughput and capacity data. The test, diagnosis and repair (TDR) costs are shown separately, as are the costs of program preparation, fixtures and ECOs. The field-service cost derived from the escape rate from the final test stage and the cost per service call can be viewed as a measure of the quality of the strategy relative to other strategies.

The key results in this analysis relate to the superior throughput rate of ICT-B relative to ICT-A. ICT-B can cope with the workload with only a 42 per cent utilization whereas ICT-A needs 104 per cent. This obviously requires two systems or a second shift. Even with two systems or two shifts there is less available capacity compared to that with ICT-B. For the calculation of the capital cost for the strategy, EVALUATE will indicate the cost of two ICT-B testers because the 'Production Data' form specified that only one shift will be operated. Since there is a need for only 4 per cent additional time on the ICT-B tester this could easily be accommodated with a small amount of overtime. However, there would be no available capacity if only one system is purchased. If the workload is fairly consistent this may not be a problem, but if the workload varies much there may be periods when this system might cause a serious bottleneck. In any event, there is a difference in annual operating costs of $93 025, which would make it easy to justify the extra $50 000 for ICT-B.

9. Evaluating automatic test equipment

Chapter 8 discussed the economic analysis of test strategies, the test tactics, and the development of analysis models to help with this work. The test 'tactic' analysis work only differs from the test strategy analysis in the level of detail required. Essentially the model will need to be more detailed in order to see the economic effects of differences in the performance of the competing testers that you may be considering. This chapter supplements the last one with additional discussion of the evaluation process for capital equipment. The suggestions made here apply equally to other types of equipment as well as ATE but the examples and the references are drawn from this specific class of equipment.

Once a decision has been made to purchase an automatic test system the next step is the evaluation of available equipment. This can be a nerve-racking task for anyone, especially if there is no previous experience of buying expensive capital equipment. There are three major reasons for this:

1. It usually involves a large investment so there will be a need to get the maximum return on this investment in order to satisfy your senior management.
2. There are many factors involved so it can be a very complex problem to analyse. As well as the ROI you also have to consider such things as capacity, programming, support, quality, diagnostics, etc.
3. Your production output will depend on the test system you choose so there is a big risk of not meeting time to market and production schedules if you make a bad decision.

All of this means that the evaluation process can be a fairly detailed and lengthy operation. The complexity of the task will depend on the purchase situation involved. Four such situations can be identified:

1. Buying a system for the first time.
2. Buying a system from a new (unfamiliar) supplier.
3. Buying a type of system that is new to the company.
4. Buying a second or subsequent system similar to one already in use.

Buying a system for the first time is the most difficult situation, especially if none of the people involved have any experience gained from other companies. Apart from the lack of experience, the problem is made worse by the fact that upper management will need a lot of convincing to part with such a large amount of cash when it has not made similar investments before. There is also a high degree of personal risk involved in this situation. Companies do not decide to buy ATE; people decide to buy ATE. Making a good decision will often have a positive effect on the career of the person involved. Making a poor decision can have the reverse effect. It is not surprising then that the evaluation of ATE is often a very long, very cautious and very costly exercise. I know of several cases where the evaluation took so long that the project the ATE was being purchased for was delayed since there was nothing to do the testing with.

Buying from a new supplier is less of a problem than buying your first system, but still quite a major undertaking. It does not happen often since there is a very high cost associated with changing suppliers, but when it does it requires almost as much care as the first-time buy situation.

Buying a kind of system that is new to the company presents its problems due to unfamiliarity with the technical and economic aspects of the system. The situation is usually complicated by the fact that you can look at all available systems on the market, since there is usually no strong reason to use the same supplier as the one who supplied your existing ATE.

A good example here is the large number of companies purchasing an in-circuit board test system when they already use functional testers. With the possible exception of considering quality data collection requirements, there is no real need to buy your ICT system from the same company that supplied your functional systems. Buying another system like the one you already have, just to increase capacity, is obviously the easiest of the four situations described. The purchase will still need to be justified properly but the evaluation phase of the purchase is not required.

9.1 The evaluation team

More often than not it will be a good idea to have a small team of people to perform the evaluation, not just because there is safety in numbers but mainly because you will need people with different expertise to do an efficient evaluation. Usually, the team leader should be the person who will ultimately decide what to buy and be responsible for making it work when it arrives. It is a good idea to include someone who will be doing the programming work and possibly someone from engineering who is familiar with new designs and their testing needs. A two- or three-person team is probably optimum.

9.2 Risk

Most business decisions are based largely upon guesswork. It may often be called something sounding a little more scientific, such as forecasting, estimating, guesstimating, deducing, budgeting, calculating, etc., but it boils down to being guesswork of some kind. Hopefully, the guesswork will be done with as much supporting evidence as can be obtained in the time available, but usually many of the relevant facts will be unavailable. The point about all this is that there will always be a degree of risk or uncertainty attached to any decision.

Strictly speaking, there is a difference between risk and uncertainty. If we know that the probability of an event occurring is 0.7 and we are relying on that event occurring, then the risk is 0.7 and is known. If we do not know the probability of occurrence then we are operating under a condition of uncertainty. For most purposes, however, including ours, the two terms can be assumed to have the same meaning and be interchangeable.

It will never be possible to eliminate totally the risk from an ATE purchase decision. It should be possible, however, to minimize the risks if we know what risks are involved. The risks can be broken down into the following five categories:

1. The cost risk.
2. The technology risk.
3. The support risk.
4. The reliability risk.
5. The vendor credibility risk.

The cost risk

This is really what the previous chapters have been all about. Here the 'cost' is the overall life cycle cost rather than the purchase price. In fact, most of the examples in the earlier chapters have shown that the cost of the equipment often pales into insignificance when compared to the cost of running an ATE system. The major factors affecting costs will be the technical features of hardware and particularly of software that lead to efficient operation in set-up, testing, diagnosis and repair. Additionally, the support capabilities of the vendor will have a big impact. System downtime is expensive in terms of lost throughput, and a test program delayed due to poor support can be equally costly.

The technology risk

The users of ATE need some assurance that their supplier can keep up with the rapidly advancing component technology. The ability of a supplier to do this will be a function of:

1. The present capability of the equipment.
2. The commitment of the supplier to the market.
3. The technical capabilities of the supplier.
4. The amount of money invested in research and development (R&D).

Probably the most important of the factors mentioned above is the R&D spending since this will illustrate the commitment and make the technical capabilities available. Most companies in electronics spend 7 to 12 per cent of their revenues on R&D. Other things being equal, the company with the bigger market share can spend more on R&D. However, spending the money is no guarantee of future capability. If its marketing people make poor decisions about what the market really needs, it will still not have the right solution to your testing problems. It is important then to look at the 'track record' of the suppliers as well as their size. Have they been innovators and leaders in terms of technology, or have they been copiers and followers?

The support risk

This is an important one. It is probably true to say that no one buying a modern, large, software-based ATE system can become independent of the supplier. Such systems are just too complex for a user to invest in having people trained well enough to handle everything themselves. The one possible exception is the servicing of the hardware, but even this only makes economic sense for a small percentage of users with large numbers of similar systems. Anything involving the software will be almost impossible for the user to handle, other than the application software. Even in this the user will still require the additional support to help with a difficult problem or to train new employees, etc.

To some degree the user will therefore be dependent on the supplier. If the supplier is unable to provide the necessary support on a local basis the full potential of the equipment, in terms of achieving its planned ROI, will never be realized.

The reliability risk

No matter how good the local support is and how fast it can respond to your needs you still want to have a reasonably reliable piece of equipment. It is important, therefore, to get some idea about the reliability of prospective supplier's equipment. It is worth bearing in mind that, once a particular product is being tested on ATE, there will probably be no other available way of testing it. Your entire production of those products tested on ATE will be dependent on the availability of the test systems.

The vendor credibility risk

To minimize your risk of purchase, it is important for the supplier to be around for the next few years so that support can be given with new developments, maintenance, spare parts, training, etc. Vendor credibility is best looked at in terms of 'How safe is it to buy from this company?' The answer will depend on the supplier's:

1. *Commitment to the ATE business.* This in turn will depend on how important it is to the supplier—is it a major part of the business or is it a small part that could easily be cut off without serious effect?

2. *Financial performance.* This will indicate how likely it is that the supplier can stay in business and spend money on resources to support customers with new developments and after-sales support. Ask prospective vendors for a copy of their annual financial report, and be suspicious if they are reluctant to supply it. It is public information—there should be no secrets.
3. *History—track record.* As with the technology risk being a leader or a follower, this can say a lot about the credibility of a supplier.
4. *Organization.* How is the supplier organized in your area and in other countries where you may need support? How many people are there in support operations, etc?
5. *Product line.* The breadth and depth of the supplier's product line will have an influence on the cost risk and the technology risk, as well as being an indication of credibility.

Having compatible equipment can save money in terms of operator training and other areas, and if the supplier has a broad knowledge of many aspects of testing he or she is more likely to understand the problem and produce effective equipment. The supplier is also less likely to 'sell' something that is less than optimum if a wide choice of alternatives is available.

These five areas of risk should therefore be the five main subjects for evaluation when purchasing ATE (see Fig. 9.1). Of these, the most important area is that of

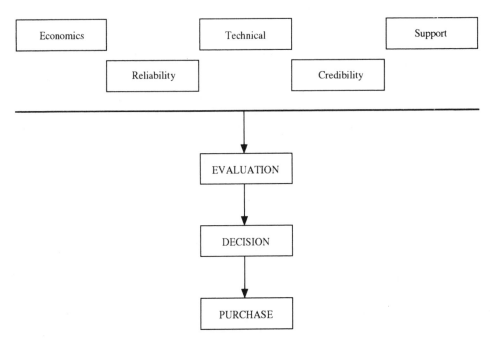

Figure 9.1 The evaluation process, considering the main areas of risk

economics. If all the others meet the minimum requirements then the system or strategy with the best cost effectiveness or ROI should be the approach taken. However, it is the technical performance, particularly of the systems software, that will determine the economic performance. Also, the economic performance expected from a system will not be achieved unless the reliability is good and the vendor stays in the business and is capable of providing all the required elements of support.

9.3 Economic evaluation

The earlier chapters have dealt with the need for economic evaluation, with examples of how to calculate costs and savings. All economic appraisal is concerned with comparing the costs of alternatives. These alternatives might be doing something or not doing something, or using system A or system B. Without some reference it is not possible to calculate 'savings'.

Having arrived al the different costs of the alternatives under evaluation, the next step will be to determine the financial performance of the alternatives. This will be of particular interest to upper management, who will have to approve the spending of the money. Chapter 11 covers the financial appraisal part of the overall economic analysis and evaluation.

9.4 Technical evaluation

It can often save a lot of valuable time if you can begin by making the suppliers familiar with your technical requirements. A written specification often helps, not only to inform the supplier but also to clarify to yourself what the needs are. It is probably best to make this specification fairly broad since different suppliers may have a different approach to solving a similar measurement problem.

Probably the best advice is 'do not try to do everything'. Separate the 'mandatory' requirements from the 'desirable' ones, and do not ask for unnecessary accuracies. A 0.01 per cent DVM may be a relatively common benchtop instrument these days, but put it in a system, under remote control, switched to the UUT via long cables and reed relays and you will be lucky if you get 0.1 per cent measurements. Most ATE systems are intended to detect and locate manufacturing process faults or defective components; with few exceptions, they are not automatic calibration systems.

Most suppliers have adequate facilities for demonstrating their products. It is usually possible to see measurements being made and to verify them manually. When looking at technical features always try to consider them in terms of the cost benefit to be derived, rather than in terms of the cleverness of the technical idea.

Going back to the comment about not trying to do everything, I have often heard a statement like: 'We cannot justify ATE because of programming costs—we have 1000 board types—it would take years to write the programs.' It probably would, but you do not need to write programs for all the boards in order to justify ATE and lower your testing costs. Keep Paretos 80/20 rule in mind at all times: 80 per cent of the problem (the cost) is often caused by 20 per cent of the boards. Looked at in a slightly different way, it may be a poor strategy to include requirements in your technical specification that are only needed for testing a few per cent of the board types. Adding 10 per cent to the cost of the ATE to solve another 1 per cent of the problem may not be good economics unless the alternative is very expensive.

Usually the brochures supplied by vendors of ATE do not contain detailed technical specifications, merely an overview of the equipment highlighting its main features and benefits. Technical specifications are normally contained in a separate technical proposal or technical description. It will normally be necessary to go over this with a representative of the supplier since it is not as easy to interpret system specifications as it is for a benchtop instrument. Measurement accuracies may be stated 'at the interface' (to the UUT) or 'at the instrument' or measurement unit. Bandwidths and cross-talk of switching matrices may be included in the accuracy statements or be stated separately. All in all it can be very difficult to compare specifications between systems.

All ATE systems are a compromise. Trade-offs have to be made between performance and price, accuracy and speed, ease of use and flexibility, etc. The important thing to consider when looking at technical performance is: 'What are we intending to achieve with this equipment?' If you want to locate gross manufacturing errors quickly, then speed is more important than accuracy. If you want to check the 'performance' of the UUT, then the reverse may be true. Most of the time, however, technical specifications and performance should be looked at in terms of their effect on the economic performance. This is particularly true in the case of the software that has no real specification as such.

9.5 Evaluation of support capabilities

Someone buying ATE for the first time is unlikely to be fully aware of the dependence he or she will have on the vendor of the equipment. The user and vendor become partners, a team, with the goal of getting production tested and shipped. If the vendor lets you down, production and shipping delays will inevitably result. Once you start to use ATE you become dependent on it—there is usually no other way of getting the job done if it is out of action or you need some new bit of software or hardware.

What kind of support is needed? The obvious one is of course *field service*. ATE systems are big and complex. They have to be repaired on site. Their complexity is such that failures will occur, especially since they may be using state-of-the-art technology. Find out about the size of the service organization. What is the vendor's field-service philosophy? Is there board-swapping or component level repair on site? What level of spares is held locally? What procedures are there for replenishing stock? What maintenance can be done by the user? Are diagnostic programs and self-test hardware available? How good are the maintenance manuals and other documentation? Are field-service training courses available?

Ideally the vendor should be able to offer whatever service arrangements best suit your needs. These could range from 'call out when you need it' to a full service contract with regular preventive maintenance visits.

Training is something that is not only required when you first buy equipment. Staff turnover and promotions will mean that training will also be required at other times. What training facilities are available locally? What language can the training be given in? How good are operation and programming manuals? Are videotape courses available?

Applications support will be required on an on-going basis but is especially important in the few months following the arrival of a new system. The training course, no matter how good, will not make your programmers 100 per cent effective in system operation and programming. It is like driving a car—you only really learn when you are out on your own after passing your driving test. During those first few months and beyond, it is important that your programmers have someone they can call upon for advice on a system operation problem or an application programming problem. In the case of a really difficult problem the vendor should have application experts available to visit your site to help in solving it.

Other forms of application support that can be useful are application notes and technical seminars provided by vendors.

Software support is essential. The system software in an ATE system is usually extremely large and complex. Quite often it is protected in some way so that users do not have access to its inner workings. If you need some non-standard capability it is necessary to get help from the vendor. Specialists who have detailed knowledge about the software will be able to explain it and what can and cannot be done on a non-standard basis.

Another important form of software support is the modelling of library elements. Many board testers using simulators for functional testing or ATG systems for ICT require a library of component descriptions or test algorithms. These libraries need updating from time to time as new devices become available. This need for updating is one of the reasons why it is difficult for a user to be independent of the vendor. At times you may require a library element that is non-standard, perhaps a custom device that only you use. You will need support from the vendor to get this device into the library. The vendor must either be willing to do it for you (probably for a fee) or be willing to train your own programmers on how to do it yourself.

Programming services is another form of support that may be needed from time to time, either to get started quickly or to help out during a programming bottleneck. It can sometimes be useful to pay for someone else to write some test programs. Some vendors offer such a service and there are also some independent companies offering programming services.

User groups can be another useful form of support. Some of the larger vendors organize user group meetings and publish news letters, which are contributed to by users of their equipment. At such meetings it is often possible to find someone who had a similar problem to one you now have, and so find a solution much more quickly. User groups also give the users an opportunity to make suggestions (hopefully constructive ones) and request certain additional features. They are also a valuable market research tool for the vendor, making it more likely that new designs or enhancements to existing equipment will meet a genuine user need.

Before making a decision to buy ATE or a specific system, you may well require some vendor assistance or support, especially if this is a first-time purchase. It can make your evaluation and selection job a lot easier if you can 'pick the brains' of the vendors. They will have product specialists who can provide detailed explanations on their specific speciality—functional testers, component testers, etc. They may well also have testing specialists who can analyse your testing problems and advise on the best approach or strategy. Also, these specialist sales people often have a lot of experience in solving testing problems simply because they talk to a lot of people (customers) who have already solved a wide range of problems.

Support is a vital element required to make an ATE operation efficient. The number of people and the amount of equipment a vendor has for support operations is a good indication of the commitment to support.

9.6   Reliability evaluation

There are two ways to obtain some ideas about the potential reliability, or otherwise, of equipment under evaluation. Probably the best way is to talk to people already using other equipment from the vendors being evaluated. There are a couple of things to watch out for if doing this. First of all, make sure that the equipment in use by any reference user is the same or similar to the system you are considering. There can be major differences in reliability between different products within one company's line, especially if they come from different, relatively autonomous, divisions. Secondly, be aware that, just as with a car, pride of ownership may sometimes colour the report you receive. Very few people will openly admit that they 'made a bad mistake by buying that so-and-so system'.

Another way to get a feel for potential reliability is to ask for details of the vendor's own testing and quality assurance programs. Does the vendor burn-in components? Does the vendor test and burn-in outside purchased peripherals? Is there a final product soak test or burn-in?

9.7   Evaluation of vendor credibility

Having the best equipment and the best support operation is of little value to a user if the vendor goes out of business the month after you bought your equipment. For the kind of long-term relationship that develops with the purchase of ATE the vendor must be in a healthy financial position with a stable organization. When buying any high-value equipment always ask to see financial reports on the company. As stated earlier, there should be no secrets here. To most companies in a healthy position the annual report is a major piece of promotional material, not only to impress potential stockholders but also to reassure customers that they are going into partnership with a winner or a leader.

The commitment the company has to the ATE business is important—many companies big and small have come and gone, come back again, gone again and so on. Each time they opt out they leave a few customers high and dry with little or no support. A measure of commitment can be seen by reading the annual report. How big a mention does the ATE division get? Is it mentioned at all?

Another important measure is whether the vendor is a leader or a follower, innovator or copier. Read any good marketing book and you will see that it is possible to run a very successful business by simply copying others. You can have lower overheads, little R&D expense and undercut prices. Many companies have a mixture of innovative products and so-called 'me-too' products, but if your prospective vendor only has 'me-too' products and always waits for someone else's innovations, you may find yourself lagging technology by a one to two year margin. If your products are mostly using established technology this may not matter, but if you are using state-of-the-art designs and devices it could well be critical.

If you take a snapshot at a point in time, the technically best product may well be from a follower. This is because it is relatively easy to add a few nice features after someone else has done all the basic design work. As with any high-technology industry competitors seek to leap-frog over each other with new products. The thing to look for is consistency in leading or innovation; after all, it will be a long-term relationship between yourself and the vendor. It takes roughly two years to develop a new system. Typically, it can be copied in less time, say one to one-and-a-half years. When the copy is announced, the original has already had one to two years of field experience, but more importantly the innovative company is already one to one-and-a-half years into the development of newer enhancements or new products.

9.8 Comparison charts

A fairly common approach used when evaluating and comparing ATE is to draw up a comparison chart. The advantages of this are that you can see at a glance the pros and cons between vendors or products: it is useful where vendors are not consistently the best at everything and it looks scientific when presented to management. The disadvantage is that the results are highly dependent on proper weighting of the importance of the factors considered, and this is often very subjective and not very scientific.

The following example, based on a real case, will highlight the potential pitfalls. The user was comparing a short-list of two functional digital board testers, and awarded a score out of 10 for each of the parameters that were felt to be important for the application (Table 9.1).

The main problem in this rather extreme example was that no weighting was given to the various parameters considered. As a result, *software* could score a maximum of 10 per cent of the total. Most people would agree that it is much more important than that, especially in the part it plays in the recurring cost areas of set-up and diagnosis. At the same time, the computer could score 10 per cent. In this case, the user was knowledgeable about computers and felt that the computer in system A was a better design than the one in system B. In fact, it is almost irrelevant since it was felt that system B had better software, faster diagnostics (without loss of accuracy) and easier program preparation than system A. System A does have a

Table 9.1 An oversimplified comparison chart

	Vendor A	Vendor B
Computer	8	6
Program storage method	7	7
Test speed	10	7
Diagnostic speed	7	8
Diagnostic accuracy	6	6
Driver/sensor capability	8	6
Ease of operation	7	7
Software	7	10
Price level	10	8
Ease of program preparation	8	9
	78	74

better test speed, but this advantage would be eliminated due to the handling time being so much longer than the test time. Another problem is that the results are not very conclusive. A result of 78 versus 74 is not a clear result when the score is a subjective rating in any case.

9.9 Tangibles and intangibles

One area where a comparison chart can come in useful is in the evaluation of intangible factors. The evaluation of ATE tends to be a mixture of evaluating tangible things such as the system itself, and less tangible elements such as 'support' and 'vendor credibility'. Since intangibles are difficult to evaluate or measure; there is often a tendency to leave them out of an evaluation and to concentrate on the more tangible elements. This is a mistake since if you leave them out of the evaluation you are in effect assigning implied values that have not been evaluated. At best you would be saying that 'all vendors are equal as far as the intangible elements are concerned' and this is unlikely to be the case.

One method of using a comparison chart to compare both tangibles and intangibles is to list all the parameters or objectives in order of importance. An importance rating is then assigned to each element on the list starting with assigning 100 to the most important element and then assigning decreasing values, proportional to their importance, to the others. The ratings can then be normalized to add up to 100 by dividing each rating by the sum of all ratings and multiplying by 100.

An effectiveness rating is then applied to each of the systems under evaluation for each of the elements on the list, and a weighted value obtained by multiplying these by the normalized importance rating. Table 9.2 will clarify this method, which can also be used for other parameters or objectives. For instance, if we use this method to look at the earlier example of the poor approach to using comparison charts we might arrive at the results given in Table 9.3.

Table 9.2 A slightly more scientific comparison chart

Parameter/ objective	Importance factor	Normalized importance	System A Effectiveness factor	System A Weighted value	System B Effectiveness factor	System B Weighted value
Cost risk	100	27.0	1.0	27.00	0.8	21.60
Technology risk	80	21.7	0.9	19.53	0.9	19.53
Support risk	70	18.9	0.8	15.12	0.9	17.01
Reliability risk	60	16.2	0.8	12.96	0.9	14.58
Vendor credibility	60	16.2	0.9	14.58	0.8	12.96
		100.0		89.19		85.68

Table 9.3 An example of poor choice of selection criteria and importance rating

	Importance	Normalized importance	System A Effectiveness factor	System A Weighted value	System B Effectiveness factor	System B Weighted value
Ease of program preparation	100	15.76	0.8	12.61	0.9	14.18
Software	95	14.96	0.7	10.47	1.0	14.96
Diagnostic speed	90	14.17	0.7	9.92	0.8	11.34
Diagnostic accuracy	80	12.60	0.6	7.56	0.6	7.56
Price level	70	11.02	1.0	11.02	0.8	8.82
Ease of operation	60	9.45	0.7	6.62	0.7	6.62
Driver/sensor capability	50	7.87	0.8	6.30	0.6	4.72
Program storage	40	6.30	0.7	4.41	0.7	4.41
Computer	30	4.72	0.8	3.78	0.6	2.83
Test speed	20	3.15	1.0	3.15	0.7	2.21
		100.00		75.84		77.65

The decision is now tipped in favour of system B; however, this is still a poor comparison chart for three reasons:

1. There are items on the list that are counted twice or that are irrelevant. 'Ease of program preparation', 'diagnostic speed' and 'diagnostic accuracy' are all determined largely by the performance of the system's software and to a lesser degree by the hardware, such as the computer and the program storage media.

'Test time' is usually irrelevant in a digital functional test system since it is dwarfed by the handling time involved.

2. For the elements on the list, the 'importance' rating is still very suspect. Since it has put ease of program preparation at the top of the list, it probably has a lot of programs to prepare. That being the case, there should probably be a greater emphasis placed on this relative to say 'driver/sensor capability' and 'program storage'.

3. The third, and most important, reason why this is a poor comparison chart is that *none of the most important selection criteria appear on the list*. Cost effectiveness appears only very indirectly under headings such as 'ease of program preparation' or 'diagnostic speed'. As was shown in earlier chapters, this can be determined in a much better way than simply using a subjective rating of a nebulous parameter. The 'price' of the equipment appears on the list but the ROI does not. Again, as will be seen in Chapter 11, this is an easy computation once the operating costs of the various alternatives have been determined. Other major areas of risk, such as *support, reliability, vendor credibility* and *technology risk*, are also missing from the list.

These examples have been included to highlight the dangers of using comparison charts. It is very easy to include relatively unimportant parameters and to get the weighting of the important issues wrong, or even leave them off entirely. It can be a highly subjective and emotional method of analysis rather than the scientific and logical approach that it sets out to be. Used wisely, however, it can be a useful technique, especially for developing a short-list of equipment for more detailed evaluation. When using such a technique, always keep in mind the overall objective of the evaluation, the five main areas of risk and the four major market forces.

9.10 The 'essentials of test'

When I was working for GenRad I came up with the concept of 'the essentials of test' for a promotional campaign. The idea behind this was to get evaluation teams to concentrate on the important issues rather than getting sidetracked by some clever features that may or may not have any real value when it comes to meeting the real objectives in hand. The remainder of this chapter is based upon a paper that I wrote on this subject and I am grateful to GenRad for allowing me to reproduce it here.

Let us assume that the decision has been made to implement a test strategy based upon in-circuit board testers. How do you decide just which one of the available systems to buy? How do you approach this very complex problem? Among all of the possible selection parameters what are the really important issues? Which of these important issues are most important for your particular set of problems?

The vendors of the various competing testers can help. They will all tell you what the key selection criteria should be. Unfortunately, human nature is such that there is a tendency for them to put the things they feel *they* are best at, at the top of *your* list of priorities. Listen to their arguments because many of them will be valid.

However, then you need to isolate yourself from all of this potentially conflicting input and take a cold hard look at just what you are trying to achieve with this tester. You need to determine your *fundamental objectives*. Why do you really need to buy this system. Basically, you need to select a tester that closely matches your chosen test strategy—one that will enable you to consistently ship your products on time, with the lowest cost and the highest quality, even if they contain all the latest component technologies and the latest manufacturing process technologies, for example ASICs, SMT, boundary scan, etc.

Be careful you do not become side-tracked. It is all too easy to lose sight of your fundamental objectives in all of the detailed discussion of bits, bytes, feeds and speeds, megahertz, millivolts, etc. At all times remain focused on the four main market forces that drive the whole evaluation process. To survive in today's globally competitive markets there is an *absolute* need to meet these potentially conflicting goals of cost, quality, time and technology. The reason that the important selection criteria *are* important is simply because they directly impact these four market forces. Within the evaluation process proper there are really two sets of issues to look at. There are the tester-related issues and there are the vendor-related issues. These two sets of issues are also linked. For example, you may not be able to utilize all of the available product features if the support capabilities of the vendor are inadequate. Poor training and poor applications support will make this impossible without a lot of work. Probably the most common mistake made in the evaluation process is to not spend enough of the available time comparing the economic merits of the various options. Tremendously detailed technical comparisons are frequently performed, but the economics is often a bit of an afterthought. It is often done to some minimal level just to satisfy the fact that 'the boss needs some numbers'.

A second mistake that is related to the first is that most of the technical evaluation is performed on the hardware elements of the tester, with relatively little effort spent on the software elements. Of course you need to have the right level of hardware performance to get the testing job done, but in any automatic system it is the performance of the software that will have the biggest impact on the overall performance. This is particularly true for the economic performance of a system that needs to be programmed and, as in the case of ATE, has to perform some diagnostics. The evaluation process is largely an analysis of risk. You are trying to select the tester that is the 'best fit' for your particular needs so that you minimize the risk of not meeting your fundamental objectives.

Once all of the competing products have been assessed, most people would tend to draw up some form of comparison chart. Even if you do not do this on paper it is still the same sort of mental process that would be applied. Comparison charts are fine, but beware. If not used properly they can produce misleading and incorrect results. Their effective use relies upon:

1. Selecting the most *important criteria.*
2. *Ranking* these criteria in the correct order of importance for your specific situation.
3. Then applying the correct *weighting* factors to the list.

If this is not done carefully then this apparently *objective* approach can end up being very *subjective* because of poorly selected, ranked and weighted criteria. How do you best determine the criteria to compare? One useful approach is to use the five-layer product model. This model is widely used in marketing and was originated by Theodore (Ted) Levitt and adapted by Phillip Kotler. These gentlemen are two of the most respected marketing gurus around. Basically the model proposes that any product (or service) can be described as having the following five distinct layers, which are shown in Fig. 9.2:

1. The core benefit.
2. The generic product.
3. The expected product.
4. The enhanced product.
5. The potential product.

For a board tester the core benefit would be the testing of boards and the diagnosis of defects. A broader view of this might be 'the shipping of quality products'. The generic product is all of the hardware and software pieces that are needed to get the job done. The expected product is all of the other things that a customer expects to get with the generic product: documentation, training, a warranty, applications support, service support, etc.

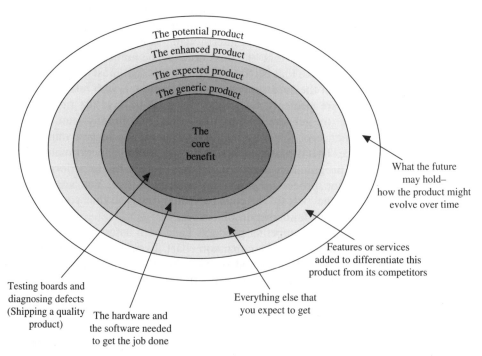

Figure 9.2 The five-layer product model

The enhanced product layer consists of the extra features and services that suppliers might add to differentiate their offerings from those of their competitors. Finally, the potential product is all that the future may hold—how the product might evolve over time. Clearly, in terms of meeting your current fundamental objectives, it is the middle three layers that are the most important. These can be regarded as the 'essentials of test'. By definition, if these are the essentials of test then everything else is not essential. Features in the enhanced layer usually fall into several categories:

1. Useful, but not essential.
2. Futureproofing, but not necessary at the moment.
3. Cosmetic.
4. Bells and whistles.
5. etc.

Any genuinely useful features or services will quickly become *expected* by the customer and so move in by one or even two layers. Promises about the future enhancement of the product, while important, clearly have to have a low weighting because the suppliers plans, or your needs, may change. Quite obviously, any differences between competing products in the middle three layers will be far more important than differences in the outer layers. As mentioned earlier, the enhanced product layer can be thought of as a *differentiation* layer, where suppliers try to differentiate their offerings from those of their competitors. As a result, a lot of the talk about the merits of a product are focused on this relatively unimportant layer. This is what I referred to earlier when I said 'don't let your fundamental objectives get lost in all of the details'. Any salesman will quite naturally try to increase the importance of some capability that he or she has that is lacked by competitors. Listen carefully because it may occasionally be true, but do not lose sight of the fundamental objectives and the market forces.

Because of this concentration on the enhanced layer there is a tendency for people to dismiss the middle three layers, implying that all testers are basically the same so that the only differences lie in the enhanced layer. *This is just not true.* There are substantial differences to be found in the three middle layers that will far outweigh any differences in the outer layers. Be suspicious of anyone that says, 'Well they are all pretty much the same basically, but our XXX model has this super feature that ...' He or she is probably trying to divert your attention away from some deficiency they have in the 'essential layers'.

What, therefore, are these essentials of test? Simply stated, you need a high capacity tester that can test large quantities of boards, with a high degree of fault coverage, for a high quality of test. It should provide accurate diagnostics so that boards can be repaired efficiently. It should have a comprehensive test program generation and debugging system and an effective test fixture preparation capability. It should be able to cope easily with new component and manufacturing technologies and have an integral quality data collection system to enable rapid yield improvements. All aspects of the system should be easy to use and the supplier should provide superior levels of support in all of the needed areas. In addition, all of this should be available for the lowest possible life cycle cost. Let us now look at these essentials in a little more detail.

The annual *testing capacity* of a test system is primarily a function of the testers throughput, in boards per hour, and the number of test hours available each year. The number of hours available will be the total number of hours worked less any lost time. Typically, productive testing time is lost to:

1. Time spent debugging test programs and test fixtures that require the testing hardware of the system. In other words this is the time for operations that cannot be performed in a 'background' mode on the tester's computer. If you are tying up the testing hardware you cannot be testing boards.
2. Productive time is lost each time the test program and test fixture are changed, during normal production test operations, to accommodate a new board type. This lost time can be quite significant when boards arrive at the tester in very small batches.
3. System housekeeping functions, routine maintenance and downtime also contribute to the loss of productive testing time on the system.

High fault coverage requires a combination of good measurement hardware and an effective test program preparation capability that can enable the rapid improvement of initial fault coverage levels. Of course the test program preparation system is of little use without the support of an accurate, robust and comprehensive device library. The fault coverage capability of the tester will, of course, have a direct impact on the quality of the products you ship to your customer.

Accurate diagnostic information also requires a combination of good measurement hardware and a software diagnostic system that not only pinpoints defects but also minimizes the diagnosis of 'defects' that are not really there. This false indictment of defects is a common problem in poorly designed in-circuit testers. It leads to unnecessary repair and re-test operations that cost money and reduce the capacity of the tester. Additional diagnostic tools are provided on some testers to more accurately identify difficult problems. Accurate dignostics lead to efficient repair and minimize incorrect diagnosis.

A good test program generation system is vital to the successful operation of any ATE. Differences in program preparation performance between alternative testers will result in major performance differences in fault coverage (quality of test), diagnostics and capacity. A comprehensive, informative and rapid debugging capability is also essential, as is an extensive and well-proven device library. There should also be the possibility to take in data from design to simplify the creation of both the physical (layout and interconnect) and the behavioural test information. The program preparation system should also output fixture information to simplify or even automate this part of the process. This test fixture preparation process is closely tied to the program preparation system. The design and the production of test fixtures should result in good electrical performance. As already mentioned it should be possible to automate much of this process. Reliability of the fixture and its interface to the tester is also important both for high-volume production and for situations where the fixture is loaded into the tester many times, for example in a JIT operation with small batch sizes.

Incorporating the latest component and manufacturing *technologies* into your new products is a necessity if you are to remain competitive in the market-place. It is one of the four major market forces. Therefore it is essential that the tester you choose be capable of coping with new technologies now and in the future. A major concern of most people making a purchase decision is 'When will it go out of date?' Take a careful look at the vendor's track records on keeping their products up to date by developing enhancements, issuing software and library updates, and adding new capabilities. Also take a look at how well a vendor supports products that have been discontinued.

Quality is possibly the single most important competitive issue today, and it will continue to be. To continuously improve your product quality you have to have accurate and timely *quality data*. Then you can continue to make the required improvements and maintain the gains that you achieve.

Ease of use is now a far more important issue than it has been in the past. Time is money and *time to market* is even more money. Anything and everything that can be done to reduce time to market should be done. A few weeks delay with a product caused by a difficult programming and debugging operation and a long learning curve in the production test will have a large impact on the profitability of the product. Ease of use features will minimize training and re-training times and speed up the learning process. They will reduce programming times, speed the product to market and generally improve efficiency and increase the tester's capacity.

When you buy a board tester you do not simply buy a set of hardware and software tools. You effectively enter into a partnership with the supplier. Your production output will depend on the tester, and the reliability of getting that production output in a consistent and constant manner will depend on the support that the supplier can provide. This support has to go a long way beyond simple maintenance. It is required in many other area such as training, applications support and software support.

The tenth essential, the lowest possible *life cycle cost*, is arguably the most important of all because the other nine are aimed at achieving this one. ATE is a solution to what is basically an economic problem. You buy ATE because it is cheaper to have it than not to have it. However, this 'cheapness' has to be viewed in terms of the life cycle cost of your products, not simply in terms of the 'cost of ownership' directly associated with the use of the ATE.

All of the costs that are directly and indirectly affected by the test system have to be considered to obtain a true life cycle cost for the various alternative testers being considered. One major problem of perception occurs in test circles because of our preoccupation with fault coverage data. It would be far better if we always thought in terms of escaping defects rather than the ones we detect. For example, is a tester capable of achieving a fault coverage of 97 per cent, 3 per cent better than one that can only achieve 94 per cent, or is it 100 per cent better? In terms of fault cover it is 3 per cent better but in terms of the shippable quality of your products it is 100 per cent better. Regardless of what the fault coverage of the following test stages is, the board tester with the 97 per cent fault cover would result in only half of the defects reaching the board tester escaping to the field to be detected by your customer, compared to the tester with a 94 per cent coverage.

In summary, your choice of evaluation criteria should be dictated by your fundamental objectives. These in turn are mainly determined by the four major market forces of cost, quality, time and technology. The ten key criteria, referred to as the essentials of test, are the variables that will have the biggest impact on your ability to meet your goals. It is the ranking of these that will be different between different production situations rather than what items appear on the list. Naturally there are bound to be some exceptions to any rules, but these are likely to be the important criteria in most situations.

10. Field-service economics

10.1 Overview

The automation of the field service of electronic products has not grown at the rate that was being predicted in the early eighties. It is difficult to be specific with reasons for this because the situation has tended to be different in different segments of the market. One area where there has been significant growth is in the automotive electronics service market. Car manufacturers have continued to add electronic control elements to their products to meet pollution standards, to extend the mileage between services and to generally remain competitive with each other. This has resulted in a large market for the supply of special-to-type testers for the thousands of service garages that each of the car companies have around the world. To date this has typically been addressed with a joint definition and design process involving the customer and the test equipment vendor, but there are signs that there is a market for a general purpose diagnostic tool that is not specific to one particular model or even to one manufacturer.

One general reason for the slow growth in the automated field-service market is the increased quality and reliability resulting from the 'quality revolution'. The field failure rates predicted in 1980 as a result of the predicted increase in complexity simply did not materialize. Another problem that affects the strategy of 'on-site' repairs and local service depot repairs is the increased use of surface mount components. These are more difficult to repair in the field and usually require specialized repair equipment as opposed to the simple soldering iron and a 'solder sucker' needed for through-hole technology. There are probably many other reasons why there is less automation in field service than was expected but this is not really the subject of this chapter. The main issues about field-service strategies remain the same as they were a decade ago so the remainder of this chapter is essentially unchanged from the first edition.

The cost of electronic hardware has decreased in real terms during the past 5 to 10 years. Some of the reasons for this decrease include such things as the greater volume production of component parts, the decreasing labour content brought about by automating production operations, the tremendous technological advances and the shift over to software intensive designs. At the same time, the cost to service this equipment, particularly field service, has risen rapidly. Labour costs have risen since the complexity of the equipment is such that highly trained and, therefore, more expensive technicians are required. Recruiting and training have become two of the biggest problems facing management, particularly in high-growth segments of the industry. A new industry, that of third-party maintenance, has emerged to capitalize on the opportunity presented by the problems faced by suppliers and users.

Service managers are under pressure to control costs while maintaining a better service support than their competitors. The use of common device technology, shorter design cycles and product lifetimes is making products look more and more alike. Companies are searching for elements other than the generic product to differentiate their overall offering from that of their competitors. One form of differentiation is field service. Often the company with the best record of servicing customer's problems is the company that gets the business.

Many companies are now recognizing that service is such a major portion of their overall business that it is now organized as a separate profit-making division. These divisions typically employ 15 to 20 per cent of the total workforce and account for 10 to 25 per cent of revenues. Some estimates show that between 5 and 10 per cent of a company's total assets can be tied up in spare board inventories.

To some degree this situation works against the quality ethic of constant improvement. So long as a substantial proportion of your revenues and profits comes from customer support activities there will not be the motivation that is needed to carry quality improvement to its logical conclusion. The cost of your field-service operation is a major part of the cost of quality, specifically the cost of non-conformance, but any revenues that result from quality problems will reduce this cost. It is very difficult to balance the service revenues against the possible loss of business caused by poor quality. However, if your competitors are taking the 'quality' route you really do not have much choice. You have to match the industry average quality levels or suffer the consequences of lost market share.

The problem of field service is enormous, and much of the problem has been caused by a general lack of automation. Traditionally, field service has been a labour-intensive operation with very little investment in capital equipment other than the ubiquitous oscilloscope. Budgets are typically 60 to 85 per cent personnel-related and only about 5 per cent is allocated to the purchase of capital equipment. This situation is now changing as many of the large companies, especially in the computer industry, have recognized the need to do something about the problem. The test and measurement companies have also responded by developing a variety of new tools using a variety of technologies. Some of these products have been developed as a result of the initiative of the end users, while others have emerged directly from test and measurement companies. They include such things as in-circuit emulators for servicing microprocessor-based designs, signature analysers and small or even portable board test systems.

The emulators and signature analysers are usually small, low-cost instruments intended to be carried on-site by a service engineer and then used for diagnosis and repair down to the component level. The low-cost board testers are usually intended to be deployed at field-service offices rather than be carried around by engineers, although this can be done with the portable versions. The cost of these systems is, however, too high to consider equipping each engineer with a unit, the term 'low-cost' being relative to a large production test system. The general strategy for using small board testers in field service is to decentralize the board repair operation, to reduce the amount of spare boards needed and to improve customer service.

More recently a new type of field-service tool has emerged. This can best be described as a complete system exerciser or tester. This type of device can be customized to the needs of a specific user or even a specific product. For optimum use, the target product should be designed with this type of tester in mind, and equipped with a field-service connector. In use, an engineer would simply connect the tester to the product and run a complete diagnostic program. If the product being serviced contains any data that needs to be retained, the tester can store it in its own memory or on magnetic disk or tape prior to loading the diagnostic programs. Such testers can be configured to diagnose to board and component level, exercise peripherals, check communications channels, etc., just by changing hardware and software options. They can communicate with a large central service database via an acoustic compiler and a telephone. The service engineer can even send reports back to the office or the central service centre using the tester's communications capabilities.

Which type of service testers are adopted for a given situation will depend largely on the overall service strategy adopted. There are two basic strategies in use today:

1. On-site repair to component level.
2. Board-swapping.

In-circuit emulators and signature analysers tend to be aimed at helping the engineer when an on-site repair strategy is adopted. Small, low-cost board testers are aimed at helping when a board-swapping strategy is adopted. The new generation of system testers can be used for either strategy or for a combination of the two.

On-site repair has the advantage of requiring small inventories since components rather than boards are stocked by the field engineer and the various local service offices and area depots. Another advantage of this approach is the better local control and lower dependence on the factory to provide good boards.

The disadvantages of this approach are that a larger staff of engineers is required to cover a given installed base since they will inevitably spend more time on-site than someone simply exchanging boards. These engineers need to be more skilled and better trained than their board-swapping counterparts and, as mentioned earlier, hiring and training such high-level people in sufficient numbers is becoming increasingly difficult. The use of logic analysers, signature analyser and in-circuit emulators does little to alleviate this problem since they require a skilled operator and will still require more on-site time than board-swapping. The customer is less happy with on-site repair due to the longer downtimes experienced and the lack of confidence in the repair relative to seeing a new board going in. There is also a risk of poor repair quality, creating reliability problems if soldering of multi-pin devices is required.

Board-swapping has virtually opposing advantages and disadvantages to the on-site repair strategy. Fewer people with less skill and training are needed to cover the installed base. The customer is happier with the lower downtime and is more confident when a replacement board is seen to be fitted rather than clouds of smoke from an engineer's soldering iron. The major disadvantage of board-swapping is the enormous stocks, or inventory, of spare boards that are required to enable rapid response to customer calls for service. In all other respects board-swapping has advantages over 'on-site' repair and is currently the predominant strategy in use. In fact, very few companies with large field-service organizations practise on-site component level repair other than in emergency situations.

The major reason for this enormous inventory of spare boards is the fact that most companies practise a centralized board-repair strategy. In this context 'centralized' might mean returning boards to a factory or to one of several repair depots around the country or the world. Here, 'centralized' is a relative term. A company with 800 field-service offices and 10 board-repair depots is still 'centralizing' board repair. If this problem of big stocks of spare boards could be solved then the only major disadvantage of board-swapping would be overcome—resulting in a significant lowering of field-service costs.

The only way to reduce the quantity of spares required is to decentralize the board test and repair operations by equipping many of the field-service offices with their own small, but powerful, automatic board test systems. To see how this can work it is necessary to understand how such large inventories of boards occur in the first place.

The remainder of this chapter describes in more detail how the field-service problem can be alleviated by adopting the philosophy of decentralized board repair. If a different field-service philosophy is more appropriate for your own situation it is hoped that the analysis method and identification of cost areas might still have some value.

10.2 The build-up of spares inventory

Many customers—geographically dispersed—using a lot of different models of PCBs, each with several revision levels, lead to a lot of inventory. In addition, the relatively long turn-around time between an engineer removing a board and that board being repaired and returned to field stock means that there are always many boards in the 'pipeline' between a field engineer and the repair depots or factories. This 'pipeline' is typically three to nine months long, with six months being about average. This number is based upon the experience of several companies with a mix of domestic and international business. Figure 10.1 shows how the pipeline stock accumulates.

In military use, the pipeline can be even longer, especially in the case of the navy, where a ship may have a long tour of duty before returning for repairs. In industrial companies, the pipeline tends to be a build-up of a number of short delays. A field-service engineer may remove a board from a customer's equipment, replacing it with one of his spares. This bad board may sit in the engineer's car for a week or two before being handed into the office. There will be a delay there before it is sent off with others to the repair depot or factory. Further delays take place there before the board is repaired and returned to the field. In the case of an international shipment there will be delays caused by generating export documentation and customs delays.

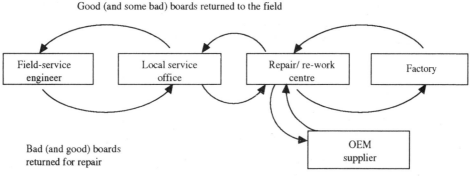

Good (and some bad) boards returned to the field

Bad (and good) boards
returned for repair

Figure 10.1   The board-repair pipeline

All of this eventually leads to the three to nine month delays when averaged out over all boards.

Due to the 'shotgun' approach sometimes used when board-swapping. or due to pressure from a customer to 'get on with it', a field engineer often removes a good board and does not replace it. As a result, a high proportion of boards going back for repair are indeed 'fault-free'. Industry averages about 30 to 50 per cent good boards returned, and for the military the number is more like 50 to 70 per cent. This higher number is often brought about by repair procedures that specify 'if this fault occurs replace boards X, Y and Z'. As a result, two good boards and one bad board are returned. Eliminating these good boards from the pipeline would save a lot of money by cutting down the inventory, but if a large number of faulty boards could also be fixed at the local office the number of boards in the pipeline would be minimized.

Example

The number of boards in the pipeline is proportional to the number of boards returned and the length of the pipeline.

N_p = number of boards in the pipeline

P = pipeline length (in months)

N_r = number of boards returned (in a year); N_r is a function of the number of system failures in a year and the proportion of good boards returned

F_s = system failures in a year

R_{gb} = ratio of good boards returned to total returned

$$N_r = \frac{F_s}{1 - R_{gb}}$$

If we assume a company has 1000 units of a particular model installed and is experiencing an MTBF due to board failures of six months, it will have 2000 failures per year, 40 per cent of all boards returned for repair are all right and its pipeline is six months long.

$$F_s = 2000$$

$$R_{gb} = 0.4$$

$$N_r = 2000/(1-0.4) = 3333 \text{ boards/year } (2000 \text{ bad} + 1333 \text{ good})$$

$$N_p = 3333 \times 6/12 = 1667 \text{ boards in the pipeline}$$

This inventory in the pipeline is often referred to as the board 'float'.

It now installs small digital board testers at its local service offices, which will eliminate sending the good boards back and will diagnose 70 per cent of the remainder (assumes some boards are analog and, therefore, not testable on a digital board tester). The turn-around (pipeline) at the local office is two weeks but additionally the service engineer still carries boards around in his or her car for a couple of weeks, so the total pipeline is one month long. As 600 boards per year are still sent down the six month pipeline, there are $600 \times 6/12 = 300$ boards in the pipeline; as 2733 boards per year go down the one month pipeline, there are $2733 \times 1/12 = 228$ in the pipeline. Thus the total pipeline inventory is now reduced from 1667 to 528 boards. At an average cost of manufacture of $300 per board, that is a saving of $342 000. This, however, is only part of the total inventory-avoidance picture. Once the pipeline is full, boards will be coming out of each end on a more or less continuous basis. However, they are not always the boards you need so it is necessary to maintain stocks at each location in the field-service organization. The field engineer has some stock in his or her car. The local office carries stock and so does the area depot, and the factory will also carry stock for replenishment of field stock. If we assume that in the example above the policy is to maintain about three month's worth of stock at the field level, and that use of board testers at the field office would allow this buffer stock to be reduced to a few weeks' worth, reducing this stock from three to one month would result in a saving of:

$$[(F_s)(3/12)] - [(F_s)(1/12)] = 333 \text{ boards}$$

or approximately $100 000 (at $300 each). This brings the inventory-avoidance savings to a total of $442 000.

10.3   Other savings

Inventory avoidance is only one of the potential savings to be made using this strategy of equipping field-service offices with board test systems. There are five major areas of cost-saving potential:

1. Repair costs.
2. Inventory carrying costs.
3. Inventory avoidance/sell-back.
4. Inventory handling and transportation.
5. Landing costs in foreign countries.

Repair costs

When the service organization returns boards to manufacturing or to an original equipment manufacturer (OEM) vendor for repair, it is normally charged with some fixed percentage of the 'manufacturing cost' or 'standard cost' of the board. Typical charges are around 30 to 45 per cent.

Repair costs using the local board tester will typically be 5 per cent or less, so savings to the field-service organization will be 25 to 40 per cent of standard cost multiplied by the number of boards returned. Here the number of boards returned includes the 'good' boards since a technician will often spend more time checking a good board than a bad one. This may sound crazy but when a bad board is encountered the fault is diagnosed, repaired and re-tested, a relatively straightforward operation, particularly if ATE is used. When a good board is encountered the technician assumes that the board was faulty when the service engineer returned it and so will spend time with hair driers, freezing sprays, rubber hammers, etc., to try to make it fail.

Of course, not all of these savings will be net corporate savings since the board-repair costs are simply being transferred from manufacturing to service. The real savings will be the OEM vendor repair charges plus those manufacturing repair costs that exceed 5 per cent. There will also be some savings on the (expensive) production testers by effectively transferring some of the workload to the less expensive field-service testers. However, if the field service organization is viewed as a 'profit centre' or a 'cost centre' then the whole of the savings are valid when looking at its performance.

Inventory (stock) carrying costs

The major part of the carrying cost is the cost to finance the inventory. Basically, money invested in inventory could as a minimum be earning interest. Better still, it could be invested in some profit-making or cost-reducing equipment (such as a few field-service board testers). If you take the simple view then the finance cost of carrying the inventory is in the region of 10 to 20 per cent depending on interest rates, but if that money could be invested in something that could give a 50 per cent return on investment, it is clear that reducing inventory could have a major impact on the profitability of a company.

A simpler way to look at this cost is to assume that if the company had less money tied up in spares inventory then its borrowing requirements would be lower.

There is also an occupancy cost associated with inventory. It costs money to store it and to insure it. There is also an obsolescence cost to write off old spares no longer needed. All in all the carrying cost is typically 25 per cent per year of the value. It is important to realize that the inventory avoidance saving is a once-off saving whereas the carrying costs are incurred each year.

Inventory avoidance/sell-back

This was covered in the example earlier. It is easy to see how the strategy would reduce the amount of inventory needed at each service location and in the pipeline when a new product is introduced. For an existing product there is often a possibility to sell back unwanted stock to the manufacturing operation for inclusion in new products. Where this can be done, typically 50 to 70 per cent of the 'standard cost' of the board will be credited to service by manufacturing. Therefore 'inventory avoidance' can save the service organization 100 per cent of the standard cost of boards and 'sell-back' can save 50 to 70 per cent of the standard cost.

There is another major financial benefit to this field-service strategy in the case of introducing a new product. Quite often the initial production rate is determined by component availability, etc. Manufacturing can only build a certain quantity of boards in the first year. Some of these boards will go into products to ship to customers and some will go into service stock and to 'fill the pipeline'. If you can reduce the need for service stock you can ship more product—and that means more profit.

Example

A company is introducing a new product, containing five boards, that will sell for $8000. First-year production of boards is limited to 1000 of each type. Field service needs one set per engineer plus a buffer stock of 90 sets, which includes an allowance for pipeline float. There are 30 service engineers; therefore, the total service requirement in the first year is 120 sets of boards. If board testers at the field offices could reduce this requirement to 50 sets then at $300 per board the inventory would be reduced by $70 \times 5 \times 300 = \$105\,000$. Carrying cost at 25 per cent would be reduced by $105\,000 \times 0.25 = \$26\,250$ for a total of $131\,250. This reduction in service inventory would permit the shipment of 70 extra units to customers, producing an additional $560\,000 in shipments, which at typical manufacturing costs of, say, 45 per cent should contribute $308\,000 towards marketing costs, overhead costs and profit.

Inventory handling and transportation

Large quanlities of spare boards having to be shipped all over the world requires a lot of administrative work, such as paperwork processing, inventory control, location, withdrawal, packing, shipping, stock-taking, etc. All this work is proportional to the volume of inventory stocked and shipped around. Shipping costs are relatively high since air freight or air parcel post is often used. Emergency shipments cost even more. For international shipments, export documents and import documents will be required.

Landing costs in foreign countries

Import duties are closely related to transportation costs but are high enough to be treated separately. They can add substantially to the cost of centralized service in some areas of the world. Some South American countries have import duties as high as 70 to 100 per cent on spare parts, and a repaired board is charged duty on its value, not the cost of the repair. Placing the board testers in foreign countries can often produce the fastest payback.

These then are some of the major cost areas associated with field service when board-swapping is the primary strategy. With other strategies, there will be other cost areas as well as some of these. With an 'on-site' strategy the inventory costs, while still there, should be lower. Personnel costs will, however, be much higher since the call rate is likely to be lower when a more detailed and lengthy service call is required.

The following example is intended to show a method for calculating the savings possible when switching from a board-swapping strategy with a central repair of boards to a decentralized board test operation using a number of small testers.

Example

The XYZ Electronics Company is introducing a new product in nine months' time. The product is expected to sell well but it will place a heavy burden on the field-service organization due to its complexity and the quantities XYZ plans to install. XYZ is a relatively small company with high-technology products. It currently operates a board-swapping repair policy with all boards going back from the field to the factory for repair. Production pressures are such that field returns are a low-priority job for the production test department. As a result its pipeline delays are fairly long. It has six field sales and service offices domestically and four international sales and service subsidiaries.

XYZ has been investigating the possibility of using some low-cost digital functional board testers at its 10 service offices as an alternative to centralized repair. This cost analysis for field service of this new product with and without the ATE follows. This analysis is based solely on using the ATE for this new product—it will, however, be possible to transfer some of its older products to the new testers also.

Since repair costs are not entirely a corporate saving, only inventory-related costs are considered in the analysis. About 40 per cent of XYZ's installations are expected to be in the international territory, and the average manufacturing cost for a board is $300.

Cost analysis with present centralized board repair (note that some figures are rounded):

		Year 1	Year 2	Year 3	Year 4
A	Units shipments (forecast)	300	600	800	900
B	Installed base	300	900	1700	2600
C	Expected MTBF (months)	4	5	6	6
D	Failures per year ($B \times 12/C$)	900	2160	3400	5200
E	Proportion of good boards returned	0.4	0.4	0.4	0.4
F	Total boards returned $[D/(1-E)]$	1500	3600	5670	8670
G	Domestic returns (0.6F)	900	2160	3400	5200
H	International returns (0.4F)	600	1440	2270	3470
I	Domestic pipeline (months)	5	5	5	5
J	International pipeline (months)	7	7	7	7
K	Domestic pipeline inventory ($G \times I/12$) boards	375	900	1417	2167
L	International pipeline inventory ($H \times J/12$) boards	350	840	1322	2022
M	Domestic buffer stock (three months) ($0.6D \times 3/12$) boards	135	324	510	780
N	International buffer stock (three months) ($0.4D \times 3/12$) boards	90	216	340	520
O	Incremental domestic inventory (boards)	510	714	703	1020
P	Incremental international inventory (boards) (additions per year)	440	616	606	880
Q	Total inventory	950	2280	3589	5489
R	Total incremental inventory (boards)	950	1330	1309	1900
S	Cost of inventory added each year ($)	285k	400k	393k	570k
T	Carrying cost total inventory (25 per cent)	71k	171k	269k	412k
U	Import duties, etc., on international inventory (20 per cent)[1]	62k	123k	173k	261k
V	Total inventory-related costs ($)	418k	649k	835k	1243k

Cost analysis with ATE at each field office:

		Year 1	Year 2	Year 3	Year 4
D	Failures per year	900	2160	3400	5200
G	Domestic returns (boards)	900	2160	3400	5200
H	International returns	600	1440	2270	3470
I_2	Domestic pipeline (months)	1	1	1	1

---

[1] Import duties, etc., based on returns (H) + incremental inventory shipped internationally. This incremental inventory is required to support the growing installed base of products—it is the total number of boards supplied by manufacturing to field service each year.

Cost analysis with present centralized board repair (note that some figures are rounded):

		Year 1	Year 2	Year 3	Year 4
J_2	International pipeline (months)	1	1	1	1
K_2	Domestic pipeline inventory (boards)	75	180	283	433
L_2	International pipeline inventory (boards)	50	120	190	290
M_2	Domestic buffer stock (one month) $(0.6D \times 1/12)$ (boards)	45	108	170	260
N_2	International buffer stock (one month) $(0.4D \times 1/12)$	30	72	113	173
O_2	Incremental domestic inventory (boards)	120	168	165	240
P_2	Incremental international inventory (boards)	80	112	111	160
Q_2	Total inventory (boards)	200	480	756	1156
R_2	Total incremental inventory (boards)	200	280	276	400
S_2	Cost of inventory added each year ($)	60k	84k	83k	120k
T_2	Carrying cost of total inventory (25 per cent)	15k	36k	57k	87k
U_2	Import duties, etc., on international inventory (20 per cent)[1]	5k	7k	7k	10k
V_2	Total inventory-related costs ($)	80k	127k	147k	217k
W	Total inventory-related savings $(V - V_2)$ (rounded) ($)	338k	567k	688k	1026k

Cost of implementing the ATE program. Eleven testers will be needed—one for each office and one for the factory to update programs to compensate for engineering changes to the boards. Ten test programs will be needed for the digital boards in the product. All the above calculations of MTBF, inventories, etc., are based upon these 10 boards. The programs will be converted from the production test programs developed for the production ATE. Interface adaptors will be required to interface between the boards and the tester. The following is an estimate of the costs involved:

11 testers plus quantities of spares	$400k
10 programs and interfaces	$100k
Training costs (mostly travel)	$40k
Total project cost	$540k

The cumulative savings over the four years are shown on the graph in Fig. 10.2. By inspection of the graph or by interpolation of the savings, the payback is achieved in 1.35 years.

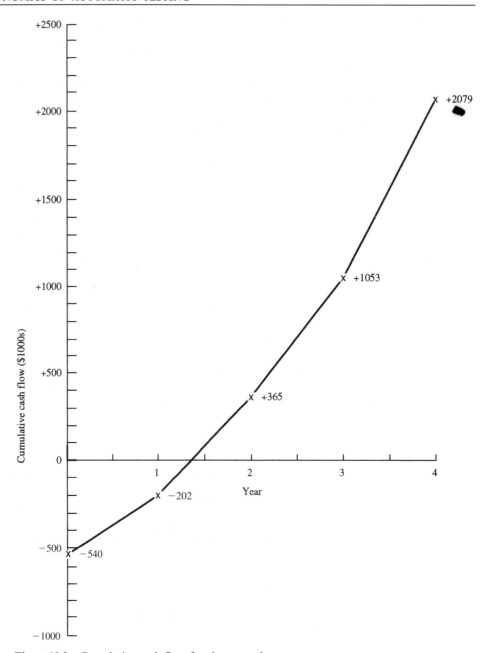

Figure 10.2 Cumulative cash flow for the example

Note. This is a very simplistic look at the payback. A more accurate method that takes in the effects of depreciation and taxation is explained in Chapter 11.

So XYZ Electronics can justify the program on the strength of a single new product. Transferring boards from existing products to be tested on the new testers will provide an even faster payback and higher ROI.

10.4 Summary/checklist

1. Field service of complex electronic products has become a major problem. Advances in semiconductor technology have lowered product costs, shortened product lifetimes and made the cost of servicing a higher proportion of the purchase price.
2. Customers demand rapid response coupled with low costs. Since customer satisfaction is not negotiable, service managers must work to make their operations more efficient.
3. Field service has traditionally been a labour-intensive operation with relatively little investment in capital equipment such as ATE. Now that new types of field service test equipment and systems are becoming available this situation is changing.
4. There are two basic field-service strategies that can be adopted:
   (a) On-site repair to the component level.
   (b) Board-swapping with bad boards returned for repair.
   These two strategies have virtually opposite advantages and disadvantages. On-site repair needs more engineers with a higher level of skill but requires relatively low investment in spare parts. Board-swapping requires fewer people with less skill and training but suffers from requiring enormous investments in spare board inventories.
     If the inventory problem could be solved, the board-swapping approach would be universally accepted due to the difficulty of recruiting and training good service engineers in sufficient numbers.
5. The major cause of the high inventories is the long turn-around time required to get boards repaired and back into circulation in the field. This, in turn, is caused by the use of centralized board repair facilities. Decentralizing board repair by using small low-cost testers at local field-service offices is one way for some companies to lower inventories and improve the efficiency of their field service operation. This is especially true for international operations where the board repair pipeline is longest if bad boards have to be shipped overseas for repair.
6. Implementing such a strategy will reduce costs in five major areas:
   (a) Repair costs will be lowered.
   (b) Inventory carrying costs will be reduced.
   (c) The build-up of large inventories will be avoided and excess inventory can possibly be 'sold' back to production.
   (d) Reduction of inventories will result in lower inventory handling and transportation costs.
   (e) Landing costs in foreign countries will be reduced.

11. Financial appraisal

Probably the single most important function common to all levels of management is that of making decisions. Most, or all, of the decisions made in industry should be based upon economic analyses. Decisions based upon other factors, such as being progressive, being modern, gut feel, rules-of-thumb or hunches, can often turn out to be poor decisions. We all envy the entrepreneur who makes a fortune by risking all on a hunch. Unfortunately, we do not always hear so much about all the failures that the entrepreneur and others may have had in the past. Once a business or an organization becomes established, with many people dependent upon it for a livelihood, it becomes irresponsible rather than glamorous to make 'gut-feel' decisions.

The major responsibility of senior management is to ensure that the owners of the company, the stockholders, receive a good return on the investment they have made in the business. This they do by making decisions on business strategy and the acquisition and use of money. It is only common sense that the investment of a large sum of money into the acquisition of an automatic test system should be carefully appraised in economic and financial terms.

It also makes sense that the person best qualified to make any economic analysis is the person most familiar with the problem that is to be solved and the capabilities of the equipment being considered to solve that problem. Simply passing a rough set of numbers on to an accountant to 'see how it looks financially' will not always give good results since he will be unfamiliar with the application. The actual process of performing the financial appraisal is fairly straightforward and in most cases will take less time than it would to familiarize a financial person with the problem. Another good reason for doing it yourself is that you will probably be responsible for making sure the project achieves its planned performance. Knowing how the economic and financial performance is arrived at will be essential to the proper management control of the project. The accounting or finance departments can always be called up to familiarize you with the company's internal criteria and to refine the financial appraisal that you have performed.

11.1 Return on investment

Most people are familiar with the expression return on investment, or ROI as it is commonly abbreviated. The concept of getting back your investment plus a little extra is a common one, and has been with us since long before the use of money. The loan to a neighbour of grain for planting would typically be repaid after the harvest with a little extra. This led to the concept of 'interest' on loans once money became more widely used than bartering. Just as you would expect to receive interest on a deposit of money in a bank so a company should expect a 'return' on any money it invests in a piece of equipment such as ATE.

Unfortunately, although the concept of ROI is very familiar it often means different things to different people. The most commonly used method of appraising an investment is to determine how long it will be before the 'payback' of the investment occurs. Many people think of payback as a measure of ROI when in fact it is not. For example, if you invest $2000 in a project with a two-year life that will return $2000 over the two years, then payback is achieved but there will be no 'return' since all you have recovered is your original $2000.

There are other methods of appraisal in common use, each with some advantages and disadvantages, so before getting too involved in the subject it may be useful to review them.

Before doing so, however, it is worth remembering that all capital investment decisions involve the comparison of alternatives. Without alternatives there can be no calculation of savings with which to develop an ROI because the savings will be the difference in the costs of the two alternatives. The most common investment situation is to compare the purchase of some new equipment against the alternative of continuing to do things the way they are currently being done. This current method might be a manual or semi-automatic test as opposed to full automation, or it might be an older piece of automatic equipment. The current costs will be compared to the predicted costs with the new equipment to determine the savings. These savings will then be used to develop some form of ROI or some other measure of the financial attractiveness of the investment. The whole process is quite simple and has three stages:

1. Determine the costs for the alternatives.
2. From the cost data determine the savings.
3. From all of the above determine an ROI.

The second most common investment situation is the comparison of two or more similar alternatives, such as comparing the ROIs for two different board testers. This comparison may not be quite as straightforward because there are several options available. You can compare each tester in turn against the common 'baseline' option of continuing with the present method. However, the present method may not be a viable alternative. You may be in a situation where a change of tester is required to cope with a technology change which the present method cannot handle. In this case it would be a nonsense to try to calculate the cost of doing things in the old way when the old way can not do the job at any cost. In this type of situation an assumption has to be made that either tester will provide a substantial return relative to the old method, so now you can simply compare the financial performance of the two testers relative to each other.

11.2 Cash flow conventions

Before getting down to the basics of the various appraisal methods there are a few simple conventions that need to be understood. The term 'cash flow' is used to refer to funds that may flow into a project (a positive cash flow such as the gross savings) or that might flow out of a project (a negative cash flow such as the operating cost). The difference between all of the positive and negative cash flows in a given period of time is referred to as the net cash flow. This net cash flow can be positive or negative depending on the relative magnitudes of all of the individual cash flows. There is a standard way of showing all of these net cash flows called the 'cash flow diagram' and this is illustrated in Fig. 11.1. The convention is to show positive cash flows as an upwards pointing arrow above the horizontal 'time line' and to show negative cash flows as downwards pointing arrows below the time line. The time line usually runs for the economic life of the project or for the period over which the analysis is taking place. The position of the arrows along the time line indicates when the cash flow is deemed to be taking place. The initial investment, a negative cash flow, is usually the first arrow on the time line, but not always. It is possible that for a large project there may be cash flows occurring for several years before the investment is made. This is not usually the case with ATE since evaluations usually take less than one year.

A convention used in investment appraisal is that any major investments occur at the beginning of a time period and all operational cash flows occur at the end of the time periods. This simplifies the calculations and also adds an element of 'worst case' to the analysis. In practice, of course, the savings will be generated continuously throughout the time periods, which are usually one year long, rather than in one lump at the end of each period.

If you use a financial calculator you will already be familiar with the 'cash flow diagram' and the convention of entering negative cash flows as a negative number.

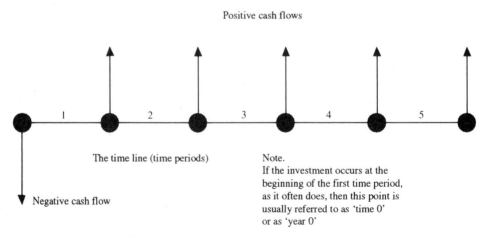

Figure 11.1 The cash flow diagram

Such calculators are themselves a good investment if you expect to do much financial appraisal because they will save a lot of calculation time.

11.3 The concept of increment investment

This concept was explained briefly in Chapter 4 and a simple example was shown in Fig. 4.1. Essentially, if the savings are calculated as being the difference between the operating costs of two alternative investments, then the investment required to obtain the savings is the difference between the two investment costs. It is not the absolute value of the investment required for the selected alternative. If one of the alternatives is to continue with current methods, with no investment needed, then the incremental investment required to introduce a new tester will indeed be the full investment cost of the tester. However, if the two alternatives under consideration

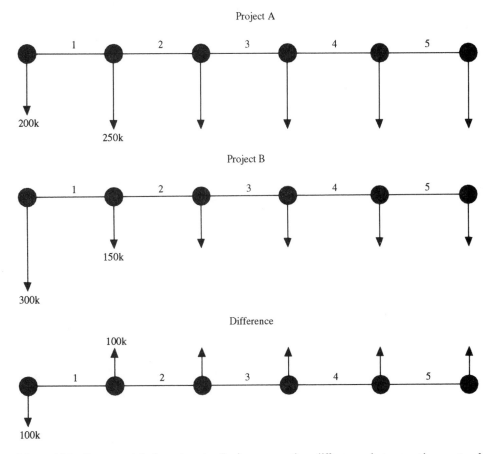

Figure 11.2 Incremental investment. Savings are the difference between the cost of two alternatives; they are generated by the difference (increment) in the investments required

are two different testers and the savings are determined as the difference in their operating costs, then the incremental investment will be the difference in the investment costs for the two testers. This becomes very obvious if we look at how the cash flow diagram is developed. Figure 11.2 shows three cash flow diagrams. The top two diagrams show the investment costs and the operating costs for the two alternatives. Since these are looking at the two alternatives individually they can only contain cost data, so all of the arrows point down. The third cash flow diagram shows the difference between the two cost diagrams which is the data we would use to determine the ROI. Here we now have some positive cash flows and a negative cash flow (the investment) that is the difference between the two investments. This is the *incremental* or the *net* investment. Obviously any ROI calculated from these figures will have resulted from the investment of the extra money rather than the absolute value of the investment needed.

This implies that you cannot develop an ROI when the cheapest alternative also has the lowest operating costs because the incremental investment would have a negative value. In cases like this all you can do is compare the costs over the life of the project and select the lowest cost option. Even this is not strictly necessary because it is completely obvious which of the two alternatives represents the best investment so long as the cost data used is correct. It is extremely important in cases like this that all of the costs that are influenced by the investment decision are included in the analysis.

This highlights why it is essential to consider the life cycle costs when comparing ATE systems rather than simply the costs directly associated with the testers. For example, a very simple in-circuit tester will usually have a lower investment cost than a more powerful ICT. It may also have lower direct operating costs in terms of programming and the use of the tester. Therefore this would fall into the above category. However, the real cost differences will lie beyond the actual board test stage. If these are not taken into account then the simple tester, which may well have the highest life cycle cost, may look more attractive than the better tester that produces better quality results.

11.4   Payback analysis

Polls conducted in the United States and in the United Kingdom show that payback analysis is the most commonly used method for rating and comparing alternative investments. It is, however, a very poor method to use for comparing alternatives and should only really be used for relatively small projects or as an additional criteria for large ones. Relying on payback analysis for making decisions about major investments is extremely dangerous. Payback analysis does not produce a measure of any return, as the example earlier indicated, but merely an indication of the time that it will take for the savings to recover the original investment. It is usually used in two different ways.

Its first use is to compare projects by ranking their payback times and selecting the alternative with the shortest time, and the second use is to define a payback time criteria that all investment projects have to meet or exceed. This second use is quite common during periods when the economy is in poor shape. It may well be used in this way in addition to some better form of selection criteria. Many companies will issue an edict that is something like the following example: 'No requests for capital funding will be considered unless the project will have a payback time of less than eighteen months.' The rationale behind this thinking is that a short payback time must mean that the project will have a good return. Unfortunately this is not necessarily going to be true as some of the examples will show.

The formula for payback is usually stated as

$$\text{Payback} = \text{net investment/average annual cash flows}$$

In practice this simply involves calculating the cash flows for as many years as is necessary to exceed the investment, dividing the cumulative cash flow by the number of years to get the annual average and then dividing the net investment by this figure.

Example

Net investment required = $1000

Year	Cash flow	Cumulative cash flow
1	400	400
2	200	600
3	300	900
4	300	1200

$$\text{Payback} = 1000/(1200/4)$$
$$= 3.33 \text{ years}$$

One of the problems of payback analysis is its inability to compare projects with different patterns of cash flow. The example shown in Fig. 11.3 illustrates the problem. Which of the two alternatives is the best investment? Most people would intuitively choose the first alternative (A) because it has the higher cash flows in earlier years. By saying that project A is preferable to project B we are in fact introducing the concept that money has a 'time value'. The earlier we receive a sum of money the greater its worth. This concept of the 'time value of money' will be covered later. Another possible reason for choosing project A might be on the grounds of risk. We can usually forecast what will happen next year with less risk than a forecast of what will happen in three years' time. Since the future cash flows are in fact forecasts, project A is inevitably less risky. So project A is better than project B—or is it?

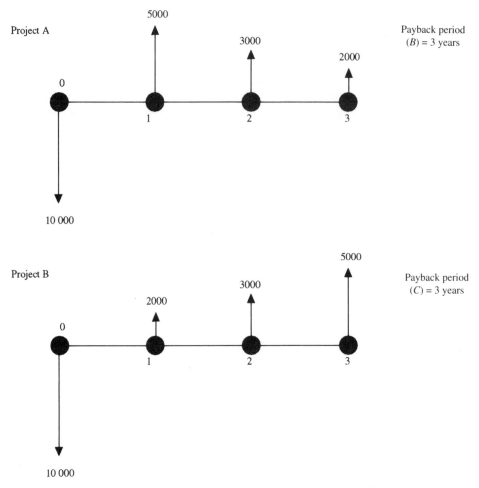

Figure 11.3 Comparing project A and project B we find dissimilar cash flows with the same payback time

Payback analysis normally only concerns itself with cash flows up to the payback time regardless of the economic life of the project. So assuming projects A and B both have five year lifetimes, what might happen in years 4 and 5? A quick look at the pattern of the cash flows would indicate that project B is more likely to have a high cash flow in year 4, whereas project A might have a very low one. This being so, project B could well have a better ROI, even allowing for risk.

The real answer, of course, is that we do not know. Without more information it is not possible to determine which is the better project in which to invest our money. With the information that is available we would probably be right to choose project A because the bigger cash flows early in the life of the project have the potential of being invested in something else, or of reducing the need for borrowing. Figure 11.4 shows what would happen if the cash flows could be invested at 15 per cent.

A. The $5000 received at the end of year 1 is reinvested at 15 per cent and becomes $5750 at the end of year 2. At this time, an additional $3000 is made on the project making a total of $8750. This is then invested at 15 per cent and becomes $10 062 by the end of year 3. At this time, a further $2000 flows out of the project, giving us a total of $12 062 at the end of year 3.
B. The $2000 received at the end of year 1 is reinvested at 15 per cent and becomes $2300 by the end of year 2. At this time $3000 flows out of the project, making a total of $5300. This is invested at 15 per cent and becomes $6095 by the end of year 3 at which time a further $5000 flows from the project, making a total of $11 095 at the end of year 3.

Clearly, project A is better than project B when looked at in this way. However, if the original $10 000 had been invested at 15 per cent it would be worth $15 209. Neither project is as profitable as investing the money—at least not over the 'payback period'.

The resulting payback period may be seen as a test of the liquidity of the investment. It does not represent a measure of profitability. It is, however, a commonly used criteria, especially in times of recession when many companies may have difficulty meeting operating expenses and limited ability to fund new investments. Under such conditions, the imposition of a short payback criterion will ensure that only quick-profit projects will be adopted, since cash availability may be more important than total potential profit at such times. Having a quick payback on something as complex as an ATE system will also guard against technological obsolescence.

Advantages and disadvantages

The major advantages of the payback method are that it is extremely simple to compute and to understand. It can also be used to eliminate unprofitable projects quickly. For instance, if the project's payback is longer than its expected life, the project is definitely uneconomical.

The major disadvantages of the payback method are its inability to differentiate between the timing of earnings and, perhaps more importantly, it concentrates its attention on the saving flows occurring only within the payback period—receipts in later years are ignored. A further defect lies in the fact that payback cannot reliably determine whether one project is better than another. It is unable to distinguish between two projects where one exhibits no savings flows outside the payback period (e.g. three years), yet one may continue to yield returns for a further three years (e.g. up to year 6). No account is taken of this possibility by the payback formula. As a result, it should only be used as an additional criteria to a more reliable method of evaluating the profitability of competing projects.

Even though the calculation of the payback time is a simple process it is frequently

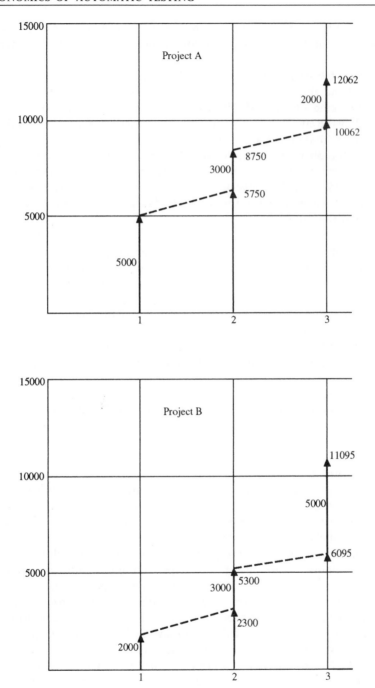

Figure 11.4   Reinvesting the cash flows at 15 per cent shows that project A is a better financial choice

performed incorrectly. In many instances the payback will be calculated using different combinations of the *full investment cost*, the *incremental investment cost*, the *gross annual savings* and the *net annual cash flows*. Each of these combinations will obviously yield different results so which is correct? If the cost savings have been calculated as the difference between the costs of two alternatives then the investment figure used should also be the difference between the two investments—in other words, the net investment. Most definitions of payback will in fact refer to the *net investment*. Similarly, it is the net annual cash flows that should be used to calculate the payback period because all of the positive and negative cash flows have to be taken into account to find the correct answer. Two major elements that need to be taken into account are the effects of depreciation of the asset and the taxation of the savings. This is actually quite straightforward.

The savings you make as a result of using the equipment under review will have the effect of increasing your profits and therefore the amount of tax you have to pay. This will therefore reduce the real amount of the gross savings and increase the payback time. It is not all bad news, however. The tax authority allows you to depreciate the capital equipment at some agreed rate, and this will reduce your tax bill and shorten the payback time. Clearly these effects have to be taken into account because they are real; you cannot legally avoid them. The example shown in Fig. 11.5 illustrates how a realistic payback analysis should be performed. Here the gross annual savings are $5000 per year. The permitted depreciation is 20 per cent of the investment per year (five year straight line) and the tax on profits is set at 50 per

Investment = $12000   Depreciation = 5 year/straight line   Tax rate = 50%

Year	Expenses	Income	Savings	Tax on savings	Depreciation	Tax saved	Net cash flow
1	2000	7000	5000	2500	2400	1200	3700
2	2000	7000	5000	2500	2400	1200	3700
3	2000	7000	5000	2500	2400	1200	3700
4	2000	7000	5000	2500	2400	1200	3700
5	2000	7000	5000	2500	2400	1200	3700

$$\text{Payback} = \frac{12000}{5000} = 2.4 \text{ years} \quad \textbf{X}$$

$$\text{Payback} = \frac{12000}{3700} = 3.2 \text{ years} \quad \checkmark$$

Figure 11.5   Calculating the payback time from the gross savings is wrong; the cash flows after tax and depreciation effects should be used to calculate the payback time

cent. If nothing else changes as a result of introducing this particular piece of equipment then the $5000 savings each year will result in extra tax to pay of $2500. The $2400 depreciation allowance will reduce the tax bill by $1200 so the net cash flow each year will be

$$5000 - 2500 + 1200 = \$3700$$

As you will see from the results in the diagram, the payback time calculated on the basis of the savings is 2.4 years. Based, as it should be, on the net cash flows, this time goes out to 3.2 years. Even though this is the correct payback time it is probably still not realistic as a measure of the financial attractiveness of the project because the analysis takes no account of the time value of money, which is just as real as tax and depreciation.

11.5 Accounting rate of return (ARR) (or average ROI)

Another commonly used method of appraising capital investment outlays is to calculate the project's simple rate of return. This is calculated by dividing the project's average annual savings (after allowing for tax and depreciation effects) by the total initial investment outlay:

$$ARR = \frac{\text{average annual cash flow}}{\text{net investment}} \times 100\%$$

This indicator attempts to determine what percentage of the initial investment the project returns each year. For instance, a project costing $10 000 while saving an average of $2000 per year during the life of the project would have 20 per cent average ROI. It is, in effect, the reciprocal of payback.

Just as with the payback method, the annual savings must be adjusted to be the net cash flows; for example, a $50 000 machine is expected to produce savings of $15 000 each year for five years. The annual maintenance costs will be $2000 and the equipment will be depreciated in a straight-line manner over its five year life. The average annual cash flow would be calculated as follows:

$$\begin{array}{lr} \text{Operating savings} & \$15\,000 \\ \text{Less maintenance} & \$2\,000 \\ \hline \text{Net savings} = & \$13\,000 \\ \hline \end{array}$$

This $13 000 would increase profits and therefore attract corporation or company income tax. However, the depreciation allowance can be deducted from the profit before tax is computed. Therefore,

$$\text{Depreciation} = \$50\,000/5 \text{ years}$$
$$= \$10\,000/\text{year}$$

and tax on the net savings would be (at a 50 per cent tax rate)

$$(13\,000 - 10\,000) \times 0.5 = \$1500$$

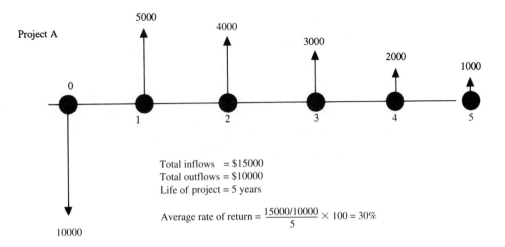

Total inflows = $15000
Total outflows = $10000
Life of project = 5 years

$$\text{Average rate of return} = \frac{15000/10000}{5} \times 100 = 30\%$$

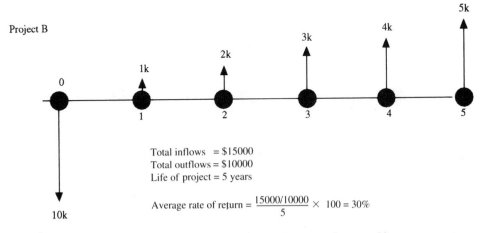

Total inflows = $15000
Total outflows = $10000
Life of project = 5 years

$$\text{Average rate of return} = \frac{15000/10000}{5} \times 100 = 30\%$$

Figure 11.6 In this example the accounting rate of return shows a 30 per cent return even though payback is only just achieved. The timing of the cash flows has no impact on the results

The net savings will be further reduced by this tax to a cash flow of $11 500 per year. The ARR would therefore be

$$\frac{11\,500}{50\,000} \times 100\% = 23\%$$

and the payback would be

$$\frac{50\,000}{11\,500} \quad \text{or 4.35 years}$$

where the reciprocal of 4.35 is 0.23.

As with payback, this measure fails to include the time value of money, and on the basis of the percentage return on the investment figure alone, we are not able to distinguish between projects resulting in the same return. It does, however, normally look at the entire expected life of the project. The example shown in Fig. 11.6 clarifies the problem. Again, as in the payback example of Fig. 11.3, there are very different patterns of cash flows but the result is the same. In this case, project A is clearly superior to B since we do know the expected cash flows over the life of the project, and A gives the bigger returns early in its life. The advantages of this method are its simplicity and the ease with which it can be used to eliminate unprofitable investments quickly. Its disadvantages are similar to those of payback, the major one being the ignoring of the time value of money. It should not be used alone to compare alternative projects or to determine ROI.

Another problem with this method is that there is no standard approach for calculating the answer. The method used above takes account of depreciation and

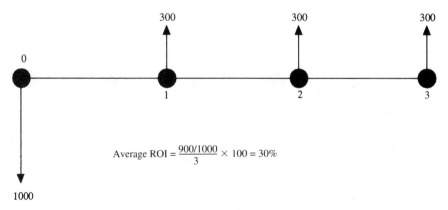

Figure 11.7 In this example the accounting rate of return shows a 30 per cent return even though payback is only just achieved. The timing of the cash flows has no impact on the results, which are not achieved

taxes, but some books on the subject do not mention the need to consider these factors. If ARR is calculated without considering these, then some very strange results can occur. Consider the example in Fig. 11.7 and assume that the cash inflows are the gross savings. The ARR computes out to 30 per cent but the project has not even achieved payback—only $900 has come in from an investment of $1000.

Despite all the disadvantages of the payback and the ARR methods, these 'rule-of-thumb' approaches are still widely used throughout industry. However, more and more they are being superseded by methods using discounted cash flow (DCF) techniques. Many companies now insist on an ROI calculation using DCF methods and possibly supplemented by a payback analysis.

11.6 DCF and the time value of money

Many people are put off by the terminology used by accountants and financial people. DCF is in fact a very simple mathematical process and is simply the inverse of compound interest, which most people are fairly familiar with.

Compounding is a method of determining the future value of a present sum of money at some given interest rate.

Discounting is a method of determining the present value of a sum of money to be received some time in the future at a given interest rate. In discounting, the interest rate is called the discount rate.

The formula for compounding is

$$FV = PV(1+i)^n$$

where

FV = future value
PV = present value (PV is sometimes also referred to as 'present worth')
i = interest (expressed as a factor)
n = number of periods

If a business can earn 10 per cent on its investments, an original sum of $100 will grow in the following way:

Year 0	Year 1	Year 2	Year 3	Year 4
$100	$110	$121	$133.1	$146.61

Alternatively, if it wants to see how much a future sum is worth today, assuming it could get 10 per cent for its money, it needs to find the PV of $1 at a 10 per cent discounting rate. Figure 11.8 shows how this could be calculated.

The formula for discounting is

$$PV = FV(1+i)^{-n}$$

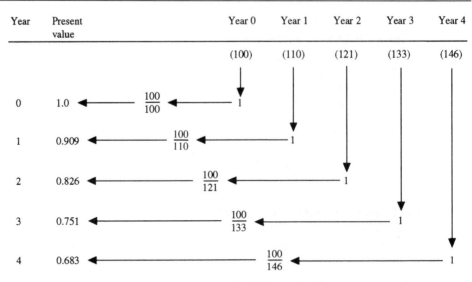

Figure 11.8 Determining the present value of a future sum of money

The PV of one unit of currency is used as a present value factor (PVF) for converting any future sum to its PV at a given rate of interest (discount). Tables of PVFs are readily available (see Appendix E) and also many pocket calculators now have the built-in capability of converting PV to FV and vice versa.

The time value of money is sometimes confused with inflation. Inflation certainly reduces the purchasing power of money, but has nothing to do with the basic concept of its time value. The time value is simply related to the fact that if you have $100 now you can invest it at some interest rate and you will have more dollars in a year's time regardless of its purchasing power. However, the effects of inflation are big enough that they cannot totally be ignored. If we can buy something today for $100 or invest it at 10 per cent, we would have $110 in a year's time. If the item we want to buy goes up in price at something higher than 10 per cent we would be better off buying it now and having the use of it for a year. This is particularly true if the item we want to buy will give us a better ROI than 10 per cent.

In general, however, interest rates and inflation are fairly closely linked, so for many purposes inflation can be ignored when performing DCF calculations. In fact, it can be argued that the whole point of bringing the value of future cash flows to some common point in time is to eliminate the effects of inflation. This is fairly valid if you consider that during the life of the project being evaluated there will be growth in the company, some of which is real and some inflationary. Conversely, some experts argue that, to offset the effects of inflation, the minimum ROI expected from an investment should be somewhat higher than the cost of money. I personally prefer to think of any addition to the minimum required ROI as an allowance for the *risk* that the project may not perform as well as expected.

Table 11.1 Net cash flows converted to 'discounted cash flows' at a present value of 10 per cent

Year	Net cash flow	PVF (10%)	DCF
1	200	0.909	182
2	500	0.826	413
3	400	0.751	301
4	300	0.683	205
		PV = 1101	

Example of use of DCF techniques

Table 11.1 shows the cash flows expected from an investment of $1000 in a piece of machinery and how these cash flows are converted to DCFs to arrive at the present value of all the cash inflows from the project.

Since the PV of the project exceeds the initial investment, it is producing a return in excess of the 10 per cent discount rate. We can now take a more detailed look at the two most common methods of using DCF techniques for financial appraisal of investment projects. These two methods are commonly called the net present value (NPV) method and the internal rate of return (IRR) method.

11.7 Minimum acceptable rate of return (MARR)

The minimum acceptable rate of return (MARR) is usually established by the financial management of the company and can also be known as the *criterion rate of return* (UK) or the *hurdle rate* (USA). There is always a lot of discussion, but very little agreement, on how this minimum rate of return should be established. There is reasonable agreement that the minimum level for MARR should be the *cost of funds* for the company, but establishing this is also open to some dispute. However, the real problems come in deciding how much to add to the cost of funds to allow for the risk involved with any major investment. The rationale here is that since the cash flows established for the ROI calculation are mainly forecasts of the future outcome of using the equipment, then some allowance for the risk that these may be in error should be included. This subject of setting the MARR is discussed in more detail later in this chapter.

11.8 Net present value (NPV)

In most American texts that I have seen on engineering economics this is called the *net present worth* (NPW) and yet financial calculators of US origin (Hewlett Packard) have PV and NPV keys. It would therefore appear that both terms are in common use within the United States. This method involves taking the minimum acceptable rate of return (MARR) and, using that rate, calculating the present value of the future cash flows. The original investment is then subtracted from the present value (PV) to arrive at the net present value (NPV). Whatever MARR is used, the NPV method requires that every cash flow is discounted in each year by the appropriate discount factor. For instance if the MARR is 12 per cent then each year's net cash flows will be discounted by multiplying them by the following present value factors (PVFs):

Year 1   0.893

Year 2   0.797

Year 3   0.712

Year 4   0.636

Year 5   0.567

and so on until the estimated lifetime of the project, or the required period for the analysis, has been reached.

Another useful index that can be computed easily once the discounted PV has been calculated is the profitability index (PI). The formula for calculation is

$$PI = \frac{PV \text{ of net cash flows}}{\text{original amount invested}}$$

The following example will help clarify both the NPV method and the PI. For an original investment or $50 000, the cash flows shown in Table 11.2 are envisaged.

Table 11.2 Expected cash flow converted to a present value (sum of the DCFs), and a net present value

Year	Cash flow	PVF (8%)	DCF
1	7 000	0.926	6 482
2	8 000	0.857	6 856
3	8 000	0.794	6 352
4	9 000	0.735	6 615
5	9 000	0.681	6 129
6	9 000	0.630	5 670
7	8 000	0.583	4 664
8	8 000	0.540	4 320
9	8 000	0.500	4 000
10	8 000	0.463	3 704
	$82 000		PV = $54 792

NPV = 54 792 − 50 000 = +$4792

The total PV sum is greater than the amount of the original investment and therefore the project earns a higher rate than 8 per cent and would therefore be acceptable. The PI can be calculated as

$$\frac{54\,792}{50\,000} = 1.096$$

When the index falls below unity then the project has failed to meet the criterion rate of return.

The PI can quite readily be used to rank projects according to their profitability. It gives an easy guide to those projects offering the best rates of return.

Another way to view NPV is—for the example above—'This project meets the criteria of earning an ROI of 8 per cent, and it also "throws off" an additional $4792, which can be invested in other projects.'

NPV advantages

The obvious advantage that this method has over those of payback and average ROI is the inclusion of the concept of the time value of money. At the same time NPV is expressed in units of currency so most people have a good feel for the relative worth of several projects.

NPV disadvantages

The only real disadvantage of the NPV method lies in the need to establish an MARR or 'hurdle rate' that is used for discounting the cash flows. As indicated earlier, there is often very little agreement on how this should be established and what factors should be included. It was primarily because of this that the internal rate of return method was developed. However, as we will see, this also requires that an MARR is established.

11.9 Internal rate of return (IRR)

The NPV method sets out to determine whether the rate of return on a project exceeds the established MARR. If the calculated NPV is positive then the project provides the minimum return and also some surplus cash (the NPV) that can be used to invest in other projects. A profitability index can then be calculated to rank the alternatives in their order of financial attractiveness. If the NPV is negative then the project has failed to meet the minimum requirements and the project proposal should be rejected. If the NPV is zero then the project just meets the requirement with no surplus. This is therefore the actual return for the project, or its *internal rate of return*. In the United Kingdom this was in fact called the *project rate of return* for a while.

The IRR method was seen as a way of overcoming the one major problem with the NPV method, which is the need to establish a discount rate (the MARR or hurdle rate) to use in the calculation. Unfortunately this was somewhat illogical because you still need some sort of MARR to determine whether a calculated IRR is acceptable. For example, you may calculate the IRR of a potential investment to be 15 per cent. Do you accept this as being a good investment or do you reject it as being a poor investment? Without some reference you do not know. The problem is easier if you are considering two or more alternatives versus the baseline of continuing with existing methods. Three alternatives may have IRRs of l0, 12 and 15 per cent. Clearly the alternative with the 15 per cent IRR is the one you should select, but is 15 per cent a good enough return? You still really need to establish an MARR even for the IRR method.

The simplest definition of IRR is 'the discount rate at which the NPV is zero'. There is no direct formula for computing IRR so it has to be determined from a series of trial and error calculations and some interpolation. Financial calculators and computer programs that can calculate IRR also have to use some sort of successive approximation approach. Most spreadsheet applications for personal computers have NPV and IRR functions available.

Example IRR calculation

The project illustrated in Fig. 11.9 requires the calculation of IRR before presentation to management. The procedure is as follows. Taking our cash flows we choose an initial discount rate and compute the NPV (Table 11.3).

Our resulting NPV is a positive figure so we must choose a higher discount factor. Performing the same computation with an 18 per cent discount factor we get the results shown in Table 11.4.

In this case we have computed a negative NPV and so we try a lower rate of discount, e.g. 15 per cent. These results are given in Table 11.5.

A 15 per cent discount factor has resulted in an NPV of 0. Therefore, the IRR on the project is 15 per cent. If the money can be obtained at a rate of interest lower than this figure then the project may be said to be profitable.

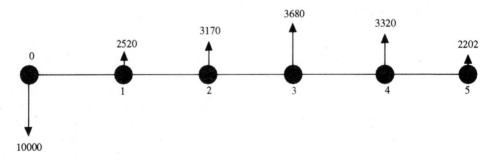

Figure 11.9 The cash flow diagram for the IRR example

Table 11.3 IRR calculation. Discount rate set too low

Year	Net cash flows	Discount factor (10%)	DCF
1	2520	0.909	2 290
2	3170	0.826	2 618
3	3680	0.751	2 763
4	3320	0.683	2 267
5	2202	0.621	1 367

PV of receipts = 11 305
Initial investment = 10 000
NPV = + 1 305

Table 11.4 IRR calculation. Discount rate set too high

Year	Net cash flow	Discount factor (18%)	DCF
1	2520	0.847	2 134
2	3170	0.718	2 276
3	3680	0.605	2 241
4	3320	0.516	1 713
5	2202	0.437	962

PV of receipts = 9 326
Initial investment = 10 000
NPV = − 674

Table 11.5 IRR calculation. Discount rate set correctly at the IRR

Year	Net cash flow	Discount factor (15%)	Present value of cash flow
1	2520	0.870	2 191
2	3170	0.756	2 397
3	3680	0.658	2 421
4	3320	0.572	1 899
5	2202	0.496	1 092

PV of receipts = 10 000
Initial investment = 10 000
NPV = 0

For a complicated analysis the trial and error method of obtaining the correct discount rate can be very time consuming; a method for pre-estimating the discount rate will save some valuable time in obtaining the correct rate. Begin by totalling the cash flows and ascertain by how much they exceed the original investment. Divide this excess by the number of years the project will extend over.

Finally, calculate the percentage that the resulting figure represents of half the amount of the original investment. For example, consider the following cash flows:

Year	Cash flow (E)
0	− 10 000
1	+ 3000
2	+ 3000
3	+ 3000
4	+ 3000
5	+ 3000

(a) Initial investment = $10 000
Cash flows total = $15 000
Excess = $5000
(b) a divided by number of years = $5000/5 = $1000
(c) b as a percentage of half the original investment =

$$\frac{1000}{5000} \times 100 = 20\%$$

Making the initial calculation at 20 per cent will require less trial and error calculations than simply guessing an initial rate.

Advantages of IRR

The usual main advantage quoted for IRR is the fact that it eliminates the need to determine a discount rate to apply to the annual cash flows. However, as indicated earlier, this is not really valid since there will still be a need for some MARR to be applied to any investment decisions. The one real advantage that IRR has is that the results of any multiple analyses are perhaps a bit more intuitive than the same results using NPV. For example, if two alternatives requiring different levels of investment are compared with a baseline situation and produce IRRs of 18 and 22 per cent, it is easy to pick a winner. The same analysis performed using the NPV method and a discount rate of 16 per cent might produce NPV figures of $22 000 and $24 000, the interpretation being that both alternatives give a return of 16 per cent and they also produce a cash surplus of $22 000 and $24 000 respectively. Since the original investment levels are different it may not be obvious as to which of these is the better investment. However, the purpose of the *profitability index* is to eliminate the effect of any differences in the investment level.

Disadvantages of IRR

In the past one of the main disadvantages of IRR calculations was the need to perform multiple manual calculations due to the trial and error approach that is required.

Today almost everybody has access to a PC or a MAC running a spreadsheet program, such as Lotus 123, Microsoft Excel, Borland Quattro Pro or a host of others, such that this is no longer a barrier to using IRR. The real problems with the IRR approach are much more fundamental and serious. In fact the method suffers from so many problems that it is surprising that it is so popular. Several new ROI calculation methods have been introduced to overcome IRR's deficiencies, and although these are gaining in popularity they are all more complex than the NPV method that IRR was supposed to simplify.

For the majority of straightforward comparisons the NPV method and the IRR method will agree in terms of the ranking of the alternatives, but there are situations where they will not agree. When this happens it will invariably be the IRR solution that is incorrect. There are three main problem areas:

1. The IRR method may result in excessively high rates of return which will not be valid because of *the reinvestment assumption* (see below).
2. The IRR method may reverse the ranking of projects relative to the results from an NPV calculation.
3. The IRR method can result in more than one mathematically correct answer.

The IRR method frequently results in very high rates of return which may sound so wonderful that you cannot believe that anyone could turn down your pet project. 'Who in their right minds would not jump at funding a project with an IRR of 78 per cent?' It sounds great, and I have seen many published examples of investment in ATE producing returns around this level and even much higher. Such levels of IRR will occur frequently if you are comparing an automatic solution with a manual solution that takes a long time. The savings will be enormous and this will lead to a very high IRR. This becomes particularly dangerous if you make the mistake of comparing an automatic solution with a manual solution, when the manual solution is not really viable. It is a nonsensical comparison that is bound to produce a high return. These high levels of IRR are actually quite misleading and meaningless. The reason for this is the *reinvestment assumption* mentioned above, so just what does this mean?

The reinvestment assumption

The obvious interpretation of IRR is that it is the return that the project is expected to make on the initial investment. This is correct up to a point but if you analyse the return and the cash flows you will get some strange results. If we take a look at the example in Table 11.5 you will see the problem. In the example the IRR worked out to be 15 per cent. Now, if we were to put the $10 000 in the bank at 15 per cent and take out the interest earned, at the end of each year we would have a cash flow of $1500 each year and a total over the five years of $7500 (before any discounting). We also still have the $10 000, so the total amount is now $17 500. The sum of the undiscounted cash flows in Table 11.5 is only $14 892. This implies that we have recovered the original $10 000 and gained another $4892. This implies that the project may actually be earning less than 15 per cent. Alternatively, if we leave the interest in the bank each year then the $10 000 would grow over the 5 years, at 15 per cent compound interest, to $20 114. Now we have gained $10 114 over and above the original $10 000, so it looks as if the IRR figure should be a lot less than 15 per cent. What is happening here? We would not expect the simple interest amounts to be the same as the compound interest amounts, but why do they both differ so much from the cash flows generated from the project, which are also supposed to be producing a 15 per cent return?

The answer is that we need to reinvest the cash flows produced by the project, at the IRR rate, in order to get agreement with the compound interest amounts. In other words, the cash flows themselves are only a part of the return from the project; the other part of the return is the additional interest we receive by reinvesting the cash flows. Another way to view this is that the 'IRR is the interest earned on the unrecovered balances of an investment such that the final investment balance is zero'.

This sounds rather complex; however, the diagram in Fig. 11.10 may clarify the meaning of this. In year 0 we invest $10 000 in the project. Had we invested this at 15 per cent we would have $11 500 at the end of year 0 (or the start of year 1). At this point our investment in the project is $11 500. We have, however, received a cash flow resulting from this project of $2520, giving us an unrecovered balance (or present investment) of $8980 at the start of year 1. If this process is continued using the IRR rate we end up with full recovery at the end of year 5. In effect the project has recovered the initial investment of $10 000 plus the interest (at the IRR) on the unrecovered balance at the end of each period. Another definition is that the IRR is the percentage earned on the amount of capital invested in each year of the life of the project after allowing for the repayment of the sum originally invested.

For the cash flows arising from the project to recover the investment and the interest on the outstanding balances (which is effectively the amount invested in the project each year) they need to be reinvested at the IRR rate for the remainder of the lifetime of the project. If we had invested our $10 000 at 15 per cent for 5 years it would have grown to $20 114, so our reinvested cash flows need to match this. The calculation in Table 11.6 shows that they do more or less match.

This then shows the flaw in the IRR approach. Since the savings or cash flows need to be reinvested at the IRR rate in order for the IRR rate to be valid, we have a problem if we cannot reinvest at this rate.

This reinvestment assumption is also true of the NPV method. However, the criterion or hurdle rate chosen usually represents a logical expectation of return on capital, so the funds generated each year by the project should have a good chance of being reinvested at the hurdle rate. This, of course, is a good argument for not making the hurdle rate too high.

The reinvestment assumption means that, for high return projects, the IRR method may produce unrealistic returns and also may rank projects in the wrong order of profitability. The NPV method will always rank projects in the correct order of profitability, assuming that the hurdle rate is a reasonable one, and so is the preferred method to use when evaluating ATE.

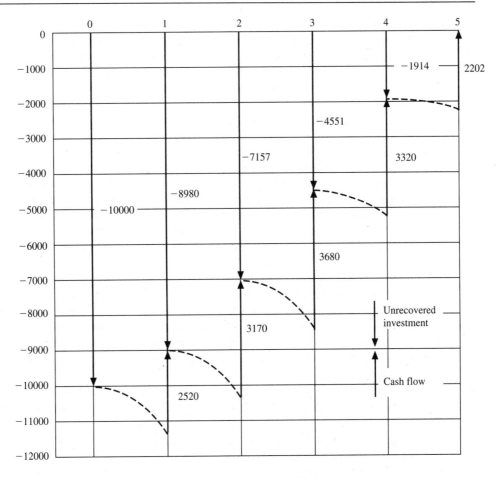

Figure 11.10 IRR is the rate at which the unrecovered balances of an investment need to be reinvested so that the NPV is equal to zero

Table 11.6 Cash flow reinvested at the IRR

Year	Cash flow	Years reinvested	Compound interest factor (15%)	Cash at end of project
1	2520	4	1.749	4 408
2	3170	3	1.521	4 821
3	3680	2	1.323	4 867
4	3320	1	1.150	3 818
5	2202	0	1	2 202
				20 116

Reversal of ranking

The second problem with the IRR method is that it may occasionally rank projects in a different order to the NPV method, and when this happens the NPV method is usually more accurate. Since the whole point of financial appraisal is to rank the alternative uses of funds, this is a fairly serious shortcoming. However, when this happens it is usually because the pattern of the cash flows is very different between the two alternatives. To illustrate how this problem occurs I have plotted the NPV curves for the two alternatives shown in Table 11.7. These curves appear in Fig. 11.11 and show the NPV versus a range of discount rates. The problem occurs because the curves for project A and project B cross each other. If you use the NPV method then project A will be preferred if the discount rate is set below the value where the curves cross. If the discount rate is set above this point then project B would be selected, as it would be if the IRR method was used. Project A has an IRR of 25 per cent whereas project B has an IRR of 33 per cent. For this example these are rather high rates which would probably be unrealistic due to the reinvestment requirement. If the cash flows have been calculated properly these are 'after tax' returns and it may be rather difficult to find other investments that can yield such high returns. In cases like this the NPV method, with a realistic discount rate, would be a better approach to use since it is unlikely that the MARR would be set as high as the point where the curves cross. The problem is that many companies only perform the IRR calculation so they would be unaware of such potential problems. My recommendation would be to plot NPV curves as I have done in Fig. 11.11. Using a financial calculator or a spreadsheet, this can be done in a few minutes once the cash flows have been entered. Such curves will immediately highlight any potential problems including the third major problem with IRR, which is the potential for multiple answers.

Table 11.7 Plotting data for the curves in Fig. 11.11

Year	Net cash flows for project A	Net cash flows for project B
0	−10 000	−10 000
1	1 000	8 000
2	3 500	4 000
3	5 000	2 000
4	6 000	2 000
5	6 000	1 000

Discount percentage	NPV for project A	NPV for project B
0	11 500	7000
5	8 083	5403
10	5 382	4068
20	1 462	1968
30	−1 167	400
Internal rate of return (IRR)	25.09%	33.05%

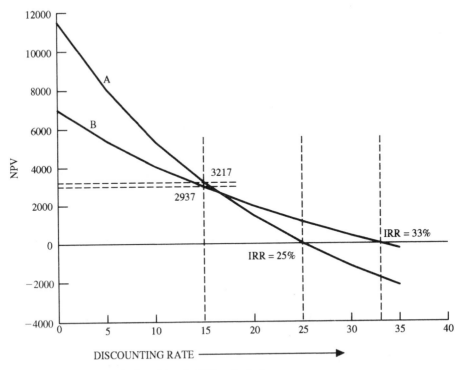

Figure 11.11   Reversal of ranking in IRR calculations

The problem of multiple IRRs

The nature of the formula for IRR is such that there will be one additional mathematically correct answer for every time that there is a sign change in the cumulative value of the cash flows. All of the examples to date have shown a single negative cash flow, the initial investment, followed by a number of positive cash flows resulting from the savings created by using the equipment. This is the most common pattern of cash flows and there is only one sign change. However, you can get a negative cash flow in any year when the outgoings exceed the savings, and this can result in another sign change in the cumulative cash flow value. If you are comparing alternative test strategies or test tactics for a major project that will run for several years it might be necessary to buy additional testers some time into the project as production volumes build up. If this happens you could well get a negative cash flow occurring one or two years into the project.

Figure 11.12 shows what happens to the NPV curve when there are two sign changes in the cumulative cash flows. Instead of continuing to fall as the discount rate is increased, the NPV starts to rise again and crosses the zero line. Since the IRR is the discount rate at which NPV equals zero, we have two values for IRR which are both mathematically correct. In cases such as this example you can ignore

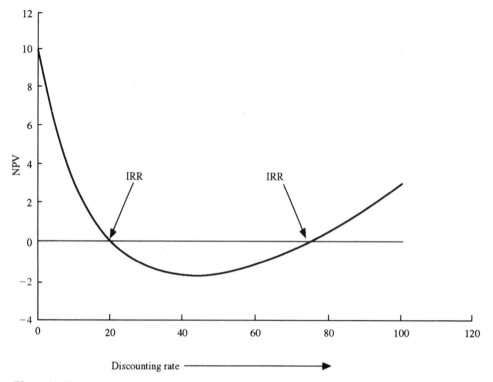

Figure 11.12 Multiple answers in IRR calculations

the higher answer because the reinvestment assumption will make this meaningless. However, you may not be aware of the fact that there are two answers because the trial and error method needed to calculate IRR will tend to converge on a single answer. Once you have obtained an answer why would you look for a different one? The important issue here is to be aware of the problem. If your financial people insist on using the IRR method, as many do, then look at the pattern of cash flows to see if there is more than one sign change in their cumulative value. If there is then it would be prudent to calculate NPVs for a range of discount percentages and plot a curve like the one in Fig. 11.12 rather than using an IRR calculation directly. Some IRR calculations will fail to compute under these conditions. A few calculations of IRR require you to enter a preliminary guess as a starting point for the trial and error iterations. By running such a program twice, once with a very low guess and once with a very high guess, you may arrive at both answers.

11.10 Modified internal rate of return (MIRR)

A number of methods have been developed to overcome the disadvantages of the IRR approach. One of the most widely used of these newer methods is the *modified internal rate of return*. This method eliminates the main problems of the reinvestment assumption and the possibility of multiple answers. It is a fairly logical approach and is quite simple to calculate. However, it does require that *two* interest rates are established. You determine a realistic rate at which any positive cash flows can be reinvested and therefore overcome the reinvestment assumption problem. You then determine a 'safe rate' for the cost of funds for any investments (any negative cash flows) required over the life of the project. All positive cash flows are converted to a *future value* at the 'end of life' point on the time line, by *compounding* their values at the established *reinvestment rate* for however many years that exist between the cash flow point and the end of life point. This sounds a lot more complex than it really is and can be better understood by looking at the diagram in Fig. 11.13, which

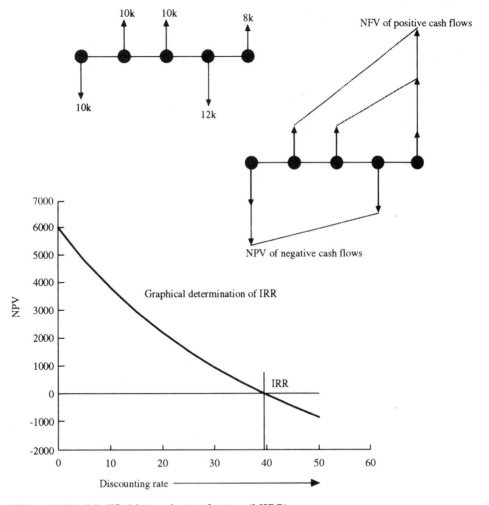

Figure 11.13 Modified internal rate of return (MIRR)

shows a project with a four year life. There are three positive cash flows at the end of years 1, 2 and 4, and two negative cash flows at the beginning of years 1 and 4. The first positive cash flow is compounded at the reinvestment rate for three years to determine its value at the end of year 4. Similarly, the second positive cash flow is compounded for two years at the reinvestment rate. The third positive cash flow needs no compounding because it occurs at the end of year 4. These future values are then summed to obtain a net future value (NFV).

Now the negative cash flows are *discounted* at the 'safe rate' (cost of funds) to bring them to a *present value*. The first negative cash flow needs no discounting because it occurs at the present time. The second negative cash flow is discounted for three years at the 'safe rate' and added to the first to obtain the net present value (NPV) of the negative cash flows. The MIRR for the project is the interest rate at which the NPV of the negative cash flows equals the NFV of the positive cash flows. Table 11.8 shows the calculation of the NPV and NFV figures. The MIRR works out to be 15.8 per cent. This is considerably less than the IRR of almost 40 per cent determined graphically from the NPV curve in Fig. 11.13, and reflects the difference between using a realistic reinvestment rate of 10 per cent rather than the 40 per cent implied by the IRR calculation. Incidentally, my trusty HP-12C financial calculator failed to compute IRR for this series of cash flows until I supplied it with an estimate. Lotus-123 requires an initial guess in its IRR function anyway, so it got the answer first time. They both agreed at 39.492 per cent.

Table 11.8   Calculation of MIRR. Interest (MIRR) at which $18 584 compounded annually for four years becomes $33 410 = 15.8%

Positive cash flows (end of periods)

Year	Cash flow	Time left, years	Compound factor (10%)	Future value (FV)
1	10 000	3	1.331	13 310
2	10 000	2	1.21	12 100
3	0	1	1.1	0
4	8 000	0	1	8 000
			Net future value	33 410

Negative cash flows (beginning of periods)

Year	Cash flow	Discount period, years	Present value factor (12%)	Present value (PV)
1	10 000	0	1	10 000
2	0	1	0.893	0
3	0	2	0.797	0
4	12 000	3	0.712	8 544
			Net present value	18 584

The MIRR method is a very logical approach in that all investments in the project (the negative cash flows) are brought to a present value (current cost) at a rate equivalent to the maximum cost (safe rate) of the funds, and the future savings (positive cash flows) are taken to a net future value at a realistically achievable reinvestment rate. Its main disadvantages are the need to define two interest rates and the fact that financial calculators and spreadsheets do not have a direct MIRR function available. The calculation is straightforward but it does take longer than NPV or IRR. In some ways it seems odd that the IRR method was introduced to overcome the need to establish a criterion rate (hurdle rate or MARR) and yet the method developed to overcome the problems with IRR requires two rates to be defined.

11.11 The external rate of return (ERR)

This is another method developed to overcome the deficiencies of IRR that is gaining in popularity. In this approach the positive cash flows are compounded at a defined realistic reinvestment rate and summed to a *net future value* (NFV) at the end of the project life, in the same way as with the MIRR method. However, in the ERR approach the negative cash flows are also compounded to a net future value at the end of the project life, instead of being discounted to a present value. The ERR is the rate at which these negative cash flows have to be compounded in order to be equal in magnitude to the net future value of the positive cash flows compounded at the reinvestment rate. Again the verbal description sounds a lot more complex than the actual process, which is shown in Fig. 11.14.

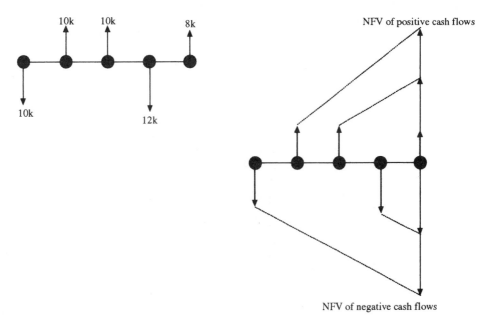

Figure 11.14 The external rate of return

If there is only one negative cash flow the ARR can be computed directly but if there are two or more negative cash flows then a trial and error approach, with interpolation, has to be used. For the example the positive cash flows have an NFV of $33.41k and we need to find the interest rate at which $10k compounded for four years plus $12k compounded for only one year will equal this.

The compound interest factor for a given interest rate and period is usually written as

$$(F/P, i, n)$$

so the equation we need to solve is

$$10k(F/P, 10\%, 3) + 10k(F/P, 10\%, 2) + 8k(F/P, 10\%, 0) = 10k(F/P, i, 4) + 12k(F/P, i, 1)$$

Tables of compound interest factors are available but they can easily be calculated since the factor is $(1+i)^n$, where i is the interest as a factor and n is the number of periods. The above equation becomes

$$(13.31 + 12.1 + 8) = 10k(1+i)^4 + 12k(1+i)$$

This reduces to

$$3.341 = (1+i)^4 + 1.2(1+i)$$

Making $i = 0.2$ (that is 20 per cent), the right-hand side of the equation becomes 3.514 so 20 per cent is too high. Making $i = 0.17$ produces an answer of 3.278 so 17 per cent is too low. Making $i = 0.18$ produces an answer of 3.355, just too high. Interpolating between these values,

$$17\% + 1\% \left[\frac{3.341 - 3.278}{3.355 - 3.278} \right] = 17.818\%$$

So the ERR for this set of cash flows is 17.818 per cent. The MIRR we calculated was 15.8 per cent, not too different considering the difference in the methods, but both very different to the 39.492 per cent IRR.

Both the MIRR and the ERR methods overcome the drawbacks of the IRR method. Of the two the MIRR is easier to calculate if there are two or more negative cash flows. However, I cannot help thinking that the NPV method with a correctly established MARR is the simplest approach. It is easy to interpret the results and the profitability index always ranks alternatives in the correct order of financial attractiveness.

A word of caution. Some of the simple examples presented in earlier chapters included the amortized cost of the equipment spread over the time or the workload of the project being analysed. This is not an unreasonable approach if you want to take a simple look at the relative *costs* of several alternatives. However, it should be obvious from the discounted cash flow examples presented here that the cost of the capital equipment should be *excluded* from the operational costs if you plan to develop an ROI. The reason is that the capital cost would get counted *twice* in the analysis. Any amortized equipment costs used in the development of the savings will obviously lower the savings by the amortized amount. Then in the DCF calculation of, say, the NPV, the capital cost will be deducted from the sum of the present values of the net cash flows which will already have been reduced by the amortized cost of the equipment. *ATE can be expensive enough without doubling its cost in your justification.*

11.12 The effects of taxation

The examples used to illustrate the different methods of financial appraisal have all referred to cash flows rather than savings. This is because the savings resulting from a project will be modified by taxation.

Assuming nothing else changes, additional savings will go right through to the bottom line of the profit and loss statement as an increase in profits or earnings. As such they will be subject to corporation tax. The investment in the equipment will, on the other hand, reduce the tax liability of a company. For an accurate assessment of the ROI on a project these important factors must be taken into account.

11.13 Depreciation

Depreciation means a decrease in value or worth. Most assets will lose value as they get older due to wear and tear. This is particularly true of mechanical equipment. The depreciation on a car, for instance, is something that affects most of us. With electronic equipment, like ATE, the main reasons for a decline in value is more likely to be a 'technological depreciation' than something wearing out. This decrease in worth is a valid cost of production in the same way that material and labour costs are. Depreciation can therefore be added to a company's expenses before calculating the profit before tax. Thus it has the effect of reducing taxation. Since in most countries company tax is about 35–50 per cent the benefit will be 35–50 per cent of the depreciation. On the negative side will be the need to pay 35–50 per cent tax on any savings from the project since these will effectively add to the overall profit of the company.

Example

A company purchases a $100 000 machine that will last for 5 years. It can be depreciated linearly over its five year life and the company expects it to save $30 000 a year in operating costs. The net cash flows will be as shown in Table 11.9.

A very beneficial refinement to this calculation of net cash flows may be possible depending on the nature of the tax laws of the country. Usually some or all of the

Table 11.9 How taxation and depreciation allowances affect the 'net cash flows'

Year	Savings	Tax on savings (at 50%)	Depreciation	Tax saved by depreciation	Net cash flows
	A	B	C	D	A − B + D
1	30 000	15 000	20 000	10 000	25 000
2	30 000	15 000	20 000	10 000	25 000
3	30 000	15 000	20 000	10 000	25 000
4	30 000	15 000	20 000	10 000	25 000
5	30 000	15 000	20 000	10 000	25 000

Table 11.10 The effect of a one year delay in tax liability

Year	Savings	Tax on savings (at 50%)	Depreciation	Tax saved by depreciation	Net cash flows
	A	B	C	D	A − B + D
1	30 000		20 000	10 000	40 000
2	30 000	15 000	20 000	10 000	25 000
3	30 000	15 000	20 000	10 000	25 000
4	30 000	15 000	20 000	10 000	25 000
5	30 000	15 000	20 000	10 000	25 000
6		15 000			(15 000)

first year's depreciation can be taken in the year that the equipment is purchased. It will, however, take a year to accumulate one year's savings from the use of the equipment. This can effectively introduce a one year delay between the tax benefit of the depreciation and the tax penalty of the savings. The calculation of net cash flows would then be as shown in Table 11.10.

A sixth year has to be added to the table to allow for the delay in paying tax on the fifth year's savings. A negative cash flow occurs in the sixth year. However, when you consider the effects of discounting, this will be small relative to the extra cash flow in year one. The result of this is that the delay creates a better ROI and a faster payback.

In the United Kingdom, a different method of compensating for investments in capital equipment is used. Here, depreciation is regarded as an internal bookkeeping exercise and is simply added to the profit before tax. The tax inspectors will then subtract something called 'capital allowances' from the total profit before tax (PBT) before calculation of tax. Therefore, unlike most countries where the percentage depreciation possible in a given year is determined by the tax authorities, in the United Kingdom it can, within reason, be determined by the company's accountant because it is not used in the determination of corporate taxes.

Table 11.11 The current (1993) UK system of capital allowances versus a straight line depreciation for five years

Year	Five year straight line	UK system	Balance of allowances
1	20 000	25 000	75 000
2	20 000	18 750	56 250
3	20 000	14 062	42 188
4	20 000	10 547	31 641
5	20 000	7 910	23 730
Total	100 000	76 269	

The system of capital allowances was originally introduced to promote investment in new and more efficient capital equipment. In its original form it was extremely beneficial because 100 per cent of the allowances (the full value of the asset) could be taken in the first year. This created a very large positive cash flow in the first year which reduced the payback time and increased the ROI relative to the usual depreciation approaches. In their infinite wisdom the UK government phased out this system and replaced it with a very poor system that never provides the full allowances over the typical life of an asset.

The new system only allows 25 per cent of the asset's value to be deducted from the profits before calculating the tax. Furthermore, the second year's allowance is 25 per cent of the 75 per cent of the original value that did not receive any allowances in the first year, not a further 25 per cent of the original value. Since you can only take 25 per cent of the remaining balance each year, you never get the full allowances. Table 11.11 shows how the present UK system compares with a system with a five year straight-line depreciation of 20 per cent per year for an asset valued at $100 000. It can be seen that the present UK system of allowances, usually called the 'writing down' allowances, is not as beneficial as most depreciation methods and nowhere near as good as the old system.

In addition to capital allowances or depreciation allowances, many countries have investment incentives for areas where the government is trying to boost the economy. Any such grants or extra tax benefits should obviously be included in the calculation of the net cash flows and hence the discounted cash flows.

The tax rules regarding the capital allowances will obviously change from time to time as government policy changes. Your finance department should be aware of any changes and should be able to advise you about the rules for equipment such as ATE.

11.14 Residual value or salvage value

If a piece of equipment is used for, say, five years and then no longer required, it is rare for it to have no value. Usually it can be sold for some price. If this is done and the tax inspector has already given you the full benefit of any depreciation or capital allowances, then he will want some of this money paid back. There are two ways that this can be handled in practice:

1. A residual value can be estimated up-front and deducted from the investment before computing the depreciation.
2. Alternatively, as is usual with the capital allowance system, the proceeds from the sale are simply deducted from any allowances available in the year the asset is disposed of.

For the purpose of comparing alternative projects in the ATE field, however, the added accuracy of trying to estimate a residual value is probably not worth the effort. It can therefore be assumed that the full allowances will be taken unless your company's procedures demand otherwise.

11.15 Inflation accounting and investment appraisal

Since capital investments in such equipment as ATE usually cover several years, managers are sometimes concerned that the effects of inflation may invalidate the assumptions and estimates upon which the decision was based. As mentioned earlier, so long as costs and revenues inflate at similar rates then inflation should have no effect on the rate of return achieved, especially if any form of accelerated depreciation or capital allowances are available. The faster payback is achieved the more likely the estimated ROI will be achieved, and the more inflation-proof will be the project.

There are a couple of situations that could cause inflation to affect the ROI, however:

1. If one or more elements of the costs of the project inflate at a substantially different rate to the general rate of inflation. This is fairly unlikely in the case of ATE since, as the examples in earlier chapters show, the major cost is labour, and it is unlikely that labour rates will change at a rate substantially different from inflation.
2. If there are reasons why prices cannot be adjusted with inflation, such as fixed-price contracts, or government prices and incomes restraints, then the ROI could be affected. However, inflation is something we have to live with, so most long-term contracts would have some built-in protection anyway.

Inflation accounting is now mandatory in some countries, at least for large companies. The effect of this from a capital equipment point of view will be that assets will be more accurately valued, and so will the depreciation charges. Historically, assets have been valued based upon their original cost and some deduction for depreciation. The depreciation has usually been an arbitrary percentage rather than a true reflection of the lost value of the asset. In addition, because of inflation the depreciation fund, intended to make it possible to replace the worn-out asset, has rarely been enough. The result for many companies has been an overstatement of profits. The new rules about inflation accounting have finally recognized these and other problems. In general, accounts will now show the 'value to the business' of assets rather than some arbitrary amount, and the depreciation charge will be based upon the current replacement cost of the asset rather than its historical cost. These measures will tend to reduce the effects that inflation might have on the forecasted ROI for an investment; so inflation can be ignored for the purpose of investment appraisal of equipment such as ATE.

11.16   The MARR revisited

There will always be arguments about which is the best method to use for the appraisal of capital investments. Each of the methods have their advocates and followers. The method you use will probably be dictated by your finance department, who will have determined what is most appropriate for the company. They will also have determined what the minimum acceptable rate of return should be for any major investment, but there are widely differing views on how to go about setting this MARR. In general most companies set the level too high. In an attempt to maximize the return on available funds this practice can actually prevent investment in equipment that is vital to the performance of the company. Some research conducted at Harvard University in the late eighties resulted in the conclusion that many industries in the United States were not as competitive as they should be because they had not invested in new capital equipment at the right time. The main reason for this lack of investment was the excessively high 'hurdle rates' or MARRs that had been established. There has been no similar research that I am aware of in the United Kingdom but I suspect that the same problem exists.

When I was selling ATE in the seventies I frequently came across companies who had set their MARRs at levels between two and three times the base lending rate. At times this would be coupled to a requirement for the project to achieve 'payback' in less than one year. Such practices may well be considered to be prudent in times of recession but if they result in an inability to purchase essential equipment then they are the corporate equivalent of shooting yourself in the foot.

The thought process that is most frequently used to develop an MARR goes something like this. 'We should set the absolute minimum at the *average internal cost of funds*, but because the anticipated cost savings are a forecast we should add a few per cent to allow for the risk that they may be less than expected.' This is not an unreasonable approach so long as the added percentage is not too high, but what if the person who developed the forecasted savings was naturally conservative? He or she may have deliberately erred on the low side to build in some safety of their own, so now there are two safety margins built in and the project may not achieve the MARR. Conversely, a naturally optimistic person might exaggerate the cost savings and so neutralize the safety built into the MARR. To avoid these potential problems there should be a set of guidelines produced about how to develop the potential cost savings so that everyone is working in the same manner. The ROI should be determined based on the 'most likely' situation and the safety built into the MARR should then be sufficient.

Another safety margin that some companies include in the MARR is an amount for inflation. This too causes many arguments in financial circles. Some people argue that the whole point of using discounted cash flow methods to bring everything to today's prices is too eliminate the effects of inflation. I tend to go along with this argument but if your financial department includes inflation in the MARR there is probably very little chance of getting it taken out. My recommendation in cases like this is to also include inflation in your calculation of the cost savings and this will then neutralize the added inflation in the MARR. The argument for this is quite logical. Most of the cost savings will be labour costs and these will increase with inflation just like anything else, so it will be difficult for the financial people to argue against this if they themselves have included inflation in the MARR.

Inflating your cost savings using the same percentage value for inflation that was added into the MARR will not quite neutralize the effect of the higher MARR because the positive cash flows that result from the depreciation or capital allowances will remain as they were. However, if inflation accounting is used then it would be quite valid to also increase these cash flows at the inflation rate. If this is done then the effect of the inflation added to the MARR is completely neutralized. This is effective proof that the inflation element should not have been added to the MARR in the first place. It may be more difficult to convince people that it is valid to add inflation to the depreciation generated savings, so there is a simple ploy that can be used to hide the fact that you have done it. Instead of showing the detailed ROI calculation which includes the savings, the tax on the savings, the depreciation, the tax saved by the depreciation and so on, simply show the net cash flow and then add the inflation to that. This may sound a bit like cheating but it really is not. If the financial people are silly enough to include inflation in the MARR then they deserve to be played at their own game.

The example that follows includes a look at the effect of including inflation in the MARR and then neutralizing this by inflating the savings only, and then inflating the savings and the depreciation.

One fundamental mistake that appears to be made quite frequently in establishing the MARR is to overlook the fact that a properly conducted ROI analysis will include the effects of taxation. The calculated return is therefore an 'after-tax' return whereas the 'cost of funds' used as the basis for MARR is a pre-tax cost. Interest expenses are fully deductible from any operating profit but this fact is rarely taken into consideration when establishing the MARR. This simple oversight widens the gap between the return that is realistically required and the return that is demanded.

11.17 Example using all of the methods

As a means of directly comparing the results for the various methods of determining the financial attractiveness of an investment, the following example will use all of the methods outlined on the same investment and savings estimates. The investment required is $80 000 and the project is expected to result in net savings of $30 000 per year for five years. At the end of the five years the project will be terminated and no

Table 11.12 NPV calculation with allowances for tax and depreciation

Initial investment = $80 000 (year 0)

Year	Savings	Tax on savings	Depreciation	Tax saved by depreciation	Net cash flows	Present value factor (10%)	Discounted cash flows
1	30 000	15 000	16 000	8000	23 000	0.909	20 907
2	30 000	15 000	16 000	8000	23 000	0.826	18 998
3	30 000	15 000	16 000	8000	23 000	0.751	17 273
4	30 000	15 000	16 000	8000	23 000	0.683	15 709
5	30 000	15 000	16 000	8000	23 000	0.621	14 283
						Present value	87 170

$$\text{Net present value (NPV)} = 87\,170 - 80\,000 \qquad \text{Profitability index} = 87\,170/80\,000$$
$$= \$7\,170 \qquad\qquad\qquad\qquad = 1.090$$

residual value for the asset is expected. Table 11.12 shows the full development of the discounted cash flows after allowances for the effects of taxation and depreciation. The tax rate is set at 50 per cent, the depreciation is 'five years straight line' and the MARR is set to 10 per cent.

Net present value (NPV)

The sum of the discounted cash flows is $87 170, so the net present value is $7170 because the initial investment required is $80 000. Hopefully the sums have been done correctly and this is really the 'incremental investment' required; i.e. this should be the difference between the investments required for the two alternatives that were used to determine the operational cost savings of $30 000 per year.

Therefore the investment will make a return of 10 per cent on the $80 000 and will additionally generate a cash flow of $7170. The profitability index for the project will be

$$\frac{87\,170}{80\,000} = 1.09$$

Internal rate of return (IRR)

Entering the 'net cash flows' into a financial calculator and computing IRR gives an answer of 13.46 per cent.

Payback

As indicated earlier, most calculations of payback time are performed incorrectly from the net savings before the effects of tax and depreciation. The payback time is also determined from the *cumulative savings*, which is the more logical approach, rather than from the average annual cash flows as defined in the generally accepted payback formula. In this example this will make no difference because the savings are the same each year and will therefore be the same as the average value. Using this approach we get:

End of year	Cumulative savings
1	30 000
2	60 000
3	90 000

Payback will obviously be achieved towards the end of year 3. By interpolation we can determine the specific point in time:

$$\frac{80\,000 - 60\,000}{90\,000 - 60\,000} = 0.667$$

So the payback time using this approach will be 2.667 years. More correctly we should calculate payback from the net cash flows as follows:

End of year	Cumulative net cash flows
1	23 000
2	46 000
3	69 000
4	92 000

Payback will occur sometime in the fourth year and again by interpolation this works out to be 3.48 years. Using the official formula we should really divide the 'net investment' by the 'average net cash flows', which will give us the same answer (80 000/23 000 = 3.48). However, as indicated above, this agreement is fortuitous since the annual cash flows are all the same. When the annual cash flows are not equal, as will usually be the case, there will be a small difference between the payback time calculated from the cumulative values and that calculated from the annual averages.

One of the major drawbacks of the payback method is that it does not take account of the time value of money. This is an easy problem to overcome because you can always calculate the payback time from the 'discounted cash flows' rather than from the 'net cash flows'. This is a simple enough concept but I have not seen any reference to doing this in any books on investment appraisal or on engineering economics.

The calculation of what I call the 'discounted payback' is shown below:

End of year	Cumulative discounted cash flow
1	20 907
2	39 905
3	57 178
4	72 887
5	87 170

By inspection the payback will occur midway through year 5. Using the same interpolation approach as in the previous calculations the discounted payback time works out to be 4.50 years.

The modified internal rate of return

The MIRR requires that all positive cash flows are compounded at a realistic reinvestment rate and summed to a 'net future value' at the end of the project's life. All negative cash flows are discounted at a safe cost of funds rate and summed to a 'net present value' at the beginning of the project. The MIRR is then the interest rate that will make the NPV of the negative cash flows equal and opposite to the NFV of the positive cash flows. In a situation like the one in this example, where there is only one negative cash flow at the very beginning of the project, it is not necessary to set a 'safe rate' for the cost of funds since there are no negative cash flows that need to be discounted. The only negative cash flow, the initial investment, is already at a 'present value'. The calculation of the NFV is performed as follows:

End of year	Years to compound	Compound interest factor	Compounded cash flow
1	4	1.464	33 672
2	3	1.331	30 613
3	2	1.210	27 830
4	1	1.100	25 300
5	0	1.000	23 000
		Total	140 415

The MIRR will now be the rate at which 80 000 compounds up to 140 415 over five years. If you have a financial calculator or a spreadsheet with FV, PV, i and n functions, then this is a simple calculation. The answer is 11.91 per cent. The calculation can also be performed using a scientific calculator that has e^x and LN keys as follows

The compound interest formula is $FV = PV(1 + i)^n$.
This can be solved for i as follows:

$$FV/PV = (1 + i)^n$$

Therefore

$$n\sqrt{\frac{FV}{PV}} - 1 = i$$

So the interest rate is the nth root of FV/PV minus one. This can be determined as follows:

Find the natural log (LN) of FV/PV.
Divide this by n (the number of periods).
Take the inverse natural log (e^x) of this.
Subtract 1 and multiply by 100 to get per cent.

On a typical scientific calculator this would appear as follows:

$FV/PV = 140\,415/80\,000 = 1.755$
$LN\ 1.755 = 0.563$
$0.563/5 = 0.113$
$e^{0.113} = 1.1191$
$(1.1191 - 1) \times 100 = 11.91\%$

The external rate of return (ERR)

The ERR method differs from the MIRR method in that the negative cash flows are also compounded to a net future value. In the ERR method it is only necessary to establish a realistic reinvestment rate for compounding-up the positive cash flows. There is no requirement to determine a 'safe rate' for the cost of funds. The ERR is the rate at which the compounding of the negative cash flows has to be performed in order to be equal (but opposite in sign) to the net future value of the positive cash flows. This being the case, there is no difference between ERR and MIRR for the case where there is a single negative cash flow at the beginning of the project. This is because there is no discounting of the negative cash flows at the 'safe rate' in the MIRR method when there is only one negative cash flow at the beginning of the project as there is in this example. The ERR is therefore also 11.91 per cent.

NPV calculation using the current UK allowances system

The calculation of the net cash flows in Table 11.12 is on the basis of a five year straight-line depreciation allowance of 20 per cent year of the investment. There are numerous other ways in which the tax authorities may allow the depreciation allowance to be taken. In the United States the other common depreciation methods are the 'double declining balance' method (DDB) and the 'sum of the years digits' method (SOYD). Functions for these are found in most spreadsheet programs of American origin. Every other country will use their own methods and these will also change from time to time depending on government policy. Many countries offer some accelerated form of depreciation allowances to promote investment. The system used in the United Kingdom at present (1993), which was described earlier, tends to do the reverse. For the benefit of UK readers Table 11.13 shows the example of Table 11.12 with the UK capital allowances system instead of the straight-line depreciation. As you can see, the reduced tax benefits for the capital allowances reduce the net present value considerably. The profitability index also falls from 1.09 to 1.015.

The effect of including inflation in the MARR

There always tends to be some confusion about the 'time value of money' and 'inflation' but the two are really quite different. The time value of money simply relates to the fact that if you have some money now you can invest it in something that will make it grow numerically over time. Inflation relates to the purchasing power of money which tends to fall with time as prices rise. Using discounting techniques to bring all future cash flows to today's value effectively eliminates inflation from the analysis, but unfortunately not everyone agrees with this concept. As a result you may find that some allowance for inflation is added to the MARR 'just to be on the safe side'. The usual way in which this gets done is to calculate a 'combined interest/inflation

Table 11.13 The example shown in Table 11.12 calculated using the current (1993) UK system of capital allowances

Initial investment (year 0) = $80 000

Year	Savings	Tax on savings	Capital allowances	Tax saved by capital allowances	Net cash flows	Present value factor (10%)	Discounted cash flows
1	30 000	15 000	20 000	10 000	25 000	0.909	22 725
2	30 000	15 000	15 000	7 500	22 500	0.826	18 585
3	30 000	15 000	11 250	5 625	20 625	0.751	15 489
4	30 000	15 000	8 438	4 219	19 219	0.683	13 127
5	30 000	15 000	6 328	3 164	18 164	0.621	11 280
						Present value	81 206

Net present value (NPV) = 81 206 − 80 000
= $1206

Profitability index = 81 206/80 000
= 1.015

rate' in the following manner:

$$\text{Combined interest/inflation rate} = (1+i)(1+f) - 1$$

where

$$i = \text{interest rate before adding inflation}$$

and $\qquad f = \text{inflation rate}$

In the above examples we used an interest rate of 10 per cent. If we now add an inflation rate of 5 per cent to this we get

$$\text{Combined interest/inflation rate} = (1 + 0.1)(1 + 0.05) - 1$$
$$= 0.155 \text{ or } 15.5 \text{ per cent}$$

Table 11.14 shows how this would change the NPV obtained in Table 11.12. The net present value drops to −$3640 from $7170. This negative NPV means that the project does not make the 15.5 per cent MARR and would normally be rejected. However, as suggested earlier, if we are going to include inflation of 5 per cent in the calculation then this 5 per cent inflation should also be applied to the estimated savings that result from the project. The result of doing this is shown in Table 11.15 where the NPV has risen up to $3541 and so the project is now acceptable. The reason it has not gone back to the original $7170 is that the tax benefit from the depreciation, which forms part of the 'net cash flows', has remained at its original values. It can be argued that if the company is using inflation accounting, which is mandatory in some countries, then it would be legitimate to also inflate the tax saved by the depreciation by 5 per cent. The effect of doing this is shown in Table 11.16 where the NPV has risen to $7373. The small difference between this figure and the original NPV of $7170 is due to an accumulation of rounding errors. This simple

Table 11.14 The example shown in Table 11.12 with a combined interest/inflation rate

Initial investment (year 0) = $80 000

Year	Savings	Tax on savings	Depreciation	Tax saved by depreciation	Net cash flows	Present value factor (15.5%)	Discounted cash flows
1	30 000	15 000	16 000	8000	23 000	0.866	19 915
2	30 000	15 000	16 000	8000	23 000	0.756	17 388
3	30 000	15 000	16 000	8000	23 000	0.649	14 927
4	30 000	15 000	16 000	8000	23 000	0.562	12 926
5	30 000	15 000	16 000	8000	23 000	0.487	11 201
						Present value	76 357

$$\text{Net present value (NPV)} = 76\,357 - 80\,000$$
$$= -\$3643$$

Table 11.15 The example shown in Table 11.14 with the savings inflated at the defined inflation rate

Initial investment (year 0) = $80 000

Year	Savings	Tax on savings	Depreciation	Tax saved by depreciation	Net cash flows	Present value factor (15.5%)	Discounted cash flows
1	31 500	15 750	16 000	8000	23 750	0.866	20 568
2	33 075	16 538	16 000	8000	24 538	0.756	18 550
3	34 729	17 364	16 000	8000	25 364	0.649	16 462
4	36 465	18 233	16 000	8000	26 233	0.562	14 742
5	38 288	19 144	16 000	8000	27 144	0.487	13 219
						Present value	83 541

Net present value (NPV) = $83 541 − 80 000
= $3541

Profitability index = 83 541/80 000
= 1.044

Table 11.16 The example shown in Table 11.15 with the tax savings also inflated at the defined inflation rate

Initial investment (year 0) = $80 000

Year	Savings	Tax on savings	Depreciation	Tax saved by depreciation	Net cash flows	Present value factor (15.5%)	Discounted cash flows
1	31 500	15 750	16 000	8 400	24 150	0.866	20 914
2	33 075	16 538	16 000	8 820	25 358	0.756	19 171
3	34 729	17 364	16 000	9 261	26 625	0.649	17 280
4	36 465	18 233	16 000	9 724	27 957	0.562	15 712
5	38 288	19 144	16 000	10 210	29 354	0.487	14 296
						Present value	87 373

Net present value (NPV) = $87 373 − 80 000
= $7373

Profitability index = 87 373/80 000
= 1.092

example proves that discounted cash flow methods really do eliminate the effects of inflation since adding inflation to all relevant elements result in the same NPV that we obtained in Table 11.12 without any inflation content.

Summary of results

The following is a summary of the results of all of the methods used to evaluate this investment opportunity:

NPV for an MARR of 10% = $7170

Profitability index = 1.09

IRR = 13.46%

MIRR = 11.91%

ERR = 11.91%

Payback time (incorrectly calculated from savings)

$$= 2.667 \text{ years}$$

Payback time (calculated with the accepted method)

$$= 3.48 \text{ years}$$

Payback time (calculated from discounted cash flows)

$$= 4.50 \text{ years}$$

NPV calculation using UK capital allowances

$$= \$1206$$

IRR calculation using UK capital allowances

$$= 10.63\%$$

NPV with inflation of 5% added to MARR

$$= -\$3643$$

NPV with 5% inflation added to MARR and savings

$$= \$3541$$

NPV with 5% inflation added to MARR, saving and depreciation

$$= \$7373$$

11.18 Example: comparing three alternative in-circuit board testers

DEF Electronics manufacture professional audio equipment. They have been using a fairly simple tester but as their production volume has grown this is rapidly running out of steam both in terms of capacity and the ability to cope with increased complexity. Having researched the available equipment they narrowed the field down to three testers with varying degrees of price and performance.

After a thorough evaluation of all three products DEF came to some conclusions about programming costs, test and diagnosis times, diagnostic accuracy, fault coverage and other important parameters. Using a commercially available test strategy cost model the DEF engineers performed a comparison of cost and quality considerations for the three short-listed testers. The results of this analysis are summarized in Table 11.17. This data was then used to develop net present values for the three alternatives using an MARR of 15 per cent, a five year straight-line depreciation and a 50 per cent tax rate. Since taxes on incremental profits resulting from the investment would be paid in the following tax year, they shifted the tax on savings by one year and added a sixth year to the calculation to account for the taxes on the year 5 savings. Since this investment will be for additional equipment, relative to continuing with present methods, the whole cost of the testers is used as the incremental investment amount. The results of the NPV calculations are shown in Table 11.18.

Table 11.17   Annual savings calculations for three alternatives

	A	B	C
Present cost/board	18.22	18.22	18.22
Expected cost per board	9.21	11.51	13.82
Savings per board	9.01	6.71	4.4
Annual volume (1500 × 12)	18 000	18 000	18 000
Therefore total savings	162 180	120 780	79 200
Average programming cost per board	3 000	4 000	5 000
Average fixture cost per board	5 000	4 000	3 500
First year set-up costs (10 boards)	80 000	80 000	85 000
Therefore first years savings	82 180	40 780	(5 800)
Set-up costs for subsequent years (5 boards per year)	40 000	40 000	42 500
At the expected growth in board volume of 20% per year			
Year 2 operational savings	194 616	144 936	95 040
Year 2 savings after set-up costs	154 616	104 936	52 540
Year 3 operational savings	233 539	173 923	114 048
Year 3 savings after set-up	193 539	133 923	71 548
Year 4 operational savings	280 247	208 708	136 858
Year 4 savings after set-up	240 247	168 708	94 358
Year 5 operational savings	336 296	250 449	164 229
Year 5 savings after set-up	296 296	210 449	121 729

Obviously system A is the winner with a profitability index of 2.04 relative to the next best alternative, system B, with a PI of 1.61. Incidentally, an IRR calculation based on the net cash flows for system A gives the result of 54.19 per cent. The reinvestment assumption would require that the cash flows generated by this investment would have to be reinvested each year in something that would return 54.19 per cent (after tax) for this IRR to be valid. Clearly the IRR method is not appropriate for high return projects. The IRR calculated for system C is 17.9 per cent. This is quite a good return even though system C is clearly the poorest use of funds of these three alternatives. This highlights a potential problem with ATE or any other investment in automation when the current method is manual or semi-automatic. Even the poorest alternative may give a high ROI when compared to the present method. If DEF Electronics were to arbitrarily set themselves a capital budget limit of $150 000 they might not even perform an ROI analysis on system A.

If we assume that the present method of testing cannot cope with the problem and that one of the commercial testers has to be purchased, then we should look at the incremental situation as well as, or instead of, comparing the three testers with the *non-viable* current method. If we do this then the system with the highest operating costs effectively becomes the baseline case. This is obviously system C, because this showed the lowest savings relative to the current method. Since we have all of the data, we can simply compare system A to system C (or system B to system C) by

Table 11.18 NPV and PI calculations for the three alternatives

Year	Savings	Tax on savings	Depreciation	Tax saved by depreciation	Net cash flows	Present value factor (15.5%)	Discounted cash flows
System A: $200 000 investment							
1	82 180	—	40 000	20 000	102 180	0.87	88 897
2	154 616	41 090	40 000	20 000	133 526	0.76	101 480
3	193 539	77 308	40 000	20 000	136 231	0.66	89 912
4	240 247	96 770	40 000	20 000	163 477	0.57	93 182
5	296 296	120 124	40 000	20 000	196 172	0.50	98 086
6	—	148 148	—	—	(148 148)	0.43	(63 704)
						Present value	407 853

Net present value (NPV) = 207 853   Profitability index = 407 853/200 000 = 2.04

Year	Savings	Tax on savings	Depreciation	Tax saved by depreciation	Net cash flows	Present value factor (15.5%)	Discounted cash flows
System B: $180 000 investment							
1	40 780	—	36 000	18 000	58 780	0.87	51 139
2	104 936	20 390	36 000	18 000	102 546	0.76	77 935
3	133 923	55 468	36 000	18 000	99 455	0.66	65 640
4	168 708	66 962	36 000	18 000	119 746	0.57	68 255
5	210 449	84 354	36 000	18 000	144 095	0.50	72 048
6	—	105 225	—	—	(105 225)	0.43	(45 247)
						Present value	289 770

Net present value (NPV) = 109 770   Profitability index = 289 770/180 000 = 1.61

Year	Savings	Tax on savings	Depreciation	Tax saved by depreciation	Net cash flows	Present value factor (15.5%)	Discounted cash flows
System C: $150 000 investment							
1	(5 800)	—	30 000	15 000	9 200	0.87	8 004
2	52 540	(2 900)	30 000	15 000	70 440	0.76	53 534
3	71 548	26 270	30 000	15 000	60 278	0.66	39 783
4	94 358	35 774	30 000	15 000	73 584	0.57	41 943
5	121 729	47 179	30 000	15 000	89 550	0.50	44 775
6	—	60 865	—	—	(60 865)	0.43	(26 172)
						Present value	161 867

Net present value (NPV) = 11 867   Profitability index = 161 867/150 000 = 1.08

using the difference between investments and the differences between the savings. If this is done using the same MARR of 15 per cent you will find that the net present value for investment in system A relative to system C is $195 986. This is the same as the difference between the NPVs for system A and system C when the two are compared with the baseline of the current method. This is only what you would logically expect, but clearly shows the concept of incremental investment as outlined in Chapter 4. This positive cash flow of $195 986 (present value) over the six years we have looked at is the result of the incremental investment of $50 000 for system A relative to system C. The correct interpretation of this takes some thought because we already determined that system C will itself generate a good return relative to continuing with current methods. The return from system A over system C is therefore incremental, or in addition, to the return from system C. This is an important point when performing an incremental analysis. You, and your financial department, must not lose sight of the fact that any return will be on top of what would be a good return even for the poorest alternative. For example, if an incremental analysis shows a negative NPV when performed with your agreed upon MARR, the project may be thrown out for not meeting the criterion or hurdle rate. This is one time when performing an IRR calculation might be useful. If the MARR is, say, 15 per cent and the IRR for an incremental analysis is 10 per cent then you have to remember that this is 10 per cent *more* than you would get if you chose the alternative used as the baseline. It is not an absolute 10 per cent because it was not an absolute analysis. The assumption here, of course, is that at least one of the alternatives has to be purchased in order to continue with production.

There is, however, a potential problem with using IRR in an incremental analysis. If the incremental investment is small relative to the baseline alternative, but there are large operating cost differences, then the calculated incremental IRR might be enormous and therefore will be totally meaningless. My own preference in cases where the purchase of a system is mandatory, because present methods cannot cope, is to assume that the return for any system will exceed the MARR and to simply compare the sum of the net cash flows (undiscounted) with the incremental investment.

For completeness let us now take a look at the payback times for the three alternative testers.

Payback calculations

The payback period can be determined by cumulating the net (undiscounted) cash flows to determine in which year the payback is reached, and then using interpolation to determine when in that year it is reached.

For system A the cumulative cash flows are:

> Year 1   102 180
> Year 2   235 706

Therefore payback is achieved in year 2. The cash flow in year 2 is $133 526; we need to find out when the difference between the $200 000 investment and the $102 180 cash flow at the start of year 2 is reached. This is estimated as follows:

$$\frac{200\,000 - 102\,180}{133\,526} = 0.73$$

Therefore payback is reached in 1.73 years.
The generic form of this interpolation is

$$\frac{\text{Investment} - \text{cash flow at start of payback year}}{\text{Cash flow during payback year}}$$

This result is then added to the number of full years prior to the payback year.

For system B the cumulative cash flows are:

Year 1	58 780
Year 2	161 326
Year 3	260 781

Therefore payback is achieved in year 3 and is given by

$$2+\frac{180\,000-161\,326}{99\,455}=2.19 \text{ years}$$

For system C the cumulative cash flows are:

Year 1	9 200
Year 2	79 640
Year 3	139 918
Year 4	213 502

Therefore payback is achieved in year 4 and is given by

$$3+\frac{150\,000-139\,918}{73\,584}=3.14 \text{ years}$$

Thus the ranking in order of payback is also A, B, C. This again is not always the case. If there are bigger differences between the investment amounts and the patterns and amounts of the cash flows it is quite possible that the three measures of NPV, PI and payback could all produce different rankings. Most of the time the project with the highest PI should be selected, but under some financial conditions the upper management might prefer a project that throws off the most cash (or minimizes the need for further borrowing) and so NPV would be the more important parameter. Other than in exceptional circumstances, however, the payback figure should not be used as the selection criteria unless the PIs and NPVs are all similar.

The calculation of payback that we have just done is the normal method, using cumulative net cash flows (undiscounted). Out of interest let us do the same thing but using the DCFs.

System A:

Year 1	88 897
Year 2	190 377
Year 3	280 289

Therefore

$$\text{Payback} = 2 + \frac{200\,000 - 190\,377}{89\,912} = 2.11 \text{ years}$$

System B:

Year 1	51 139
Year 2	129 074
Year 3	194 714

Therefore

$$\text{Payback} = 2 + \frac{180\,000 - 129\,074}{65\,640} = 2.8 \text{ years}$$

System C:

Year 1	8 004
Year 2	61 538
Year 3	101 321
Year 4	143 264
Year 5	188 039

Therefore

$$\text{Payback} = 4 + \frac{150\,000 - 143\,264:}{44\,775} = 4.15 \text{ years}$$

These discounted payback times tell a slightly different story. System C now goes part of the way into the fifth year of operation before payback is achieved. This then would be a very risky project since a change in one of the assumptions might mean that payback, when viewed in this way, might never be achieved.

Discounted payback analysis is not in common use. I have found no references to it in any financial textbook. I personally feel, however, that it is a more realistic way of looking at payback than the traditional method, provided that the discount rate selected is a reasonable one.

11.19 Summary

This chapter has discussed the alternative methods of financial appraisal that are in common use. Their major advantages and disadvantages have been covered with the conclusion that the NPV method is probably the most suitable and most accurate for ranking alternative projects. The reinvestment assumption common to DCF analysis technique means that the IRR method is inappropriate for ATE since the rates of return are very high for this type of equipment.

1. Effective investment appraisal is vital to business decision making.
2. Simple rule-of-thumb techniques such as 'payback analysis' or ARR are of no value when trying to rank alternative investment projects. Their use should be restricted to the quick elimination of totally unprofitable alternatives and as additional data to support a DCF analysis.
3. The best method to use for ATE is the NPV method with a calculation of the PI.

Note. The examples used throughout this chapter assume that the cash flows generated by a project occur at the end of each year. This is the normal method used even though, in practice, the cash flows will be generated gradually, throughout each year. This being so, it is a little unfair to discount them all as if they appeared magically on the last day of each year.

A more accurate analysis for the situation where savings are generated continuously can be performed by using PVFs for continuously received amounts rather than for amounts received at the end of each period. Such a table appears in Appendix E. It is generated by taking the average of the PVFs for the beginning and end of each period, e.g. PVFs for the sum received at the end of each period (10 per cent):

$$
\begin{array}{ll}
\text{Year 0} & 1.0 \\
\text{Year 1} & 0.909 \\
\text{Year 2} & 0.826 \\
\text{Year 3} & 0.751, \text{ etc.}
\end{array}
$$

PVFs for sums received continuously:

$$
\begin{array}{ll}
\text{Year 0} & 1.0 \text{ (base year)} \\
\text{Year 1} & \dfrac{1+0.909}{2}=0.955 \\
\text{Year 2} & \dfrac{0.909+0.826}{2}=0.868 \\
\text{Year 3} & \dfrac{0.826+0.751}{2}=0.789
\end{array}
$$

The use of this method rather than the traditional 'end of period' tables may raise some eyebrows with your financial people. It is, however, a more accurate representation of the ATE situation where the savings are made on a daily basis. It will, of course, result in a higher PV, NPV and PI, compared to those obtained using the normal tables.

12. Presentation to management

Once the evaluation of the short-list of vendors and equipment is complete and the evaluation team have reached a decision, the next step will be to present its findings to upper management. The purchasing department will probably have been involved during the evaluation to help with the vetting of the vendors being considered.

The exact form of any presentation, or written proposal, to convince upper management to allocate funds for purchase of ATE will obviously vary a lot from company to company. The size of the company will determine how formal or informal the procedure is and who might be involved in any final purchase decision. The size of typical investments in ATE is such that even in very large companies the final decision gets to very high levels—certainly to divisional director level, and more often than not up to corporate headquarters level.

In medium to large sized companies, the procedures involved are probably well documented and previous examples will be available as a model. Something worth remembering is that having made your decision you now take on the role of salesman—to sell the project to management. It is possible, of course, that management needs no convincing about the need for some form of ATE, in which case it is easier. Now all you have to do is convince them that you have done a good evaluation job and have made a good decision on what to buy based upon good reasoning.

It should go without saying that any presentation or proposal should be in the language that management understands. More often than not this is the language of money. Juran in Section 4 of the *Quality Control Handbook* (Juran and Gryna, 1988) talks about there being two basic languages in a company. Lower management and non-supervisory staff talk in the language of things. To communicate at this level you need to talk in terms of tons, man-hours, units of production, microfarads, resistors, etc. Upper management talks in the language of money—sales, costs, profits, ROI, value of the shares, the economy in Djibouti, etc. Middle management needs to communicate with upper and lower management so they need to be bi-lingual. As an aside, Juran points out that there are also numerous local dialects peculiar to certain functions and understood by few others, e.g. accounting, marketing research, quality control.

At this point it may be worth while reviewing the comments made in the first chapter regarding the responsibilities of upper management to manage the funds to

355

give the shareholders a good return on their investment in the company. The usual procedure for the allocation of funds is a two-part process. First of all, a primary allocation of funds is made at board or chief executive officer level, to ensure that there is the right balance of resources required to achieve the long-term goals of the organization. Following this primary allocation comes the normal process of formulating new investment proposals, appraisal of them by strategic and profitability criteria, and their approval by management.

This approval process consists of allocating funds to some projects and turning down others. The usual criteria for selection is the relative financial performance of the projects competing for funds. Apart from the need to maximize the return on the shareholder's funds, there is another good reason why the selection criteria is financial. It is often the only way to compare alternative projects since they may be so different from each other. For example, a proposal to purchase an ATE system may be competing for funds with a proposal to purchase a numerically controlled punch-press. Both may be very necessary, but if there is only money available for one of them then the project with the highest return should be the one selected.

Having selected a bunch of profitable projects there are a couple of other criteria that the board of directors or upper management should apply before final selection. One criteria is to check the effect that the group of projects will have on the overall cash flows of the company. They need to check whether the projects selected will create a serious cash shortage at some point in their expected lifetimes. If they do, they may well select some less profitable projects for inclusion in the overall mix to even out cash flows.

A second criteria applied will be to look at the effect the group of projects will have on reported profits. If a number of projects only become profitable in the later years of their life there may be an adverse effect on profits for the next couple of years. This might affect the share price and the company's ability to raise new capital via loans or share issues. It might also affect the directors personally since they are judged—among other things—on the 'return on assets', so the more the assets and the lower the return the worse they look.

Much of the so-called 'asset-stripping' that takes place from time to time is, in some cases, nothing more than a few clever people telling shareholders that their companies are being mis-managed. Assets are being undervalued either through ignorance or to make the performance of upper management look good. This undervaluing of assets usually applies to things like buildings and land that do not depreciate by much. In fact they would normally appreciate. With equipment such as ATE it has been more likely that the asset would have been valued incorrectly due to the usual method of valuing at cost less depreciation. Depreciation, or at least the method of applying it, is usually an arbitrary process having little resemblance to the real market value of the asset. The new methods of inflation accounting, which are coming into use in most countries, take a more realistic view of the valuation of assets and the depreciation of them in order to set aside money for their eventual replacement.

Fortunately, investment in ATE will score heavily in all of the financial criteria that management may apply:

1. The ROI is very hgh.
2. The NPV and the Pl are usually very high.
3. Payback is very short.
4. There are usually no major negative cash flows following the initial purchase.
5. Contribution to reported profits will start in the first year of operation.
6. ATE is a very marketable asset should it become necessary to sell it.

As a result, if you do a reasonable job of presenting the facts, it cannot fail to be one of the accepted projects.

12.1 Sensitivity analysis

The potential profitability of a good investment project will be lost if it is poorly managed when it is operational. Management will therefore want to know which of the assumptions made are the ones most critical to the project's ability to meet its predicted ROI. An analysis of the sensitivity of the ROI to the major elements of the project is therefore well worth including. Some of the more important elements are likely to be:

1. Volume predictions.
2. Fault spectrum.
3. Yields from manufacturing (predicted FPB).
4. Estimated test, diagnosis and repair times.
5. Estimated set-up costs.
6. Estimated start-up time.
7. Labour costs.

Labour cost changes can really only make the project look better since if they go up, the savings will increase. One of the major advantages of automating is the stabilizing effect it has. Once purchased a large part of the cost is controllable. Labour costs, however, are relatively outside the control of management since they are affected by the general economy, inflation, etc.

12.2 Other factors

It is always worth while to remind management about the general benefits of ATE:

1. Staffing. ATE will reduce the number of people required and the skill level of the operators. This has a potential negative side to it if there is any danger of upsetting the unions. In practice, ATE rarely puts anyone out of a job. The main problem

is finding enough people with the right skill to get the job done manually. For this reason, there is rarely any union problems caused by the introduction of ATE.

2. Better control of operations due to more feedback and management information from the test systems. Test systems are becoming much more sophisticated in their ability to generate feedback to management about the production process. The capabilities can vary from simple reports on fault data, test times, diagnose times, etc., to highly sophisticated networked systems.

3. More consistent quality.

4. Fewer unnecessary repairs—less re-work.

5. Faster inventory turnover (work in process is in the factory for less time).

6. Better documentation.

7. Less floor space required. This benefit is often overlooked but can be very important. An automatic board tester can take the place of 20 or more manual test stations for the same volume. Apart from the fact that manual testing could tie up as much capital as the ATE (20 scopes, 20 DVMs, 20 logic analysers, 20 test jigs), it also takes a lot more floor space. The biggest potential benefit here, however, is the possible delay that ATE might create, in the need to extend the building or even to move to larger premises.

8. Prestige. This may sound a bit funny, but most companies are in the business of selling something. To a degree this means selling the company image as well as products. Factory tours take place frequently—sometimes it is customers, sometimes men from the ministry, local dignitaries or a party of school leavers from whom you hope to recruit new staff. Whoever it is the objective is the same—you want to impress them. As such, a modern ATE installation often becomes one of the stopping-off points on a tour. You cannot put a value on it, but it could tip the balance in your favour when bidding for a large contract.

12.3 Brevity is key

Upper management usually includes extremely busy people. If your written proposal to justify the purchase of ATE ends up being two inches thick, with every piece of supporting documentation included and all calculations presented in great detail, it will not be read. At best it will probably be glossed over by a possibly irate manager searching in vain for a synopsis or a few major facts.

A short proposal containing brief descriptions of the problem, the alternative solutions considered, the evaluation methodology, and your proposed purchase, along with relevant financial data, is much more likely to be read. Having more detailed information available at your fingertips, if and when requested, is likely to impress the management more than any two-inch pile of paper will. The temptation to say to yourself 'We have spent nine man-months on this evaluation—let's give them something to show for it' should be resisted.

12.4 Keep it simple

As mentioned at the beginning of the chapter most upper management people talk a more financial rather than a technical language. Unless your management is an exception to this rule, try to keep the language used as understandable as possible.

Avoid the use of technical jargon and abbreviations wherever possible. If you do use abbreviations make sure you always explain the meaning of them the first time you use them. You will not insult anyone's intelligence by explaining that UUT means 'unit under test', but it is unlikely that the financial director would want to show his ignorance by asking what it means.

12.5 Emphasize important issues

Throughout the written proposal, wherever the opportunity arises, emphasize the key issues that are likely to be uppermost in the management's minds. There are four that will almost certainly apply to every company or organization, and others that will depend on the individual situation at the time. The four common concerns will be:

- Cost
- Quality
- Time to market
- Technology

12.6 A checklist for the proposal to management

1. Be brief and to the point.
2. Use the 'language' of management.
3. Minimize jargon and abbreviations.
4. Emphasize key concerns.
5. Concentrate on economic and financial performance of the proposed purchase.
6. A typical proposal might include:
   (a) Definition of the problem.
   (b) Alternative solutions considered.
   (c) Solution chosen—and why.
   (d) Equipment selected.
   (e) Brief details about the chosen vendor (organization, support, credibility, etc.).
   (f) Effects on staffing and facilities.
   (g) Effects on other departments.
   (h) Cost of the equipment, training, site preparation, etc.
   (i) Estimated start-up time.
   (j) Financial analysis, ROI, payback, etc.
   (k) Major assumptions, risk, sensitivity analysis.
   (l) Additional benefits (intangibles).

As appendices or to be produced upon request:

1. Details of the evaluation program methodology and results.
2. Reasons for selection of chosen vendor in favour of others considered.

3. More detailed cost savings analysis.

4. Capacity estimates and when additional capacity will be needed.

12.7   The presentation

In a few cases a written proposal will be passed on to management who will read it, discuss it and pass judgement on it. This is not really a very satisfactory approach so more often than not the recommendations of the evaluation team will be personally presented to management at a meeting attended by all of the interested parties. Presentations of this type take place many times each day in large companies—design engineers presenting an idea to their boss, test engineers presenting an idea to the production manager, someone presenting an idea to the quality team or an evaluation group presenting their recommendation to buy a 'Genpackardynetron Mk 111 board tester'.

All of these internal presentations have a number of things in common and a knowledge of these can be put to good use in preparing your presentation. I know of many instances where many months of evaluation effort have been wasted because of a poorly planned and poorly executed presentation. You are usually only given a few short hours to get your material across in a positive manner, so careful preparation is essential.

The nature of a presentation

The first thing to understand about successful presenting is that a 'presentation' is not a 'lecture'. The two are very different. Most business presentations, whether they are internal or external (i.e. to another company), are usually *persuasive* in nature. You are usually trying to persuade a group of people to accept your product or your ideas and recommendations. The salesman presenting to an ATE evaluation team is trying to persuade them that the product on offer is the best for their needs. The evaluation team presenting their recommendation to purchase this ATE to senior management is trying to persuade them to fund their proposal. In contrast a 'lecture' is usually the transferring and the explanation of information by a lecturer passing on his or her knowledge and experience to the students, usually without any need for persuasion.

Presentations also differ from lectures in many other ways. In a lecture situation the lecturer is usually superior in status, rank, knowledge, experience and age to the audience. In most presentations the reverse is usually the norm. Most presentations are made to superiors by their subordinates and many of these are the result of a task given to the subordinate by the superior. The speaker-to-audience relationship is therefore the reverse relative to the lecture, which has serious implications about the need for planning the presentation. A lecturer can decide on how much time to allocate to any specific subject but the presenter is usually told how much time they can have to get their message across. This will usually be less time than the presenter feels is needed but there will usually be very little that can be done to get additional time. In a lecture most of the audience will be present because they want to be. They want to learn as much as they can from the lecturer. In many presentation situations a portion of the audience may well be there under duress. They would rather be somewhere else doing something else and as a result they may be somewhat hostile towards the presenter. The lecturer can command the attention of the audience by virtue of his or her superior status and knowledge. The best the presenter can do is to 'deserve' the attention of the audience by performing a well-prepared and professionally delivered presentation. I chose the word 'performing' quite deliberately because a well-executed presentation is something of a performance. You only have to watch some of the better politicians, most of whom have been trained in presentation skills, to see that this is true.

The simple advice that I can give in this chapter is no substitute for attending a course on presentation skills. If you get the opportunity to attend such a course then take advantage of it. Practically everyone in industry has to make a presentation now and then and having good presentation skills is a definite advantage in the promotion stakes.

Overcoming nerves

Most people, even experienced presenters, get nervous about a forthcoming presentation. A small amount of nervousness can actually be a good thing. It keeps you on your toes and makes you do a better job. Being a little nervous usually means that you will practice the presentation whereas the 'old hand' who is confident that just about anything can be 'winged' without any preparation at all may well make a mess of it. When this happens self-confidence is so great that the presenter will not recognize failure and will simply blame the audience or the lack of time. 'Those dummies didn't want to listen—they have already decided what they want to do.' It is very noticeable at trade conferences that a speaker who is obviously nervous but still does a good job will often get more applause from the audience than a more experienced and polished presenter. The audience have a lot of empathy for the situation, and I mean empathy as opposed to sympathy. Many people in the audience will have been in a similar position and the ones that have not been are probably terrified at the thought of it. You only have to witness the reluctance for people in the audience to ask questions to understand this.

Too much nervousness, however, is a bad thing because it can cause you to miss a vital point or to gabble on rapidly because you want to get the ordeal over with. The first step to overcoming nerves is to understand what we get nervous about. This usually boils down to one or more of the following fears:

1. 'Will I say the right things?'
2. 'Will I look a fool?'
3. 'Will they know more than I do?'
4. 'Will they ask awkward questions?'
5. 'Will I leave something important out?'
6. etc.

The answers to these questions are usually:

1. 'Yes—if you are well prepared'.
2. 'Not if you are well prepared'.
3. 'Probably'.
4. 'Probably'.
5. 'Not if you are well prepared'.
6. etc.

The solution obviously lies in good preparation. If you are well prepared you will reduce the nervousness down to a level that is helpful rather than counter-productive.

Preparation

The five key questions to ask yourself about any presentation are:

1. Why are you making this presentation?
2. What are you going to say?
3. Who are you saying it to?
4. Where will you be saying it?
5. How will you say it?

WHY?

Every presentation has an objective. If several presentations are required before the final decision is made then the objective of the earlier presentations may simply be to get to the next presentation. This is certainly the case in a selling situation but even with internal presentations you are often 'selling' your ideas in competition with other groups so the two situations are not very different. If the process does involve a need for multiple presentations you have to resist the temptation to do too much in the early stages. A carefully planned strategy is needed to move from stage to stage sensibly. The objectives should be agreed by the team and written down.

WHAT?

Make notes of all the information, illustrations and the arguments that you may need. Also list the possible objections that might be raised and how you will answer them. It may be necessary to prepare a short section, with supporting visuals, that you only present if certain objections crop up or if something requires a more in-depth explanation. If you are fairly sure that certain objections will be raised, possibly by someone who you know is favouring an alternative project, then refer to them yourself and then minimize their effect with your answer. This technique has several advantages:

1. Bringing up possible objections yourself immediately reduces their importance by showing that you do not consider them to be a problem. It also shows that you have covered 'all angles'.
2. If you wait for a 'competitor' in the audience to bring up the objection it gives them credibility and may appear to put you on the defensive in the eyes of any unbiased members of the audience.

3. Raising the objections yourself, and then answering them, eliminates the need to 'shoot the objector down in flames'. The more you have to do this the more they will dig their heels in and work against you. Remember that the objector is probably your superior. He or she could even be your boss one day.

WHO?

Find out all you can about the people who will attend your presentation. Obtain names, job titles, responsibilities, previous areas of responsibility, likes and dislikes, hot buttons, etc. What are their concerns likely to be? What are their problems and how might your proposal affect their problems? Will they feel threatened by what you are proposing? How can you eliminate or minimize this problem? Which of the attendees is likely to exert the most influence? Are there any attendees that you should 'break the ice with' before the presentation begins?

WHERE?

The 'where' is usually less important than the other questions but if it is on unknown territory it may be advisable to check on a few things ahead of time. What is the room layout? What projection equipment exists for your visual aids? Will you need extension cables? Can the room be darkened? This is only important if you plan to use 35 mm slides.

HOW?

Having set your objectives, researched your audience and checked on the venue, you can now turn your attention to how you say what you want to say. You now put yourself into the minds of your audience. What objectives do they have? What anxieties will your proposal relieve? What needs and wants can you identify? etc. This will lead you to identify a suitable introduction that will catch their attention. Once the introduction is clear you can then sort out the most logical order in which to present your facts and your arguments. The most logical order is the one that is best for the audience's understanding and the best for your persuasive arguments. You can then select or create the visual aids needed to support and clarify your factual content and your arguments.

Once all of the support material is assembled you can make your notes. If you have time to write out the entire presentation do so because this will help to clarify things in your own mind and also help to determine the 'best way' to say things and the 'best order' to say them in.

The most important segments of any presentation are the beginning and the end. Make sure that you have a very clear understanding of what you want to say in these segments and commit it to memory. Most definitely write these segments down and go over them again and again. Your opening and your close are too important to leave to chance.

Research has shown that most people's attention span is about 15 to 20 minutes. Ideally a presentation should be broken down into time slots of this length or less, especially if you have to get the message across to those people who do not really want to be there. With a lecture, a conference or a seminar, you can go a little longer because the audience is usually a willing one, especially if they have paid to get in.

During these 15 to 20 minute sections the attention level will also vary. It is usually highest in the first and the last five minute segments, which is why it is important to get these parts of your presentation right. It is high at the beginning because curiosity, and the hope of solving some problems, is high. As the speaker continues attention drops and people start to look around the room, doodle on their notepads or get ' distracted. As they sense that the speaker is coming to some conclusion, attention picks up again because they are interested to hear the proposed course of action and the summary. Obviously you should take this into account when preparing your presentation. You also need to give some clues, either verbally or by the tone of your voice, to let the audience know when you are coming to the end of your presentation to trigger the increase in attention mentioned above. If you cannot avoid having to talk for longer than the attention span there is a trick that can be employed to keep attention at a high level. You can have a 'false finish'. In other words, you can make the audience think that you are coming to a close about five minutes before you really are. The attention level will increase and just as it is flagging you hit them with the real finish so that the attention rises yet again. This is something that you can usually get away with only once during a presentation. There are a number of ways in which you can break up the presentation into a series of 15 to 20 minute sections, most of which simply involve adding some variety. You can change speakers, change the visual media, pass something around the audience, have a tea break and so on.

Finally, you should rehearse the presentation. Depending on its nature this may range from going over the talk mentally to a full scale 'dry run'. Under-rehearsed presentations are very common but over-rehearsed presentations are very rare. Good preparation and good rehearsal are effective ways to reduce the fear element. If you do not know what you are going to say, who you are going to say it to and how you are going to say it, then you have every reason to be afraid.

The structure of the presentation

Athough every presentation is different there are some common structures that most effective presentations will fit. Following an introduction, the main body of the presentation is then followed by a summary. The middle part, the presentation proper, is itself formed of three or four main sections. In many ways the structure is similar to that of a symphony, a play or a film:

EXPOSITION	DEVELOPMENT	RECAPITULATION
FIRST MOVEMENT	SECOND MOVEMENT	THIRD MOVEMENT
ACT 1	ACT 2	ACT 3
TELL THEM WHAT YOU ARE GOING TO TELL THEM	TELL THEM	TELL THEM WHAT YOU TOLD THEM
SITUATION	COMPLICATION	RECOMMENDATION

There are many ways that the structure can be described but the model I like best is a four part model:

POSITION   PROBLEM   POSSIBILITIES   PROPOSAL

Adding the 'introduction' and the 'summary' results in a six part model (the six Ps).

PREFACE	Introduction
POSITION	Current situation
PROBLEM	Current problems
POSSIBILITIES	Alternative solutions
PROPOSAL	Recommendations, our solution, etc.
POSTSCRIPT	Summary and questions, etc.

PREFACE

This introductory section, often very brief, should contain several elements:

1. Welcoming courtesies. Thank people for attending, giving up their time, 'I hope you will find it beneficial', etc.
2. Self-identification. Your name, your job, your background (if relevant), details about other colleagues present, etc. Some of this is not needed if you yourself have already been introduced by someone else.
3. The intention. What you are proposing to explain, describe or demonstrate at this presentation. Present this in terms of your audience's interest not your own, e.g. 'What I thought you would like to know ...' not 'What I'm going to tell you'.
4. The agenda. How long the presentation will take, will there be a break, will there be a video, will there be any documentation handed out, will there be a demonstration, etc.
5. The ground rules. Do you want questions during the talk or would you prefer them all at the end? (This will depend on the nature of the presentation. For an informal, interactive kind of presentation you should welcome questions at any time. For a more formal event such as a seminar or a conference questions should normally be at the end so that the schedule can be controlled.) Will other people be helping to field questions, etc.?

POSITION

At the start of a presentation it is important that everyone has the same knowledge about the current situation that has led to the need for the presentation in the first place. Do not assume that each member of the audience has been briefed beforehand. It is also important that you show the audience that you understand the situation and the background that led up to it. The 'position' stage also focuses attention on the particular part of the current situation that you are addressing (assuming there are other presentations to be made).

This 'position' stage, where you are establishing common ground, may only require a sentence or two, or it may require an in-depth treatment. It all depends on the nature of the presentation. However, some statement on the position needs to be presented and agreed upon. Some interaction at this stage can be very beneficial; it helps you to tailor the remainder of your talk more precisely. Also, a little two-way communication at this early stage is a valuable 'ice-breaker'.

PROBLEM

Here the problems with the current situation are exposed and the need for change is identified. You need to show why the present situation cannot continue. 'Technology is changing, competition is increasing, lead-times need shortening, margins are declining, etc.' There has to be some opportunity, change, danger, concern, etc., or else you are wasting your time making this presentation.

At this stage you dig the hole in which you intend to plant your idea.

POSSIBILITIES

This stage and the 'problem' stage are often combined. Here you look at the alternative solutions to the problems you have outlined. You weigh up the pros and cons for each alternative and lead into your 'proposal' stage.

PROPOSAL

This is usually the main part of the presentation, often taking most of the available time. This is where your ideas, suggestions, product solution, etc., are presented. This is what many people think of as 'the presentation'. They often omit the other Ps.

Here you make your recommendations to your audience. You also offer supporting references, proof statements, comparisons, pre-empt objections, etc. In short, you offer your solutions with reasons why they are the best solutions to the problems outlined earlier.

POSTSCRIPT

The ending is a vital part of the overall presentation. It should not be left to chance but should be carefully worked out in advance. It is usually quite short so it is important to get the key points across in a few words.

To work out your ending you have to go back to the objectives of the presentation. This is what dictates the 'postscript' which will normally include:

1. A summary of the salient facts and arguments plus a reprise of the key visuals.
2. A recommended course of action.
3. A proposal for the next step, if the recommendation is accepted, with target dates.
4. A description of the handouts that you are about to pass out.
5. Thanks for their time and attention.
6. Thanks to anyone who helped set up the meeting.
7. An invitation to ask questions.

QUESTION TIME

The question session at the end of the presentation should give you a feel for how well the presentation was understood and accepted. It also lets you know more about individual's views and reactions. Even a reluctance to ask questions can be a clue to someone's feelings (usually negative). Some questions may hide objections so you may have to ask a few questions yourself to uncover these.

Inevitably you will occasionally meet some 'smart Alec' who tries to test your knowledge (or show off his or her own). The golden rule here is not to bluff your way out of it. Refer the question to a colleague who is more knowledgeable on that subject or promise to get back with an answer. The audience will usually respect your honesty.

Sometimes a question will be more of a statement aimed at impressing the questioner's colleagues. You will make him or her very happy if you commend his or her expertise, so commend the cleverness of the question, but not in a condescending manner. 'Of course you're absolutely right. I didn't mention it because it's too technical for most people, but our proposal handles it very well....' If you can cite a benefit of your offering that applies to the question then he or she is effectively endorsing your solution.

Some members of the audience may feel challenged or threatened by your proposal and ask a defensive question. This is something that you should have planned for so your presentation should have addressed and minimized this problem. However, if it comes as a surprise it is best to retreat immediately, concede full territorial rights to the questioner and possibly ask him or her to comment (consult his or her wisdom). If it is a question relating to a concern over changes, it can be useful to throw this out for discussion or comment by the audience.

If you need more time to think of an answer, ask the questioner to repeat or elaborate the question. In general the options you have in the question and answer session are:

1. Answer the question.
2. Admit ignorance and promise to find out.
3. Defer it to a private discussion after the meeting.
4. Refer it to an expert colleague.
5. Throw it back to the questioner.
6. Throw it to another member of the audience.
7. Put it up for general discussion.

PRESENTATION TECHNIQUES

It is not possible in a short chapter like this to make someone into a good presenter. Some people are naturally more comfortable with presenting than others. Practice is probably the best way to improve. There are, however, some hints and tips that can be useful.

Most people are quite comfortable when speaking to a small group in a round table discussion. Learning to present to a larger group is little more than developing the ability to transfer this conversational style to a different situation. Many people fall into the trap of being too formal, using words and phrases they would not normally use. This comes across to the audience as being unnatural (because it is) and usually quite tedious. So be yourself. Simply talk to the group as if they were one person because from their individual points of view they are. Just as there is no need to develop a 'telephone voice' there is no need for a 'presentation voice'.

Getting the audience 'on your side' at the beginning can be very beneficial. This is why many books and courses suggest starting with a joke. But a joke that fails can be a major disaster. Think carefully before using jokes. Unless you really are regarded as an expert in your field it is often a good idea to start with some comment that makes it clear that you are not setting yourself up above the audience. For example, stress that you know less about their business than they do and then ask for their sympathetic indulgence rather than risk stimulating resentment.

THE DELIVERY

Some of the main faults to try to recognize and avoid during the actual delivery of your presentation are:

1. *Reading. Never never ever* read your script to the audience. Very few people can do this and make it sound natural. It usually comes across in a rather stilted, hesitant and extremely tedious manner. The whole point of your being there is to *talk* to the audience not to *read* to them. Some authorities on presentation go so far as to say that reading is an insult to the audience. You might as well be saying, 'I know that you are too stupid to be able to read a report if I mailed it to you, so I have come here to read it to you.'

   For the same reasons do not read out the text on your overheads word by word—paraphrase, or say it in a slightly different manner. An exception here would be to deliberately read something out to emphasize it.
2. *Mumbling.* It is better to be too loud than too quiet.
3. *Hesitancy.* Excessively long and frequent pauses, usually interspersed with lots of '...er...ummm...er...umm...', etc., usually indicates a lack of preparation and rehearsal.
4. *Gabbling.* Speaking too quickly, especially to a mixed nationality audience, usually comes from nerves and a desire to 'get it over with'. Less frequently it comes from overenthusiasm for the subject. Either way it is easily controlled once you recognize the problem.
5. *Catch phrases.* Repeating the same catch phrases over and over quickly irritates the audience to the point where they concentrate on counting the number of times they are repeated, rather than listening to the talk.
6. *Poor eye contact.* Staring at the floor, ceiling, walls, screen, projector, etc., can be disturbing. Look at the audience, but not just one area or one person or in an aggressive or hypnotic manner. Look at them as you would in the course of a normal conversation and look all round the audience.
7. *Mannerisms.* Physical mannerisms are only worth worrying about if they cause a real distraction. We all have mannerisms just as we have our favourite catch

phrases. You can sometimes do more harm by trying to eliminate mannerisms than by leaving them alone.

8. *Dropping the voice* at the end of sentences becomes rather tedious to the listener and may cause them to miss an important piece of information. It can also make it sound as though you are coming to the end of your talk. Try to keep the volume and the pitch up at the end of sentences.

12.8　Visual aids

I am a firm believer in the fact that good visual aids are absolutely vital to the effectiveness of a presentation. The emphasis here, however, is on the word 'good'. All too often the 'visuals' are not visuals but 'words' and they are there for the benefit of the presenter rather than the audience. I have seen far too many presentations where the content of the slides or the overhead transparencies (OHTs) were the presenter's notes. Constantly checking the screen to see what to talk about next is disturbing to the audience and it wastes valuable time when one of the key reasons for using visuals is to save time.

Visual aids save time because they enable you to explain a complex subject quickly without any loss of the audience's understanding. Trends can be seen more easily on a graph or a chart than as a list of numbers, and appropriate clip-art can liven up the presentation and help the audience to remember the points you made. There is an excellent series of training films produced in the United Kingdom by a company called Video Arts, many of which star John Cleese. The success of these films has been based heavily on using humour to get the message across, but the main benefit of the humour is the retention levels that it creates. Most people can remember most of the key points made in these films after just one viewing because of the humorous way in which the points are made. It is just the same with visual aids. They do not necessarily have to be humorous, but the more pictorial and graphical they are the easier it is for the audience to remember them. Incidentally, Video Arts and other training companies have films available on presentation subjects. In fact the six P's model for structuring a presentation is described in one of the Video Arts productions.

Problems with the word slide

'Words are not visuals' is a statement that is emphasized in many texts, videos and training courses on presentation skills and the creation of visual aids. There can be the occasional exception to this rule, however, and a typographer or someone skilled in using a desktop publishing system will argue strongly that good typography is a graphic art that enhances the readability of a document. I agree with this concept. You only have to compare some of the manuals that come with computer software to see this. The well-designed manuals practically invite you to read them whereas the poor ones just put you off. The occasional word slide that makes a statement can sometimes work just as well as a visual, but its use should be very occasional. It is also all right to read this kind of slide to the audience just for impact, it can often be done in an effective and theatrical manner. A simple example might be 'Buying this tester is the only effective solution to our problem'. Such a slide, however, should use a big and bold typeface that almost fills the frame rather than the standard text that might be used on a bullet slide. The word slides to definitely avoid are the one or two word platitudes such as 'integrity'.

There are times when we cannot easily avoid using a simple bullet slide, either due to a lack of time to develop a pictorial version or because it is difficult to show the subjects visually. The main problem with bullet slides is that they frequently contain too many bullet points and each point is fully spelled out as full sentences. In other words, there are too many words. There is a tendency for the audience to read ahead when such slides are on the screen, so they are reading point 8 while you are still talking about point 2. Few people can read and listen at the same time so something gets lost. Usually it is the important point that you are trying to make. Bullet slides should be restricted to a maximum of four or five points, and these should be brief resumés of what you will be saying when the slide is on view. Thus, if the audience do read ahead it will only take a few seconds and then they should be listening again. For example, you might be saying, 'We have estimated that we will require enough capacity to test 44 000 complex boards per year by 1995 and the recommended tester will still have capacity to spare at this level.' The bullet point may simply state 'adequate capacity margins'. It sounds like common sense but you would be surprised at the number of times I have seen the equivalent of the spoken sentence spelled out fully on the OHT. A better approach to bullet slides can be to use a 'reveal' technique where one slide is broken down into several. The first slide shows only the first bullet point, the second slide shows the first and the second points but the first point is now in a subdued colour relative to the newly revealed point 2. As each point is revealed the previous points are still shown as a reminder but there can be no reading ahead by the audience.

One situation where text slides can be useful is when you are presenting to an audience in something other than their natural language. Many people can read a foreign language much better than they can 'hear' it so the text can fill in for bits of your talk that they may miss as they translate it 'on the fly'. However, a well-designed visual slide will get the point across without the need for much text and is still to be preferred in this situation.

Presentation media

There is a wide variety of media available for visual aids, including whiteboards, blackboards, flip-charts, videos, notepads, etc. All have their uses but by far the most common media are 35 mm slides and overhead transparencies. I personally believe that all presentations of a persuasive nature should be supported by OHTs and not by 35 mm slides. A 35 mm presentation is a fairly formal format where the room lighting is dimmed, the sequence of events is fixed and eye contact with the audience may be difficult. In effect the format says 'Sit down. Be quiet. I am going to talk to you.' In contrast the 'overhead' presentation is much less formal, you can see the audience's reaction to what you are saying and you can vary the sequence of events 'on the fly' if necessary. It is a much more interactive format than a slide-based presentation. The audio-visual industry often refers to the choice as 'slides are for *telling*, overheads are for *selling*'. In a persuasive presentation it is the presenter who does the selling not the visuals. The visuals simply support and clarify what is being said. In order to sell your recommendation you have to be able to judge people's responses and react to them accordingly. You cannot do this when you cannot see their faces.

The 35 mm slides are more appropriate for informative presentations, lectures and conference papers. They can also be used in seminars, even though many seminars contain a mixture of telling and selling. The large size of the audience at a seminar and the need to keep to a fairly strict timetable prevent a truly interactive approach. However, my own preference is to use overheads, even for a seminar.

Creating the visual aids

The main arguments from engineers and managers against preparing pictorial or graphical visuals have been that it takes too much time and that they are not artists. These were valid arguments in the past but today we have an absolute wealth of easy to use tools to create presentations rapidly and professionally. Some industry analysts claim that graphics and presentation packages for the PC are the fastest growing segment of the applications software market. The market is so competitive that the capabilities of these packages has grown out of all recognition in the past few years. Today there are specialized business, scientific and statistical graphing and charting applications. There are high-performance drawing packages for the graphics artist. There are photo retouching and image enhancement packages and there are packages that are dedicated to the complete task of developing presentations. Spreadsheet applications have had basic business graphing for some time but the latest releases now include sophisticated presentation capabilities as the developers recognize the need to *present* the information developed in a spreadsheet in a professional manner. Even the humble word-processor has grown out of recognition. Current versions of the leading packages include drawing, charting and equation generation, and they are supplied with large quantities of 'clip-art' for inclusion in your documents. In case you are not familiar with 'clip-art' this is professionally drawn pictures, icons, maps and symbols, which can easily be incorporated into documents or overhead transparencies.

Of most interest to the presenter are the 'presentation' packages that are available. The leading products in this category at present are 'Lotus Freelance', 'Aldus Persuasion', 'Microsoft Powerpoint' and 'Harvard Graphics'. These are characterized by having the capability to create everything you need for a complete presentation. With these products you can draw, generate graphs and charts, and incorporate clip-art to create your slides. In addition you can generate hand-outs for the audience that show a number of reproductions of your slides on each page and 'speaker notes' that incorporate a picture of your slide with space below for your presentation notes.

They work on the principal of a set of 'templates' that define each slide type. You simply enter the data and your text and the style and formatting are taken care of automatically. This saves an enormous amount of time and also forces a high degree of consistency in the appearance of the slides. As well as being supplied with a wide variety of template designs the packages give users the ability to create their own 'corporate style'. These products also provide the capability to run a presentation directly on the PC screen or to project from an overhead projector using a liquid crystal display (LCD) panel on top of the projector's light box. This is becoming an increasingly popular way to visually support a presentation because it offers the capability for some animation and even for sound clips.

A trend seems to have started to include all of the capabilities of the different types of graphics packages into a single product. The current version of 'Corel Draw' includes a high end drawing capability, a charting capability, a photo-retouch/image enhancement capability and an on-screen presentation capability. It comes supplied on a CD-ROM with 14000 pieces of clip art, and all at a very reasonable price. It is not quite there as a fully integrated system because these capabilities are run as separate applications and it lacks the ability to produce hand-outs and speaker notes automatically; however, it does show the direction that things are moving in.

There is no longer any excuse. Anyone can now develop a professional looking set of presentation support visuals in a very short time. The availability of reasonably priced colour inkjet printers means that all companies can produce the end product in-house. Even the 'high end' colour thermal transfer printers, which can give projected results that come close to matching a 35 mm slide, are coming down in price considerably.

12.9 Summary

Preparation

1. Know the facts—the facts your audience will want, as well as the ones you want to give them.
2. Find out about the people, including their past experiences, present situation and future needs.
3. Always have notes, with key phrases as well as subject headings.

Structure

1. Preface—opening courtesies, the purpose of the presentation, its duration and shape and the ground rules.
2. Position—a brief outline of the present situation.
3. Problem—a description of the audience's needs which can be met by accepting your proposal.
4. Possibilities—a look at the main alternatives your audience will want to consider.

5. Proposal—your recommended course of action.
6. Postscript—summary of the proposal, the next step, a description of supporting documents, thanks, invitation to ask questions.

Technique

1. Do not read your presentation or your overheads.
2. Make good notes and use them during the presentation.
3. Convert statistics into charts and graphs wherever possible.
4. Relegate details to supporting documents.
5. Use short words and short sentences.
6. Avoid or explain jargon.
7. Sectionalize the presentation to make it easier for the audience to take it all in.
8. Never mumble or gabble.
9. Look at the audience—all of them.
10. Keep your voice up at the end of sentences.
11. Write out the introduction and memorize it.
12. Rehearse thoroughly.

Visual aids

1. Words are *not* visuals.
2. Use graphs, charts, pictures wherever possible.
3. Never have more than five bullet points on one slide.
4. OHTs are best for persuasive presentations.
5. Use a 'Presentation' application on a PC or a MAC for speed and professional results. Provide handouts of your slides for the benefit of the audience.

Reference

Juran, Joseph M. and Frank M. Gryna (1988) *Juran's Quality Control Handbook,* 4th edn, McGraw-Hill, New York.

Appendix A: Loaded or unloaded labour rates?

Whenever an economic analysis is to be performed to justify the purchase of a piece of equipment based upon saving time and hence labour costs, there is always a debate about overhead ratios. Should the direct labour rates be loaded (burdened) or not? If so, what loading factor should be used? Two times? Four times? Six times?

Unfortunately, there is no easy answer. Each situation will be different and therefore will require some investigation to determine a suitable loading factor.

It is fairly certain, however, that the manufacturing overhead multiplier used for internal costing will be too high a rate to use for justifying a purchase. Many of the elements in the overhead will be unaffected by the introduction of the equipment under consideration. To justify a major capital investment, we must attempt to calculate the real savings that will be generated by using the equipment. Using too high a loading factor will falsify the savings. Using too low a loading factor may understate the savings and lead to a poor decision.

One approach to the problem is to separate fixed overheads from variable overheads and then try to determine which might be affected by the introduction of the proposed equipment.

Financial people might argue that there will be no effect at all on fixed overheads such as buildings, etc. However, I know of several situations where the purchase of ATE has considerably delayed the need to extend a factory or move to a new site. ATE can easily perform the same amount of throughput as 10 to 30 manual testing set-ups, but if the saved space cannot be utilized it is still there and forms part of the overhead. If a genuine saving in space or building costs can be achieved by the introduction of ATE, it is probably better to include this as a one-time saving resulting from the acquisition rather than to account for it in the labour loading factor.

Most of the examples in this book have used either direct labour costs or partially loaded rates.

If you want to see the effect that a change in testing strategy might have on the internal costing and pricing of a particular product then it would be valid to use the fully loaded labour rates that are normally used for such calculations.

Appendix B: Calculation of equipment costs

For many economic analyses, it is often necessary to calculate a cost per hour or cost per year for a piece of equipment. The simple approach to this problem, which is adequate for many 'first-cut' comparisons, is to take the cost of the equipment and divide this by the number of years or hours that it is expected to be used.

Example

A $100 000 tester amortized over four years would be $25 000 per year. If it is expected that each year there would be 1680 hours of use, then the cost per hour is simply

$$\$25\,000 \div 1680 = \$14.88 \text{ per hour}$$

This is a very simplistic approach since it makes no allowance for any tax benefits created by depreciation of the equipment. In some parts of some countries there may even be some form of investment grant available that should also be accounted for.

The tax laws governing the depreciation of fixed assets vary from country to country, but the principles remain much the same. A couple of examples will show how this might be taken into account.

Example

Assuming the piece of equipment can be depreciated linearly over four years, there will be a depreciation of $25 000 per year. This will lower the tax bill by $12 250 assuming a 50 per cent rate of company or corporation tax. It might therefore be more accurate to base the hourly cost of the equipment on a cost of $12 500 per year rather than $25 000. However, to be more accurate still we should consider the time value of money as described in Chapter 11. The tax benefits will come in over four years so each year's $12 500 benefit should be discounted to a present value in order to arrive at an actual cost for the equipment. This cost can then be distributed on a per year or per hour basis as required: $100 000 invested in equipment, 20 per cent discount rate (see Table B.1).

Table B.1

Year	Depreciation	Tax saved	PVF (20%)	DCF
1	25 000	12 500	0.833	10 413
2	25 000	12 500	0.694	8 675
3	25 000	12 500	0.579	7 238
4	25 000	12 500	0.482	6 025
				32 350

Table B.2

Year	Depreciation	Tax saved	PVF (20%)	DCF
1	25 000	12 500	1	12 500
2	25 000	12 500	0.833	10 413
3	25 000	12 500	0.694	8 675
4	25 000	12 500	0.579	7 238
				38 826

Thus the depreciation allowance will save us $32 350 over the four years at PVs assuming a 20 per cent discount rate. Therefore, the equipment will cost

$$\$100\,000 - 32\,350 = \$67\,650$$

or $16 913 per year, which for 1680 hours per year usage is $10.07 per hour.

The timing of the tax benefits will determine which years should be discounted. The example above assumes that the benefit will occur one year after the equipment was purchased. If the benefit occurs more rapidly than that, as it would if the purchase takes place near the end of a tax year, then it might be more appropriate to delay the discounting by one year, as shown in Table B.2.

The equipment cost is now

$$\$100\,000 - 38\,826 = \$61\,175$$

for an annual cost of $15 294 and an hourly cost of $9.1.

Note. These approaches for computing the equipment costs are most likely to be used when looking at the overall cost of ownership or making a quick calculation of operating costs, etc. When using this approach to calculate savings it should be remembered that the savings will be reduced by taxation also.

It is, however, incorrect to include the equipment costs when calculating savings for inclusion in an ROI calculation using DCF techniques. If you do so then the equipment costs will effectively be counted twice. They will be counted in the savings calculations, thus lowering the savings, and they will be counted in the DCF computation of NPV or IRR.

Appendix C: ICT—calculation of yields, PFB and ideal diagnosis/repair loop numbers

An ICT system can theoretically detect all faults with one test operation. In practice, however, it is normal to set up the system so that it stops testing at the end of the shorts test portion (usually the first portion) of the test program, if any shorts have been detected. The reason for this is that the presence of a short will make it difficult or even impossible to measure or test accurately some of the components connected to the shorted nodes. This can result in inaccurate diagnosis messages and possibly cause the system to indicate the presence of faults where none exist.

Any boards that contain both a short and some other 'component-related' fault will therefore pass round the diagnosis/repair loop a minimum of two times. Therefore, the average number of loops for ICT, even if all diagnosis and repair is 100 per cent correct, will be something greater than 1.0. For any test stage where all fault location is sequential (such as functional test) then the number of loops, assuming 100 per cent accurate diagnosis and repair, will be equal to the average number of FPB. In addition, there will be a loop number multiplier that adjusts for any inaccuracy of diagnosis or repair action. It is this loop number multiplier that appears in the various examples in the book, since the basic number of loops will be accounted for in the formulae used within the EVALUATE program.

If you are monitoring loop numbers as part of your test statistics, it will be necessary to determine the theoretical loop number for ICT before any judgement on the effectiveness of the diagnosis/repair loop can be made, based on the measured loop number. Since in perfect operation only those boards containing both shorts and component-related faults will pass around the loop twice, the theoretical loop number can be determined from the proportion of boards containing both kinds of fault relative to the total proportion of faulty boards:

$$Y = e^{-FPB}$$

where

Y = yield
e = 2.718
FPB = average number of FPB
        detected at this test stage

$$PFB = 1 - Y$$

where

PFB = probability of a board being faulty—or
        the proportion of boards that contain
        one or more faults

Now let:

Y_s = yield for a 'shorts' only test
Y_c = yield for a 'component-related' test with no shorts present
PFB_s = probability of a board failing at the 'shorts' test stage
PFB_c = probability of a board failing at the 'component testing' stage of the in-circuit
        test (no shorts present)
PFB_{so} = probability of a failing board having only 'shorts'
PFB_{co} = probability of a failing board having only 'component-related' faults
PFB_b = probability of a failing board having both 'shorts' and 'component-related'
        faults

The various probabilities are calculated as follows:

$PFB = 1 - Y$
$PFB_s = 1 - Y_s$
$PFB_c = 1 - Y_c$
$PFB_b = (PFB_s + PFB_c) - PFB$
$PFB_{so} = PFB - PFB_c$ (or $PFB_s - PFB_b$)
$PFB_{co} = PFB - PFB_s$ (or $PFB_c - PFB_b$)

A couple of examples should clarify how these probabilities are arrived at.

Example 1

Shorts per board   = 0.6
Component-related FPB = 0.9
           Total FPB = 1.5

$$Y = e^{-1.5} = 0.22$$

$$\text{Therefore} \quad \text{PFB} = 0.78$$
$$Y_s = e^{-0.6} = 0.55$$
$$\text{Therefore} \quad \text{PFB}_s = 0.45$$
$$Y_c = e^{-0.9} = 0.41$$
$$\text{Therefore} \quad \text{PFB}_c = 0.59$$
$$\text{PFB}_b = (0.45 + 0.59) - 0.78$$
$$= 0.26$$
$$\text{PFB}_{so} = 0.78 - 0.59$$
$$= 0.19$$
$$\text{PFB}_{co} = 0.78 - 0.45$$
$$= 0.33$$

As a check $\text{PFB}_b + \text{PFB}_{co} + \text{PFB}_{so} + Y$ should equal 1.00, which they do.

The distribution of faulty and fault-free boards is shown diagrammatically in Fig. C.1.

Example 2 (Fig. C.2)

$$\text{Shorts per board} = 0.2$$
$$\text{Component-related FPB} = 0.5$$
$$\text{Total FPB} = 0.7$$

$$Y = e^{-0.7}$$
$$= 0.50$$
$$\text{Therefore} \quad \text{PFB} = 0.50$$
$$Y_s = e^{-0.2}$$
$$= 0.82$$
$$\text{Therefore} \quad \text{PFB}_s = 0.18$$
$$Y_c = e^{-0.5}$$
$$= 0.61$$
$$\text{Therefore} \quad \text{PFB}_c = 0.39$$
$$\text{PFB}_b = (0.18 + 0.39) - 0.5$$
$$= 0.07$$
$$\text{PFB}_{so} = 0.5 - 0.39$$
$$= 0.11$$
$$\text{PFB}_{co} = 0.5 - 0.18$$
$$= 0.32$$

As a check $\text{PFB}_{so} + \text{PFB}_{co} + \text{PFB}_b + Y = 1.0$. This distribution is shown in Fig. C.2.

Calculation of the ideal loop number

If the diagnosis and repair actions are 100 per cent correct and no additional faults are added during the repair process, then the average number of times each faulty board passes round the diagnosis/repair loop will be a function of PFB and PFB_b

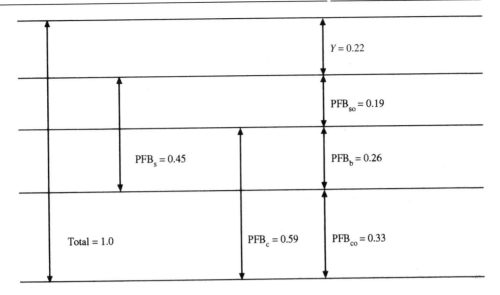

Figure C.1   Distribution of faulty and fault-free boards for Example 1

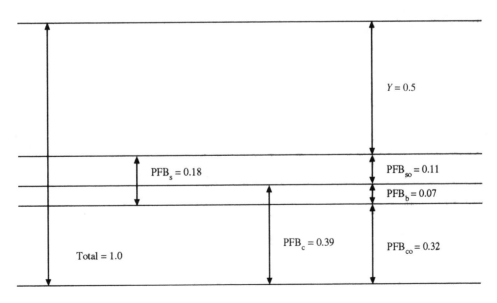

Figure C.2   Distribution of faulty and fault-free boards for Example 2

as follows:

$$\text{Ideal loop number} = \frac{\text{PFB} + \text{PFB}_b}{\text{PFB}}$$

For our two examples this works out at:

1.33 for example 1

1.14 for example 2

Assuming that your records show that for these board types each faulty board passes around the loop 1.46 and 1.31 times respectively, it is possible to determine how effective your diagnosis/repair operation is.

The loop number multiplier in each case will be

$$\frac{1.46}{1.33} = 1.1 \text{ for board 1}$$

and

$$\frac{1.31}{1.14} = 1.15 \text{ for board 2}$$

These multipliers indicate directly the percentage of incorrect repair actions (or the addition of faults) as being 10 and 15 per cent respectively. These figures may or may not be acceptable but in any case the reasons should be investigated since it is a possible area of process improvement.

Appendix D:
The relationship between average FPB and average FPFB

Throughout the book there are numerous examples where the average number of FPB is used to determine such things as yield, or the time spent on the test system, etc.

FPB is a useful parameter to know since it affects just about everything within the testing and production process. Only if FPB and the fault spectrum is known can you hope to make improvements to your manufacturing process. In practice, FPB is measured by monitoring the number of diagnosis or repair actions relative to the number of boards produced over the same period of time. By comparing the measured FPB with the FPB computed from the initial yield data, it is possible to develop a measure of the diagnosis/repair loop multiplier and hence the effectiveness of the diagnosis/repair process (see Appendix C for details).

Depending on how your statistics are gathered you may end up with the average number of faults contained on each *faulty* board rather than the average FPB across all boards tested.

The faults per faulty board or FPFB figure is related directly to FPB in the following way: since FPB determines the yield and the PFB factor then all the faults (FPB) will be contained on the proportion of boards that are faulty (PFB).

Example

For an FPB figure of 1.0 the PFB will be

$$1 - e^{-FPB} = 0.63$$

Therefore the average of 1.0 FPB will all be contained on 63 per cent of the boards. The average number of FPFB will be

$$\frac{1}{0.63} \quad \text{or} \quad 1.59$$

Therefore when FPB = 1, FPFB will be $\dfrac{FPB}{PFB}$ or 1.59

This relationship is shown graphically in Fig. D.1 for values of FPB up to 3.0.

You will notice that as FPB increases the FPFB figure tends towards FPB since PFB tends towards 1.0 (yield tends towards 0). As FPB tends towards zero FPFB tends towards 1.0. It may seem odd that FPFB is 1.0 when FPB is 0; however, if you have one faulty board in one million good ones, that faulty board must have at least one fault on it.

The chart can be used to convert between FPB and FPFB to enable calculations to be made regardless of which figure is maintained in your database.

Note. Since this is a statistical relationship (as is yield versus FPB) the sample size (number of boards) must be big enough to make the relationship valid.

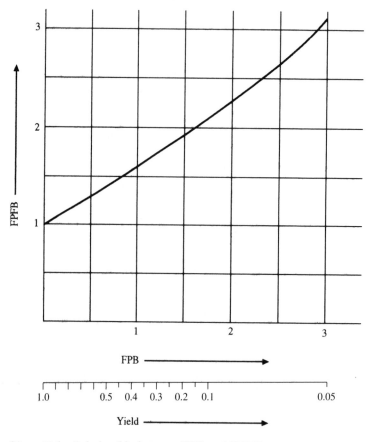

Figure D.1 Relationship between FPB and FPFB

Appendix E: Useful tables and charts

Table E.1 Cumulative binomial probabilities

p = probability of success in a single trial; n = number of trials. The table gives the probability of obtaining r *or more* successes in n independent trials, i.e.

$$\sum_{x=r}^{n} \binom{n}{x} p^x (1-p)^{n-x}$$

When there is no entry for a particular pair of values of r and p, this indicates that the appropriate probability is less than 0.00005. Similarly, except for the case $r=0$, when the entry is exact, a tabulated value of 1.0000 represents a probability greater than 0.99995.

	$p =$	0.01	0.02	0.03	0.04	0.05	0.06	0.07	0.08	0.09
$n=2$	$r=$ 0	1.0000	1.0000	1.0000	1.0000	1.0000	1.0000	1.0000	1.0000	1.0000
	1	0.0199	0.0396	0.0591	0.0784	0.0975	0.1164	0.1351	0.1536	0.1719
	2	0.0001	0.0004	0.0009	0.0016	0.0025	0.0036	0.0049	0.0064	0.0081
$n=5$	$r=$ 0	1.0000	1.0000	1.0000	1.0000	1.0000	1.0000	1.0000	1.0000	1.0000
	1	0.0490	0.0961	0.1413	0.1846	0.2262	0.2661	0.3043	0.3409	0.3760
	2	0.0010	0.0038	0.0085	0.0148	0.0226	0.0319	0.0425	0.0544	0.0674
	3		0.0001	0.0003	0.0006	0.0012	0.0020	0.0031	0.0045	0.0063
	4						0.0001	0.0001	0.0002	0.0003
$n=10$	$r=$ 0	1.0000	1.0000	1.0000	1.0000	1.0000	1.0000	1.0000	1.0000	1.0000
	1	0.0956	0.1829	0.2626	0.3352	0.4013	0.4614	0.5160	0.5656	0.6016
	2	0.0043	0.0162	0.0345	0.0582	0.0861	0.1176	0.1517	0.1879	0.2254
	3	0.0001	0.0009	0.0028	0.0062	0.0115	0.0188	0.0283	0.0401	0.0540
	4			0.0001	0.0004	0.0010	0.0020	0.0036	0.0058	0.0088
	5					0.0001	0.0002	0.0003	0.0006	0.0010
	6									0.0001
$n=20$	$r=$ 0	1.0000	1.0000	1.0000	1.0000	1.0000	1.0000	1.0000	1.0000	1.0000
	1	0.1821	0.3324	0.4562	0.5580	0.6415	0.7099	0.7658	0.8113	0.8484
	2	0.0169	0.0599	0.1198	0.1897	0.2642	0.3395	0.4131	0.4831	0.5484
	3	0.0010	0.0071	0.0210	0.0439	0.0755	0.1150	0.1610	0.2121	0.2666
	4		0.0006	0.0027	0.0074	0.0159	0.0290	0.0471	0.0706	0.0993
	5			0.0003	0.0010	0.0026	0.0056	0.0107	0.0183	0.0290
	6				0.0001	0.0003	0.0009	0.0019	0.0038	0.0068
	7						0.0001	0.0003	0.0006	0.0013
	8								0.0001	0.0002

Table E.1 (*contd*)

Table E.1 (*contd*)

	$p =$	0.01	0.02	0.03	0.04	0.05	0.06	0.07	0.08	0.09
$n = 50$	$r = 0$	1.0000	1.0000	1.0000	1.0000	1.0000	1.0000	1.0000	1.0000	1.0000
	1	0.3950	0.6358	0.7819	0.8701	0.9231	0.9547	0.9734	0.9845	0.9910
	2	0.0894	0.2642	0.4447	0.5995	0.7206	0.8100	0.8735	0.9173	0.9468
	3	0.0138	0.0784	0.1892	0.3233	0.4595	0.5838	0.6892	0.7740	0.8395
	4	0.0016	0.0178	0.0628	0.1391	0.2396	0.3527	0.4673	0.5747	0.6697
	5	0.0001	0.0032	0.0168	0.0490	0.1036	0.1794	0.2710	0.3710	0.4723
	6		0.0005	0.0037	0.0144	0.0378	0.0776	0.1350	0.2081	0.2928
	7		0.0001	0.0007	0.0036	0.0118	0.0289	0.0583	0.1019	0.1596
	8			0.0001	0.0008	0.0032	0.0094	0.0220	0.0438	0.0768
	9				0.0001	0.0008	0.0027	0.0073	0.0167	0.0328
	10					0.0002	0.0007	0.0022	0.0056	0.0125
	11						0.0002	0.0006	0.0017	0.0043
	12							0.0001	0.0005	0.0013
	13								0.0001	0.0004
	14									0.0001
$n = 100$	$r = 0$	1.0000	1.0000	1.0000	1.0000	1.0000	1.0000	1.0000	1.0000	1.0000
	1	0.6340	0.8674	0.9524	0.9831	0.9941	0.9979	0.9993	0.9998	0.9999
	2	0.2642	0.5967	0.8054	0.9128	0.9629	0.9848	0.9940	0.9977	0.9991
	3	0.0794	0.3233	0.5802	0.7679	0.8817	0.9434	0.9742	0.9887	0.9952
	4	0.0184	0.1410	0.3528	0.5705	0.7422	0.8570	0.9256	0.9633	0.9827
	5	0.0034	0.0508	0.1821	0.3711	0.5640	0.7232	0.8368	0.9097	0.9526
	6	0.0005	0.0155	0.0808	0.2116	0.3840	0.5593	0.7086	0.8201	0.8955
	7	0.0001	0.0041	0.0312	0.1064	0.2340	0.3936	0.5557	0.6968	0.8060
	8		0.0009	0.0106	0.0475	0.1280	0.2517	0.4012	0.5529	0.6872
	9		0.0002	0.0032	0.0190	0.0631	0.1463	0.2660	0.4074	0.5506
	10			0.0009	0.0068	0.0282	0.0775	0.1620	0.2780	0.4125
	11			0.0002	0.0022	0.0115	0.0376	0.0908	0.1757	0.2882
	12				0.0007	0.0043	0.0168	0.0469	0.1028	0.1876
	13				0.0002	0.0015	0.0069	0.0224	0.0559	0.1138
	14					0.0005	0.0026	0.0099	0.0282	0.0645
	15					0.0001	0.0009	0.0041	0.0133	0.0341
	16						0.0003	0.0016	0.0058	0.0169
	17						0.0001	0.0006	0.0024	0.0078
	18							0.0002	0.0009	0.0034
	19							0.0001	0.0003	0.0014
	20								0.0001	0.0005
	21									0.0002
	22									0.0001

	$p =$	0.10	0.15	0.20	0.25	0.30	0.35	0.40	0.45	0.50
$n = 2$	$r = 0$	1.0000	1.0000	1.0000	1.0000	1.0000	1.0000	1.0000	1.0000	1.0000
	1	0.1900	0.2775	0.3600	0.4375	0.5100	0.5775	0.6400	0.6975	0.7500
	2	0.0100	0.0225	0.0400	0.0625	0.0900	0.1225	0.1600	0.2025	0.2500
$n = 5$	$r = 0$	1.0000	1.0000	1.0000	1.0000	1.0000	1.0000	1.0000	1.0000	1.0000
	1	0.4095	0.5563	0.6723	0.7627	0.8319	0.8840	0.9222	0.9497	0.9688
	2	0.0815	0.1648	0.2627	0.3672	0.4718	0.5716	0.6630	0.7438	0.8125
	3	0.0086	0.0266	0.0579	0.1035	0.1631	0.2352	0.3174	0.4069	0.5000
	4	0.0005	0.0022	0.0067	0.0156	0.0308	0.0540	0.0870	0.1312	0.1875
	5		0.0001	0.0003	0.0010	0.0024	0.0053	0.0102	0.0185	0.0313
$n = 10$	$r = 0$	1.0000	1.0000	1.0000	1.0000	1.0000	1.0000	1.0000	1.0000	1.0000
	1	0.6513	0.8031	0.8926	0.9437	0.9718	0.9865	0.9940	0.9975	0.9990
	2	0.2639	0.4557	0.6242	0.7560	0.8507	0.9140	0.9536	0.9767	0.9893
	3	0.0702	0.1798	0.3222	0.4744	0.6172	0.7384	0.8327	0.9004	0.9453

Table E.1 (*contd*)

	p =	0.10	0.15	0.20	0.25	0.30	0.35	0.40	0.45	0.50
	4	0.0128	0.5000	0.1209	0.2241	0.3504	0.4862	0.6177	0.7430	0.8281
	5	0.0016	0.0099	0.0328	0.0781	0.1503	0.2485	0.3669	0.4956	0.6230
	6	0.0001	0.0014	0.0064	0.0197	0.0473	0.0949	0.1662	0.2616	0.3770
	7		0.0001	0.0009	0.0035	0.0106	0.0260	0.0548	0.1020	0.1719
	8			0.0001	0.0004	0.0016	0.0048	0.0123	0.0274	0.0547
	9					0.0001	0.0005	0.0017	0.0045	0.0107
	10							0.0001	0.0003	0.0010
$n = 20$	$r = 0$	1.0000	1.0000	1.0000	1.0000	1.0000	1.0000	1.0000	1.0000	1.0000
	1	0.8784	0.9612	0.9885	0.9968	0.9992	0.9998	1.0000	1.0000	1.0000
	2	0.6083	0.8244	0.9308	0.9757	0.9924	0.9979	0.9995	0.9999	1.0000
	3	0.3231	0.5951	0.7939	0.9087	0.9645	0.9879	0.9964	0.9991	0.9998
	4	0.1330	0.3523	0.5886	0.7748	0.8929	0.9556	0.9840	0.9951	0.9987
	5	0.0432	0.1702	0.3704	0.5852	0.7625	0.8818	0.9490	0.9811	0.9941
	6	0.0113	0.0673	0.1958	0.3828	0.5836	0.7546	0.8744	0.9447	0.9793
	7	0.0024	0.0219	0.0867	0.2142	0.3920	0.5834	0.7500	0.8701	0.8423
	8	0.0004	0.0059	0.0321	0.1018	0.2277	0.3990	0.5841	0.7480	0.8684
	9	0.0001	0.0013	0.0100	0.0409	0.1133	0.2376	0.4044	0.5857	0.7483
	10		0.0002	0.0026	0.0139	0.0480	0.1218	0.2447	0.4086	0.5881
	11			0.0006	0.0039	0.0171	0.0532	0.1275	0.2493	0.4119
	12			0.0001	0.0009	0.0051	0.0196	0.0565	0.1308	0.2517
	13				0.0002	0.0013	0.0060	0.0210	0.0580	0.1316
	14					0.0003	0.0015	0.0065	0.0214	0.0577
	15						0.0003	0.0016	0.0064	0.0207
	16							0.0003	0.0015	0.0059
	17								0.0003	0.0013
	18									0.0002

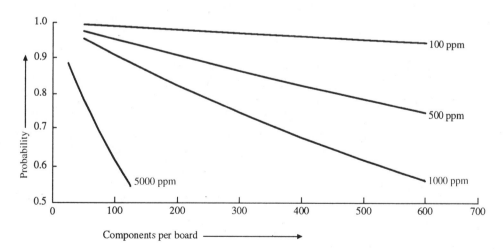

Figure E.1 Probability of boards being free from defective components, for typical levels of component quality

Table E.2 Cumulative Poisson probabilities

The table gives the probability that r *or more* random events are contained in an interval when the average number of such events per interval is m, i.e.

$$\sum_{x=r}^{\infty} e^{-m(m^r/x!)}$$

Where there is no entry for a particular pair of values of r and m, this indicates that the appropriate probability is less than 0.00005. Similarly, except for the case r = 0 when the entry is exact, a tabulated value of 1.0000 represents a probability greater than 0.99995.

m =	0.1	0.2	0.3	0.4	0.5	0.6	0.7	0.8	0.9	1.0
r = 0	1.0000	1.0000	1.0000	1.0000	1.0000	1.0000	1.0000	1.0000	1.0000	1.0000
1	0.0952	0.1813	0.2592	0.3297	0.3935	0.4512	0.5034	0.5507	0.5934	0.6321
2	0.0047	0.0175	0.0369	0.0616	0.0902	0.1219	0.1558	0.1912	0.2275	0.2642
3	0.0002	0.0011	0.0036	0.0079	0.0144	0.0231	0.0341	0.0474	0.0629	0.0803
4		0.0001	0.0003	0.0008	0.0018	0.0034	0.0058	0.0091	0.0135	0.0190
5				0.0001	0.0002	0.0004	0.0008	0.0014	0.0023	0.0037
6							0.0001	0.0002	0.0003	0.0006
7										0.0001

m =	1.1	1.2	1.3	1.4	1.5	1.6	1.7	1.8	1.9	2.0
r = 0	1.0000	1.0000	1.0000	1.0000	1.0000	1.0000	1.0000	1.0000	1.0000	1.0000
1	0.6671	0.6988	0.7275	0.7534	0.7769	0.7981	0.8173	0.8347	0.8504	0.8647
2	0.3010	0.3374	0.3732	0.4082	0.4422	0.4751	0.5068	0.5372	0.5663	0.5940
3	0.0996	0.1205	0.1429	0.1665	0.1912	0.2166	0.2428	0.2694	0.2963	0.3233
4	0.0257	0.0338	0.0431	0.0537	0.0656	0.0788	0.0932	0.1087	0.1253	0.1429
5	0.0054	0.0077	0.0107	0.0143	0.0186	0.0237	0.0296	0.0364	0.0441	0.0527
6	0.0010	0.0015	0.0022	0.0032	0.0045	0.0060	0.0080	0.0104	0.0132	0.0166
7	0.0001	0.0003	0.0004	0.0006	0.0009	0.0013	0.0019	0.0026	0.0034	0.0045
8			0.0001	0.0001	0.0002	0.0003	0.0004	0.0006	0.0008	0.0011
9							0.0001	0.0001	0.0002	0.0002

m =	2.1	2.2	2.3	2.4	2.5	2.6	2.7	2.8	2.9	3.0
r = 0	1.0000	1.0000	1.0000	1.0000	1.0000	1.0000	1.0000	1.0000	1.0000	1.0000
1	0.8775	0.8892	0.8997	0.9093	0.9179	0.9257	0.9328	0.9392	0.9450	0.9502
2	0.6204	0.6454	0.6691	0.6916	0.7127	0.7326	0.7513	0.7689	0.7854	0.8009
3	0.3504	0.3773	0.4040	0.4303	0.4562	0.4816	0.5064	0.5305	0.5540	0.5758
4	0.1614	0.1806	0.2007	0.2213	0.2424	0.2640	0.2859	0.3081	0.3304	0.3528
5	0.0621	0.0725	0.0838	0.0959	0.1088	0.1226	0.1371	0.1523	0.1682	0.1847
6	0.0204	0.0249	0.0300	0.0357	0.0420	0.0490	0.0567	0.0651	0.0742	0.0839
7	0.0059	0.0075	0.0094	0.0116	0.0142	0.0172	0.0206	0.0244	0.0287	0.0335
8	0.0015	0.0020	0.0026	0.0033	0.0042	0.0053	0.0066	0.0081	0.0099	0.0119
9	0.0003	0.0005	0.0006	0.0009	0.0011	0.0015	0.0019	0.0025	0.0031	0.0038
10	0.0001	0.0001	0.0001	0.0002	0.0003	0.0004	0.0005	0.0007	0.0009	0.0011
11					0.0001	0.0001	0.0001	0.0002	0.0002	0.0003
12									0.0001	0.0001

m =	3.1	3.2	3.3	3.4	3.5	3.6	3.7	3.8	3.9	4.0
r = 0	1.0000	1.0000	1.0000	1.0000	1.0000	1.0000	1.0000	1.0000	1.0000	1.0000
1	0.9550	0.9592	0.9631	0.9666	0.9698	0.9727	0.9753	0.9776	0.9798	0.9817
2	0.8153	0.8288	0.8414	0.8532	0.8641	0.8743	0.8838	0.8926	0.9008	0.9084
3	0.5988	0.6201	0.6406	0.6603	0.6792	0.6973	0.7146	0.7311	0.7469	0.7619

Table E.2 (*contd*)

$m=$	3.1	3.2	3.3	3.4	3.5	3.6	3.7	3.8	3.9	4.0
$r=$ 4	0.3752	0.3975	0.4197	0.4416	0.4634	0.4848	0.5058	0.5265	0.5468	0.5665
5	0.2108	0.2194	0.2374	0.2558	0.2746	0.2936	0.3128	0.3322	0.3516	0.3712
6	0.0943	0.1054	0.1171	0.1295	0.1424	0.1559	0.1699	0.1844	0.1994	0.2149
7	0.0388	0.0446	0.0510	0.0579	0.0653	0.0733	0.0818	0.0909	0.1005	0.1107
8	0.0142	0.0168	0.0198	0.0231	0.0267	0.0308	0.0352	0.0401	0.0454	0.0511
9	0.0047	0.0057	0.0069	0.0083	0.0099	0.0117	0.0137	0.0160	0.0185	0.0214
10	0.0014	0.0018	0.0022	0.0027	0.0033	0.0040	0.0048	0.0058	0.0069	0.0081
11	0.0004	0.0005	0.0006	0.0008	0.0010	0.0013	0.0016	0.0019	0.0023	0.0028
12	0.0001	0.0001	0.0002	0.0002	0.0003	0.0004	0.0005	0.0006	0.0007	0.0009
13				0.0001	0.0001	0.0001	0.0001	0.0001	0.0002	0.0003
14									0.0001	0.0001

$m=$	4.1	4.2	4.3	4.4	4.5	4.6	4.7	4.8	4.9	5.0
$r=$ 0	1.0000	1.0000	1.0000	1.0000	1.0000	1.0000	1.0000	1.0000	1.0000	1.0000
1	0.9834	0.9850	0.9864	0.9877	0.9889	0.9899	0.9909	0.9918	0.9926	0.9933
2	0.9155	0.9220	0.9281	0.9337	0.9389	0.9437	0.9482	0.9523	0.9561	0.9596
3	0.7762	0.7898	0.8026	0.8149	0.8264	0.8374	0.8477	0.8575	0.8667	0.8753
4	0.5858	0.6046	0.6228	0.6406	0.6577	0.6743	0.6903	0.7258	0.7207	0.7350
5	0.3907	0.4102	0.4296	0.4488	0.4679	0.4868	0.5054	0.5237	0.5418	0.5595
6	0.2307	0.2469	0.2633	0.2801	0.2971	0.3142	0.3316	0.3490	0.3665	0.3840
7	0.1214	0.1325	0.1442	0.1564	0.1689	0.1820	0.1954	0.2902	0.2233	0.2378
8	0.0573	0.0639	0.0710	0.0786	0.0866	0.0951	0.1040	0.1133	0.1231	0.1334
9	0.0245	0.0279	0.0317	0.0358	0.0403	0.0451	0.0503	0.0558	0.0618	0.0681
10	0.0095	0.0111	0.0129	0.0149	0.0171	0.0195	0.0222	0.0251	0.0283	0.0318
11	0.0034	0.0041	0.0048	0.0057	0.0067	0.0078	0.0090	0.0104	0.0120	0.0137
12	0.0011	0.0014	0.0017	0.0020	0.0024	0.0029	0.0034	0.0040	0.0047	0.0055
13	0.0003	0.0004	0.0005	0.0007	0.0008	0.0010	0.0012	0.0014	0.0017	0.0020
14	0.0001	0.0001	0.0002	0.0002	0.0003	0.0003	0.0004	0.0005	0.0006	0.0007
15				0.0001	0.0001	0.0001	0.0001	0.0001	0.0002	0.0002
16									0.0001	0.0001

$m=$	5.2	5.4	5.6	5.8	6.0	6.2	6.4	6.6	6.8	7.0
$r=$ 0	1.0000	1.0000	1.0000	1.0000	1.0000	1.0000	1.0000	1.0000	1.0000	1.0000
1	0.9945	0.9955	0.9963	0.9970	0.9975	0.9980	0.9983	0.9986	0.9989	0.9991
2	0.9658	0.9711	0.9756	0.9794	0.9826	0.9854	0.9877	0.9897	0.9913	0.9927
3	0.8912	0.9052	0.9176	0.9285	0.9380	0.9464	0.9537	0.9600	0.9656	0.9704
4	0.7619	0.7867	0.8094	0.8300	0.8488	0.8658	0.8811	0.8948	0.9072	0.9182
5	0.5939	0.6267	0.6579	0.6873	0.7149	0.7408	0.7649	0.7873	0.8080	0.8270
6	0.4191	0.4539	0.4881	0.5217	0.5543	0.5859	0.6163	0.6453	0.6730	0.6993
7	0.2676	0.2983	0.3297	0.3616	0.3937	0.4258	0.4577	0.4892	0.5201	0.5503
8	0.1551	0.1783	0.2030	0.2290	0.2560	0.2840	0.3127	0.3419	0.3715	0.4013
9	0.0819	0.0974	0.1143	0.1328	0.1528	0.1741	0.1967	0.2204	0.2452	0.2709
10	0.0397	0.0488	0.0591	0.0708	0.0839	0.0984	0.1142	0.1314	0.1498	0.1695
11	0.0177	0.0225	0.0282	0.0349	0.0426	0.0514	0.0614	0.0726	0.0849	0.0985
12	0.0073	0.0096	0.0125	0.0160	0.0201	0.0250	0.0307	0.0373	0.0448	0.0534
13	0.0028	0.0038	0.0051	0.0068	0.0088	0.0113	0.0143	0.0179	0.0221	0.0270
14	0.0010	0.0014	0.0020	0.0027	0.0036	0.0048	0.0063	0.0080	0.0102	0.0128
15	0.0003	0.0005	0.0007	0.0010	0.0014	0.0019	0.0026	0.0034	0.0044	0.0057
16	0.0001	0.0002	0.0002	0.0004	0.0005	0.0007	0.0010	0.0014	0.0018	0.0024
17		0.0001	0.0001	0.0001	0.0002	0.0003	0.0004	0.0005	0.0007	0.0010
18					0.0001	0.0001	0.0001	0.0002	0.0003	0.0004
19								0.0001	0.0001	0.0001

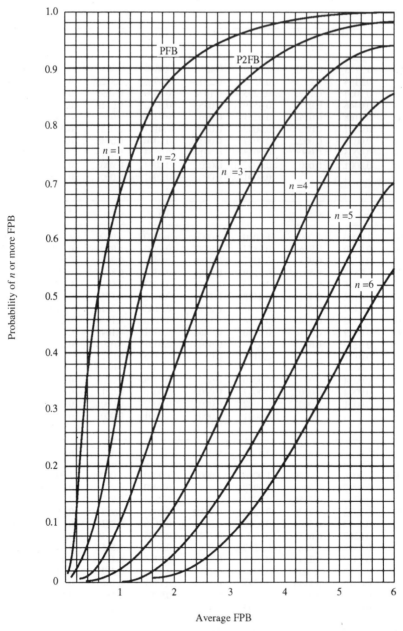

Figure E.2 Probability of n or more FPB versus the average number of FPB

Table E.3 Present value factors (PVFs). This table shows the present value of a single payment received *n* periods in the future discounted at *x* per cent per period

Period	1%	2%	3%	4%	5%	6%	7%	8%	9%	10%	11%	12%	13%	14%	15%	16%	17%	18%	19%	20%	Period
1	0.9901	0.9804	0.9709	0.9615	0.9524	0.9434	0.9346	0.9259	0.9174	0.9091	0.9009	0.8929	0.8850	0.8772	0.8696	0.8621	0.8547	0.8475	0.8403	0.8333	1
2	0.9803	0.9612	0.9426	0.9246	0.9070	0.8900	0.8734	0.8573	0.8417	0.8264	0.8116	0.7972	0.7831	0.7695	0.7561	0.7432	0.7305	0.7182	0.7062	0.6944	2
3	0.9706	0.9423	0.9151	0.8890	0.8638	0.8396	0.8163	0.7938	0.7722	0.7513	0.7312	0.7118	0.6931	0.6750	0.6575	0.6407	0.6244	0.6086	0.5934	0.5787	3
4	0.9610	0.9238	0.8885	0.8548	0.8227	0.7921	0.7629	0.7350	0.7084	0.6830	0.6587	0.6355	0.6133	0.5921	0.5718	0.5523	0.5337	0.5158	0.4987	0.4823	4
5	0.9515	0.9057	0.8626	0.8219	0.7835	0.7473	0.7130	0.6806	0.6499	0.6209	0.5935	0.5674	0.5428	0.5194	0.4972	0.4761	0.4561	0.4371	0.4190	0.4019	5
6	0.9420	0.8880	0.8375	0.7903	0.7462	0.7050	0.6663	0.6302	0.5963	0.5645	0.5346	0.5066	0.4803	0.4556	0.4323	0.4104	0.3898	0.3704	0.3521	0.3349	6
7	0.9327	0.8706	0.8131	0.7599	0.7107	0.6651	0.6227	0.5835	0.5470	0.5132	0.4817	0.4523	0.4251	0.3996	0.3759	0.3538	0.3332	0.3139	0.2959	0.2791	7
8	0.9235	0.8535	0.7894	0.7307	0.6768	0.6274	0.5820	0.5403	0.5019	0.4665	0.4339	0.4039	0.3762	0.3506	0.3269	0.3050	0.2848	0.2660	0.2487	0.2326	8
9	0.9143	0.8368	0.7664	0.7026	0.6446	0.5919	0.5439	0.5002	0.4604	0.4241	0.3909	0.3606	0.3329	0.3075	0.2843	0.2630	0.2434	0.2255	0.2090	0.1938	9
10	0.9053	0.8203	0.7441	0.6756	0.6139	0.5584	0.5083	0.4632	0.4224	0.3855	0.3522	0.3220	0.2946	0.2697	0.2472	0.2267	0.2080	0.1911	0.1756	0.1615	10
11	0.8963	0.8043	0.7224	0.6496	0.5847	0.5268	0.4751	0.4289	0.3875	0.3505	0.3173	0.2875	0.2607	0.2366	0.2149	0.1954	0.1778	0.1619	0.1476	0.1346	11
12	0.8874	0.7885	0.7014	0.6246	0.5568	0.4970	0.4440	0.3971	0.3555	0.3186	0.2858	0.2567	0.2307	0.2076	0.1869	0.1685	0.1520	0.1372	0.1240	0.1122	12
13	0.8787	0.7730	0.6810	0.6006	0.5303	0.4688	0.4150	0.3677	0.3262	0.2897	0.2575	0.2292	0.2042	0.1821	0.1625	0.1452	0.1299	0.1163	0.1042	0.0935	13
14	0.8700	0.7579	0.6611	0.5775	0.5051	0.4423	0.3878	0.3405	0.2992	0.2633	0.2320	0.2046	0.1807	0.1597	0.1413	0.1252	0.1110	0.0985	0.0876	0.0779	14
15	0.8613	0.7430	0.6419	0.5553	0.4810	0.4173	0.3624	0.3152	0.2745	0.2394	0.2090	0.1827	0.1599	0.1401	0.1229	0.1079	0.0949	0.0835	0.0736	0.0649	15
16	0.8528	0.7284	0.6232	0.5339	0.4581	0.3936	0.3387	0.2919	0.2519	0.2176	0.1883	0.1631	0.1415	0.1229	0.1069	0.0930	0.0811	0.0708	0.0618	0.0541	16
17	0.8444	0.7142	0.6050	0.5134	0.4363	0.3714	0.3166	0.2703	0.2311	0.1978	0.1696	0.1456	0.1252	0.1078	0.0929	0.0802	0.0693	0.0600	0.0520	0.0451	17
18	0.8360	0.7002	0.5874	0.4936	0.4155	0.3503	0.2959	0.2502	0.2120	0.1799	0.1528	0.1300	0.1108	0.0946	0.0808	0.0691	0.0592	0.0508	0.0437	0.0376	18
19	0.8277	0.6864	0.5703	0.4746	0.3957	0.3305	0.2765	0.2317	0.1945	0.1635	0.1377	0.1161	0.0981	0.0829	0.0703	0.0596	0.0506	0.0431	0.0367	0.0313	19
20	0.8195	0.6730	0.5537	0.4564	0.3769	0.3118	0.2584	0.2145	0.1784	0.1486	0.1240	0.1037	0.0868	0.0728	0.0611	0.0514	0.0433	0.0365	0.0308	0.0261	20

Period	21%	22%	23%	24%	25%	26%	27%	28%	29%	30%	31%	32%	33%	34%	35%	36%	37%	38%	39%	40%	Period
1	0.8264	0.8197	0.8130	0.8065	0.8000	0.7937	0.7874	0.7813	0.7752	0.7692	0.7634	0.7576	0.7519	0.7463	0.7407	0.7353	0.7299	0.7246	0.7194	0.7143	1
2	0.6830	0.6719	0.6610	0.6504	0.6400	0.6299	0.6200	0.6104	0.6009	0.5917	0.5827	0.5739	0.5653	0.5569	0.5487	0.5407	0.5328	0.5251	0.5176	0.5102	2
3	0.5645	0.5507	0.5374	0.5245	0.5120	0.4999	0.4882	0.4768	0.4658	0.4552	0.4448	0.4348	0.4251	0.4156	0.4064	0.3975	0.3889	0.3805	0.3724	0.3644	3
4	0.4665	0.4514	0.4369	0.4230	0.4096	0.3968	0.3844	0.3725	0.3611	0.3501	0.3396	0.3294	0.3196	0.3102	0.3011	0.2923	0.2839	0.2757	0.2679	0.2603	4
5	0.3855	0.3700	0.3552	0.3411	0.3277	0.3149	0.3027	0.2910	0.2799	0.2693	0.2592	0.2495	0.2403	0.2315	0.2230	0.2149	0.2072	0.1998	0.1927	0.1859	5
6	0.3186	0.3033	0.2888	0.2751	0.2621	0.2499	0.2383	0.2274	0.2170	0.2072	0.1979	0.1890	0.1807	0.1727	0.1652	0.1580	0.1512	0.1448	0.1386	0.1328	6
7	0.2633	0.2486	0.2348	0.2218	0.2097	0.1983	0.1877	0.1776	0.1682	0.1594	0.1510	0.1432	0.1358	0.1289	0.1224	0.1162	0.1104	0.1049	0.0997	0.0949	7
8	0.2176	0.2038	0.1909	0.1789	0.1678	0.1574	0.1478	0.1388	0.1304	0.1226	0.1153	0.1085	0.1021	0.0962	0.0906	0.0854	0.0806	0.0760	0.0718	0.0678	8
9	0.1799	0.1670	0.1552	0.1443	0.1342	0.1249	0.1164	0.1084	0.1011	0.0943	0.0880	0.0822	0.0768	0.0718	0.0671	0.0628	0.0588	0.0551	0.0516	0.0484	9
10	0.1486	0.1369	0.1262	0.1164	0.1074	0.0992	0.0916	0.0847	0.0784	0.0725	0.0672	0.0623	0.0577	0.0536	0.0497	0.0462	0.0429	0.0399	0.0371	0.0346	10
11	0.1228	0.1122	0.1026	0.0938	0.0859	0.0787	0.0721	0.0662	0.0607	0.0558	0.0513	0.0472	0.0434	0.0400	0.0368	0.0340	0.0313	0.0289	0.0267	0.0247	11
12	0.1015	0.0920	0.0834	0.0757	0.0687	0.0625	0.0568	0.0517	0.0471	0.0429	0.0392	0.0357	0.0326	0.0298	0.0273	0.0250	0.0229	0.0210	0.0192	0.0176	12
13	0.0839	0.0754	0.0678	0.0610	0.0550	0.0496	0.0447	0.0404	0.0365	0.0330	0.0299	0.0271	0.0245	0.0223	0.0202	0.0184	0.0167	0.0152	0.0138	0.0126	13
14	0.0693	0.0618	0.0551	0.0492	0.0440	0.0393	0.0352	0.0316	0.0283	0.0254	0.0228	0.0205	0.0185	0.0166	0.0150	0.0135	0.0122	0.0110	0.0099	0.0090	14
15	0.0573	0.0507	0.0448	0.0397	0.0352	0.0312	0.0277	0.0247	0.0219	0.0195	0.0174	0.0155	0.0139	0.0124	0.0111	0.0099	0.0089	0.0080	0.0072	0.0064	15
16	0.0474	0.0415	0.0364	0.0320	0.0281	0.0248	0.0218	0.0193	0.0170	0.0150	0.0133	0.0118	0.0104	0.0093	0.0082	0.0073	0.0065	0.0058	0.0051	0.0046	16
17	0.0391	0.0340	0.0296	0.0258	0.0225	0.0197	0.0172	0.0150	0.0132	0.0116	0.0101	0.0089	0.0078	0.0069	0.0061	0.0054	0.0047	0.0042	0.0037	0.0033	17
18	0.0323	0.0279	0.0241	0.0208	0.0180	0.0156	0.0135	0.0118	0.0102	0.0089	0.0077	0.0068	0.0059	0.0052	0.0045	0.0039	0.0035	0.0030	0.0027	0.0023	18
19	0.0267	0.0229	0.0196	0.0168	0.0144	0.0124	0.0107	0.0092	0.0079	0.0068	0.0059	0.0051	0.0044	0.0038	0.0033	0.0029	0.0025	0.0022	0.0019	0.0017	19
20	0.0221	0.0187	0.0159	0.0135	0.0115	0.0098	0.0084	0.0072	0.0061	0.0053	0.0045	0.0039	0.0033	0.0029	0.0025	0.0021	0.0018	0.0016	0.0014	0.0012	20

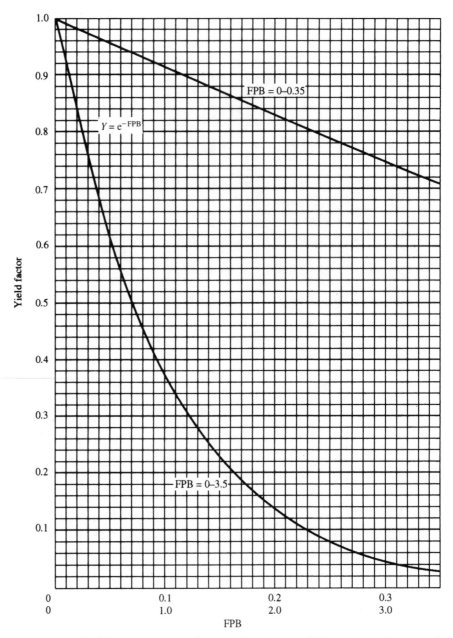

Figure E.3   Yield factor for various levels of FPB. (*Note*: yield in % = yield factor × 100)

Table E.4   Present value factors for one unit of currency receivable continuously

Period	2%	4%	6%	8%	10%	12%	14%	15%	16%	18%	20%	22%	24%	26%	28%	30%	35%
1	0.990	0.981	0.971	0.962	0.954	0.945	0.937	0.933	0.929	0.922	0.914	0.907	0.900	0.893	0.886	0.880	0.864
2	0.971	0.943	0.916	0.891	0.867	0.844	0.822	0.812	0.801	0.781	0.762	0.743	0.726	0.709	0.692	0.677	0.640
3	0.952	0.907	0.865	0.825	0.788	0.754	0.721	0.706	0.691	0.662	0.635	0.609	0.585	0.562	0.541	0.520	0.474
4	0.933	0.872	0.815	0.764	0.717	0.673	0.633	0.614	0.595	0.561	0.529	0.500	0.472	0.446	0.423	0.400	0.351
5	0.915	0.838	0.769	0.708	0.652	0.601	0.555	0.534	0.513	0.475	0.441	0.409	0.381	0.354	0.330	0.308	0.260
6	0.897	0.806	0.726	0.655	0.592	0.536	0.487	0.464	0.442	0.403	0.367	0.336	0.307	0.281	0.258	0.237	0.193
7	0.879	0.775	0.685	0.607	0.538	0.479	0.427	0.404	0.382	0.341	0.306	0.275	0.248	0.223	0.202	0.182	0.143
8	0.862	0.745	0.646	0.562	0.489	0.428	0.375	0.351	0.329	0.289	0.255	0.225	0.200	0.177	0.157	0.140	0.106
9	0.845	0.717	0.610	0.520	0.445	0.382	0.329	0.305	0.284	0.245	0.213	0.185	0.161	0.141	0.123	0.108	0.078
10	0.829	0.689	0.575	0.481	0.405	0.341	0.288	0.265	0.244	0.208	0.177	0.151	0.130	0.112	0.096	0.083	0.058
11	0.812	0.662	0.543	0.446	0.368	0.304	0.253	0.231	0.211	0.176	0.148	0.124	0.105	0.089	0.075	0.064	0.048
12	0.797	0.637	0.512	0.413	0.334	0.272	0.222	0.201	0.182	0.149	0.123	0.102	0.085	0.070	0.059	0.049	0.032
13	0.780	0.612	0.483	0.382	0.304	0.243	0.195	0.174	0.157	0.127	0.103	0.084	0.068	0.056	0.046	0.038	0.024
14	0.766	0.589	0.455	0.354	0.276	0.217	0.171	0.152	0.135	0.107	0.085	0.068	0.055	0.044	0.036	0.029	0.018
15	0.750	0.555	0.430	0.328	0.251	0.194	0.150	0.132	0.116	0.091	0.071	0.056	0.044	0.035	0.028	0.022	0.013
16	0.736	0.545	0.405	0.303	0.228	0.173	0.131	0.115	0.100	0.077	0.059	0.046	0.036	0.028	0.022	0.017	0.010
17	0.721	0.523	0.383	0.281	0.208	0.154	0.115	0.100	0.085	0.065	0.049	0.038	0.029	0.022	0.017	0.013	0.007
18	0.707	0.504	0.360	0.260	0.189	0.138	0.101	0.087	0.075	0.055	0.041	0.031	0.023	0.018	0.013	0.010	0.005
19	0.694	0.484	0.341	0.241	0.172	0.123	0.089	0.075	0.064	0.047	0.034	0.025	0.019	0.014	0.010	0.008	0.004
20	0.679	0.465	0.321	0.223	0.156	0.110	0.078	0.066	0.055	0.040	0.029	0.021	0.015	0.011	0.008	0.006	0.003
21	0.667	0.448	0.303	0.206	0.142	0.098	0.068	0.057	0.048	0.034	0.024	0.017	0.012	0.009	0.006	0.005	0.002
22	0.653	0.430	0.285	0.192	0.129	0.088	0.060	0.050	0.041	0.029	0.020	0.014	0.010	0.007	0.005	0.004	0.002
25	0.615	0.382	0.240	0.151	0.070	0.062	0.040	0.033	0.026	0.017	0.012	0.008	0.005	0.004	0.002	0.002	0.001

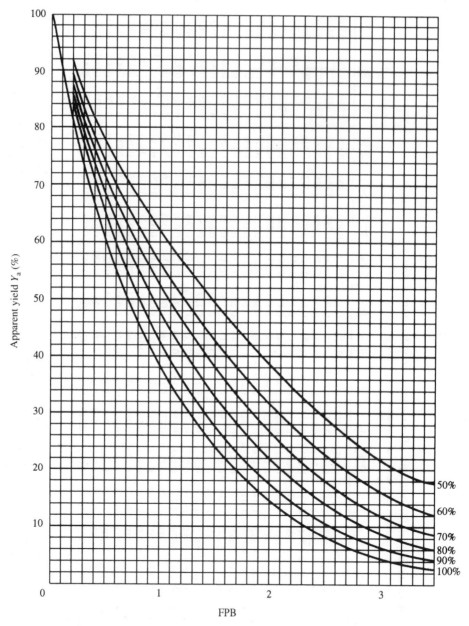

Figure E.4   Apparent yield (Y_a) versus levels of fault coverage

Appendix F: The fault spectrum

Throughout the book there are numerous references to the *fault spectrum* which is simply a listing of different types of defect with some indication of the number, or more often the percentage, of each that are present. The concept of the fault spectrum is an important one in that it is one of the main determinants of the 'optimum test strategy'. It is also one of the most widely abused concepts in testing because it can be used to prove almost anything in terms of which strategy, or more likely which tactic, will be the best to adopt. If you do not know what your fault spectrum is for your current production then you are unlikely to be able to select the best strategy or to improve your quality. It is often said that if you do not know where you are going then any direction will be all right. Equally, if you do not know where you are then you cannot possibly know which direction to take.

The big problem occurs when we need to perform some analysis for a future situation where the various determinants of the fault spectrum will be different from those with which we are familiar. There are so many variables that predicting the fault spectrum will be prone to a lot of error. Ideally we would like to know what other manufacturers, with a similar set of conditions, are experiencing. Unfortunately very little information has been made available about 'industry typical' spectra and for this reason a lot of companies have to rely on the experiences of the equipment vendors. In general, the ATE vendors have a lot of useful information available because they see many different manufacturing operations in many different industry segments. Unfortunately there can be a tendency in some cases for the fault spectrum to get biased in a manner that shows a particular tester in a good light. Probably the most blatant example of this that I have seen was in early 1993 at a conference in Switzerland where a presenter from a company producing manufacturing defects analysers (MDAs) showed his version of the fabled 'industry typical fault spectrum'. In this he had grouped open and short circuits together and referred to them as *connection problems*. He then made the tacit assumption that his product could detect all of these and proceeded to address other parts of the spectrum. In this way he neatly side-stepped the problem that many MDAs have in detecting open circuits and showed that his product could detect about 90 per cent of the defects present in this industry typical spectrum.

Wouldn't it be nice then if we could have some unbiased data on fault spectra that was gathered from real world manufacturing operations and made available to anyone with a need to know? Well, just as the final editing touches were being made to this book, this very information came to light. The TEST'93 conference held in Brighton, England, in October 1993 saw the official launch of an organization called the 'Industry Test Forum' (ITF). Following a gestation period of about twelve months this launch took place to broaden the membership base and the scope of the ITF's activities. The ITF had been formed 'to bring together practising test professionals in a forum that encourages the interchange of ideas and experiences, that will improve the efficiency and effectiveness of the test process.' During the first twelve months the activities had been limited somewhat but the founder members had undertaken a study of the fault spectra experienced for a variety of different manufacturing technologies. The results of this initial study were presented and showed a few surprises. A summary of these results is shown in Fig. F.1 by kind permission of the ITF. The first surprise was that there was less variation in the spectra for the three technologies studied than might have been expected. For example, it is widely accepted that shorts usually dominate the spectra for 'through hole technology' (THT) boards, and that open circuits are more of a problem with 'surface mount technologies' (SMT). The results show that the spectrum shifts in the expected direction but not to the degree that is

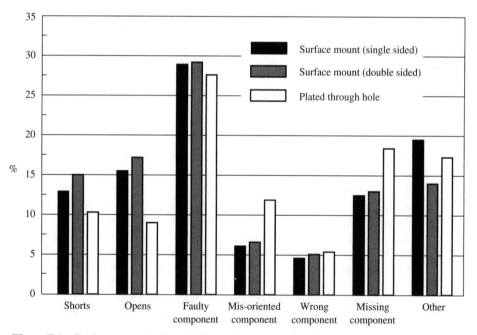

Figure F.1 Fault spectra for three different board technologies. Data averaged from a large volume of boards, from a number of companies, and compiled by the 'Industry Test Forum' in the UK

usually expected. Much of the board volume that the data was taken from came from a few large companies with high volumes and well controlled manufacturing processes. This obviously has to be borne in mind when interpreting these results and may well account for this apparent anomaly relative to what might have been expected. The second surprise is that component problems are the biggest category for all three board technologies. Again this is probably due to the fact that the manufacturing processes are under tight control and this results in some other problem bouncing up to the top of the pile. This does, however, support a number of things that I have discussed in this book. Firstly, the usual approach to quality improvement is the Pareto approach of attacking the biggest problems first. Once you have successfully got these under control then something else becomes the biggest problem and you then attack that. This results in biasing the fault spectrum towards defects that are more difficult to detect and this in turn will mean that your test strategy, but more particularly your test tactics, should be selected to address these more difficult defects. In more simple terms, a fault spectrum such as this will result in an economic analysis showing that a more powerful board tester, with a higher price tag, will probably outperform a simpler but cheaper tester. Even if many of the 'component' problems cannot be detected at the board test stage the more we can detect there the better. Obviously there will be a point where it might be cheaper to detect these defects at the 'functional verification' or 'subsystem' test stages but without the detailed economic analysis we would never know.

The second point that I make in the book that is supported by these results also concerns these component related defects. I am frequently told by clients that 'we don't get bad components any more'. My usual response is to ask how many components are consumed by manufacturing in a year. The resultant numbers, even for many small companies, imply that even for defect rates in the 50 ppm to 200 ppm they should statistically be getting several hundred to several thousand defective components per year. Of course, many of these will be defective by degree rather than totally useless but the point is made. The problem I feel is that we can easily fool ourselves about what is the norm. Read the technical press and review the contents of conference proceedings over the past couple of years and you might assume that the whole industry and his brother is manufacturing multi-chip modules (MCMs). In fact there are very few being built at present (late 1993). In a similar manner we can be misled into thinking that all component types have very high quality levels just because some of them do. Compare the typical defect rates for a high volume low complexity digital device with those for a programmable logic device and you will see my point. At the same time the use of testers and test strategies that have a poor fault coverage for component defects will simply confirm that there are no defective components! The usual lack of any failure analysis then means that component defects detected at final test or in the field get classified as 'early life failures' so there is no attempt to see if the defect could have been detected in a more cost effective manner.

The ITF in the UK are to be applauded for performing this analysis and publishing the results. I hope that they continue to do more of the same and that similar groups are formed in other countries and do their own research. At the time of writing, the full results of this study have not been published so I am grateful to the ITF for allowing me to publish this preliminary data. The data shown in Fig. F.1 is, of course, an average of data collected from a number of companies of varying size and in different segments of the industry. The results will obviously be biased by the higher volume manufacturers who will also probably be the ones with better process control. There is also likely to be some errors caused by a lack of consistency in the definition of defect types and I understand that the ITF plans to standardize the defect classifications among its members. There was no differentiation made in the component defect category between problems with the components as received and problems due to handling. It may also be that some defects in this category are really design problems. I am not aware of the degree of variability of the fault spectra among the contributing companies and I hope that some measure of this is included when they publish the results. However, it is a great start and I hope that the work continues.

Appendix G: Abbreviations

AOQ	Average outgoing quality
AOQL	Average outgoing quality level (limit)
AQL	Acceptance quality level
	Acceptable quality level
ARR	Accounting rate of return
ASIC	Application specific integrated circuit
ATE	Automatic test equipment
ATG	Automatic test generation (general term)
ATPG	Automated test pattern generation (digital)
BILBO	Built-in logic block observation
BIST	Built-in self-test
CAD	Computer aided design (drafting, drawing)
CAE	Computer aided engineering
COGS	Cost of goods sold
COMB	Combinational board tester (testing)
DCF	Discounted cash flow
DDB	Double declining balance
DFPB	Detected faults per board
DFT	Design for test
DIRFT	Do it right first time
DPDU	Defects per defective unit
DPU	Defects per unit
DUT	Device under test
ECO	Engineering change order
EDA	Electronic design automation
EFPB	Escaping faults per board
ERR	External rate of return
ESS	Environmental stress screening
e	The exponential coefficient (2.718)

FC	Fault coverage
FNT	Functional board tester (testing)
FPB	Faults per board (average across all boards)
FPFB	Faults per faulty board
FV	Future value (of a sum of money)
GIGO	Garbage in garbage out
IC	Integrated circuit
ICA	In-circuit analyser
ICT	In-circuit tester (testing)
IRR	Internal rate of return
JIT	Just in time
JTAG	Joint test action group
KGB	Known good board (as applied to comparator based test)
LN/P	Loop number multiplier
LTPD	Lot tolerance per cent defective
MARR	Minimum acceptable rate of return
MCM	Multi-chip module
MDA	Manufacturing defects analyser
MIRR	Modified internal rate of return
MST	Multi-strategy tester
MTBF	Mean time between failures
NFV	Net future value
NPV	Net present value
NPW	Net present worth
NRE	Non-recurring engineering
NW	Net worth
OC	Operating characteristic
OEM	Original equipment manufacturer
PBT	Profit before tax
PCB	Printed circuit board
PFB	Probability of there being a faulty board $(1 - Y)$
PI	Profitability index
PLC	Product life cycle
PV	Present value
PVF	Present value factor
PW	Present worth

PWA	Printed wire assembly
PWB	Printed wiring board
PWF	Present worth factor
ROCE	Return on capital employed
ROI	Return on investment
SMT	Surface mount technology
SOYD	Sum of the year digits
SPC	Statistical process control
SST	Sub-system test
SYS	Final system test
TDR	Testing, diagnosis and repair
TQC	Total quality control (commitment)
TTM	Time to market
TTT	Total time on the tester
UUT	Unit under test
Y	Yield
Ya	Apparent yield
Yp	Perceived yield

Appendix H: Bibliography

Ambler, A. P., M. Abadir and S. Sastry (1992) *Economics of Design and Test — For Electronic Circuits and Systems,* Ellis Horwood Limited, Chichester, UK.

Blank, Leland T. and Anthony J. Tarquin (1989) *Engineering Economy,* 3rd edn, McGraw-Hill, New York.

Crosby, Philip B. (1979) *Quality Is Free,* McGraw-Hill, New York.

Deming, W. Edwards (1982) *Quality, Productivity and Competitive Position,* Massachusetts Institute of Technology, Cambridge, MA, USA.

Feigenbaum, Armand (1983) *Total Quality Control,* 3rd edn, McGraw-Hill, New York.

Juran, Joseph M. and Frank M. Gryna (1988) *Juran's Quality Control Handbook,* 4th edn, McGraw-Hill, New York.

Pynn, Craig (1986) *Strategies for Electronics Test,* McGraw-Hill, New York.

Riggs, James L. and Thomas M. West (1986) *Engineering Economics,* 3rd edn, McGraw-Hill, New York.

Scheiber, Stephen F. (due spring 1994) *Building a Successful Board-Test Strategy,* Butterworth-Heinemann, Stoneham, MA, USA.

Index